High Performance Printed Circuit Boards

Other Books of Interest from McGraw-Hill

High Performance Printed Circuit Boards

Charles A. Harper

McGraw-Hill

New York   San Francisco   Washington, D.C.   Auckland   Bogotá
Caracas   Lisbon   London   Madrid   Mexico City   Milan
Montreal   New Delhi   San Juan   Singapore
Sydney   Tokyo   Toronto

Library of Congress Cataloging-in-Publication Data

High performance printed circuit boards / [edited by] Charles A. Harper.
    p.   cm.
    Includes bibliographical references and index.
    ISBN 0-07-026713-8
    1. Printed circuits. I. Harper, Charles A.
TK7868.P7 H545   1999.
621.3815'31—dc21

                                           99-054066
                                             CIP

McGraw-Hill

A Division of The McGraw-Hill Companies

1 2 3 4 5 6 7 8 9 0 AGM/AGM 0 6 5 4 3 2 1 0

ISBN 0-07-026713-8

The sponsoring editor for this book was Steve Chapman, the editing supervisor was Frank Kotowski, Jr., and the production supervisor was Pamela A. Pelton. It was set in Century Schoolbook by Michele Pridmore of McGraw-Hill's Professional Book Group Composition Unit in Hightstown, N.J.

Printed and bound by Quebecor/Martinsburg.

McGraw-Hill books are available at special quantity discounts to use as premiums and sales promotions, or for use in corporate training programs. For more information, please write to the Director of Special Sales, McGraw-Hill, Two Penn Plaza, New York, NY 10121-2298. Or contact your local bookstore.

This book is printed on recycled, acid-free paper containing a minimum of 50% recycled, de-inked fiber.

Contents

Chapter 3. Substrates for RF and Microwave Systems   3.1

Stephen G. Konsowski and John W. Gipprich

Chapter 4. Advanced Ceramic Substrates for Microelectronic Systems   4.1

Jerry E. Sergent

Chapter 5. Flexible Printed Circuits

Marty Jawitz and Michael Jawitz

Chapter 6. Design of High Performance Circuit Boards

Sam Shaw and Alonzo S. Martinez, Jr.

Chapter 7. Soldering and Cleaning of High Performance Circuit Board Assemblies   7.1

Charles G. Woychik

Chapter 8. Environmentally Conscious Printed Circuit Board Materials and Processes   8.1

John W. Lott

Chapter 9. Reliability and Performance of Advanced PWB Circuits   9.1

Thomas J. Stadterman and Michael D. Osterman

Preface

In recent years, the almost explosive technological advances in electronic systems have demanded ever-higher functionality, ever-faster circuit speeds, and always increasing interconnection density. Semiconductor technology has well kept up with, or even led, advancing system requirements. However, the baseline interconnection technology of the total system, namely, printed circuit boards and substrates, has been hard pressed to keep up. Fortunately, with much hard, dedicated, and brilliant work, the critical and limiting printed circuit board and substrate technologies are making almost phenomenal advances. These exciting advances are being made in all areas of printed circuit board and substrate technology, including materials, processes, design, fabriction, substrate form, soldering and cleaning, reliability, and also environmentally conscious materials and processes. Note that the term "substrates" is used herein, since this level of interconnection technology includes not only basic rigid printed circuit boards, but also such important forms as flexible circuits, special substrates for RF and microwave systems, and ceramic substrates for microelectronics. The tremendous achievements in all of these advanced printed circuit boards and substrate areas form the basis for this book. It is hoped that whatever your role in the total scheme of electronic systems and interconnection technologies, this book will provide you not only with up-to-date knowledge, but also useful information, data, and guidelines for your use and application areas. As will be evident by a review of the subject and author listings, I have had the good fortune to be able to bring together a team of outstanding chapter authors, each with a great depth of experience in his or her field. Together, they offer the reader a base of knowledge as perhaps no other group could. Hence, I would like to give special credit to these authors in this preface.

This book has been prepared as a thorough sourcebook of data, information, and guidelines for all ranges of interests. It offers an extensive array of property and performance data as a function of the most important product variables. The chapter organization and coverage are well suited to reader convenience for the wide range of product categories. The first chapter covers the important groups of high performance rigid circuit boards which are of increasing importance in modern electronic systems. The second chapter is an excellent presentation of perhaps the fastest growing and important area of high performance circuit boards, namely, microvias, built-up multilayers, and high-density circuit boards. Next come three chapters on substrates of special importance in modern electronic systems, namely, substrates for RF and microwave systems, advanced ceramic substrates for microelectronics, and flexible circuits for advanced electronic systems. The sixth chapter is an outstanding presentation on perhaps the most critical but often poorly understood subject of design of high performance circuit boards. The seventh chapter is yet another most excellent presentation on the basic and always critical subject of soldering and cleaning high performance circuit boards. Next comes a very thorough and comprehensive chapter on the increasingly important subject of environmentally conscious circuit board materials and processes—an area which must be understood and applied. Last, and also of vital importance, is another outstanding chapter by leading experts with great depth of experience on the subject of reliability and performance of high performance circuit boards.

It is my hope, and expectation, that this book will serve its readers well. Any comments or suggestions will be welcomed.

Charles A. Harper

Contributors

Carl T. Brooks Northrop Grumman Corporation, Baltimore, MD (CHAP. 1)

Rolf E. Funer Consultant, Funer Associates, Greenville, SC (CHAP. 2)

John W. Gipprich Northrop Grumman Corporation, Baltimore, MD (CHAP. 3)

Charles A. Harper President, Technology Seminars, Inc., Lutherville, MD (APP.)

Martin W. Jawitz Consultant, Las Vegas, NV (CHAP. 5)

Stephen G. Konsowski Technology Seminars, Inc., Lutherville, MD (CHAP. 3)

John W. Lott DuPont *i*Technologies, Research Triangle Park, NC (CHAP. 8)

Alonzo S. Martinez, Jr. Dell Computer, Round Rock, TX (CHAP. 6)

Michael Osterman University of Maryland, College Park, MD (CHAP. 9)

Jerry E. Sergent TCA, Inc., Williamsburg, KY (CHAP. 4)

Sam Shaw AMP/TYCO Electronics, Harrisburg, PA (CHAP. 6)

Thomas J. Stadterman U.S. Army Material Systems Analysis Activity, Aberdeen, MD (CHAP. 9)

Charles G. Woychik IBM Corporation, Endicott, NY (CHAP. 7)

High Performance Printed Circuit Boards

Carl T. Brooks

Northrop Grumman
Baltimore, Maryland

1.1 Introduction

The rapid move toward electronic system miniaturization, which has been driven in large part by consumer electronics, has forced electronic material manufacturers to rethink their approach to the development and marketing of new laminates. Material suppliers have been forced to think more globally about the ultimate systems which will employ their materials. These systems are more and more becoming a collection of advanced packaging techniques with increasing inputs and outputs (I/Os) and increased levels of integration. Rapidly changing integrated circuit technology is driving interconnect designers to deal with ever increasing die sizes. These advanced components dissipate larger amounts of heat and operate at ever increasing frequencies to meet the demands of today's electronics system consumer.

Today's electronics market is diverse, growing, and functionally demanding. It can be characterized by a generic set of requirements that encompasses the prevailing trend of "cheaper, faster, better":

Reduced weight and size

Increased functionality

Higher performance

Lower cost

Greater reliability

Decreased time to market

These requirements have placed a tremendous demand on the printed wiring board (PWB) industry. Significant challenges in resin/reinforcement (laminate) choices have emerged. Computer, telecommunications, automotive, medical, and military markets are creating demands for innovative solutions to electronic interconnection problems.

Radio frequency (RF) components in some telecommunications applications can operate anywhere from 900 MHz to 4 GHz. Base stations, which constitute the infrastructure for global network communications, can operate anywhere from 4 to 10 GHz, requiring PWBs with low dielectric loss (tan δ). Increased logic chip performance in high speed digital applications where propagation delay and signal rise times are critical require PWBs with lower dielectric constants (ε_r).

In many military applications where high reliability cannot be compromised by extreme temperature excursions, multiple soldering operations or thermal stress testing, higher T_g (glass transition temperature), and chemically resistant PWBs are required. Increased packaging densities, which are required to meet the needs of the move toward miniaturization, has meant that CTE (coefficient of thermal expansion) must match or be tailored to various packaging schemes and materials.

These formidable electrical, mechanical, and chemical requirements imposed on PWBs by today's (and tomorrow's) electronic systems have made conventional FR-4 epoxy/glass solutions virtually impossible in many cases. The demand for high performance PWB material alternatives will continue to grow in locked step with the demand for more sophisticated industrial and personal electronics.

This chapter will present PWB material data with an emphasis on high performance options to meet the needs of emerging electronic system requirements. Emphasis will be placed on the PWB attributes mentioned previously:

Higher T_g

Lower tan δ

Lower ε_r

Increased chemical resistance

Matched or tailorable CTE

Manufacturing considerations related to materials displaying these characteristics will be discussed as well as observed advantages and disadvantages for particular applications.

1.2   PWB Material Systems

The choice of a material system for a designated PWB application has historically been driven by cost: minimum material cost for the stated application. However, in today's world of high speed devices, advanced material packages, and high reliability systems, in many cases, *performance* must also be a key component in the material selection process. Trade-off analyses involving cost, performance, reliability, and manufacturability must be made. This analysis process must not be performed by any *one* of the stakeholders in the success of the ultimate deliverable product, but rather jointly by an interdisciplinary team comprised of the customer and representatives from design, manufacturing, purchasing, quality, and testing.

PWB material systems, by definition, are complex. We refer to *laminates* as the designed result of the combination of a resin system and a reinforcement material. These engineered laminates display a variety of mechanical, electrical, thermal, and chemical properties. We will consider in this discussion of material systems only those laminates considered high performance candidates by virtue of the attributes mentioned earlier in this chapter.

1.2.1   High T_g material systems

Glass transition temperature (T_g) is the temperature at which resins begin to behave more like a gel or high viscosity liquid rather than a solid. Near this temperature a softening of the resin begins to occur because of a weakening of molecular bonds. This temperature is normally determined by measuring the change in slope of the volume expansion of the resin with temperature rise.

High T_g materials are required for certain applications in virtually every market segment where products are exposed to extreme temperature variations and/or severe thermal shocks. High T_g laminates are able to maintain their dimensional stability over higher temperature ranges, thereby improving overall performance and reliability. Since these materials are able to withstand higher temperatures, designers are able to allow for much broader temperature operating windows, thus increasing the functionality of the PWB.

One of the key characteristics of high T_g materials is resistance to plated-through hole failures which are related to cracked-hole barrel plating. Since thermal expansion increases at a much more rapid rate above the glass transition temperature, the z axis expansion, and the resulting cracking, is less likely to occur in higher T_g laminates. This is an especially attractive attribute in high power analog applications. Reduced plated-through hole cracking is also a key reliability component in computer applications where high layer count (>20 layers), finer lines and spaces (<0.005 in), and microvias (<0.010 in) are becoming more prevalent.

Reworkability of PWBs has become an important attribute in applications where components may be routinely replaced because of planned improvements, replacement of a defective component, or the incorporation of engineering changes. In the case of high dollar products, such as multichip modules (MCMs) where the assembly is not a "throw away" after failure, the ability to replace defective devices is key. This kind of rework or repair, in most cases, requires multiple soldering and/or reflow operations.

The component or device must first be removed from the PWB, the area where the device is to be reattached must be prepared, and finally the device/component must be resoldered and evaluated. Designers must choose laminates that are capable of withstanding these extreme thermal conditions. High T_g laminates have been developed to fill this need. Some high T_g laminates are also ideally suited for harsh environments such as those encountered in the automotive industry where exposure to temperature extremes, wide humidity ranges, fluxes, and solvents used in rework and repair are common. Solvent- and moisture-resistant properties displayed by the highly cross-linked resin systems used in high temperature laminates are effective choices in many applications. Although moisture absorption is not the strong suit of most of the high T_g materials, creative combinations of high temperature resins and low water-absorbing reinforcements may mitigate some of the risks encountered in humid environments.

Applications for high temperature laminates are numerous. The choice of the appropriate laminate will depend on a very rigorous examination of system performance, cost, and reliability requirements as defined by the customer. As stated previously, a high level of attention should be paid to the manufacturability of the design in question.

A word of caution regarding reported T_g data: note the method of measurement used. The IPC (Institute for Interconnecting and Packaging Electronic Circuits) uses thermal mechanical analysis (TMA) when reporting T_g values in its report IPC-4101, "Specification for Base Materials for Rigid and Multilayer Printed Boards." Other documents may report T_g using digital scanning calorimetry (DSC) or digital mechanical analysis (DMA). It is important to note that the TMA method may yield lower values (4 to 6%) for identical specimens. All T_g data presented in this chapter were measured using TMA.

1.2.1.1 Bismaleimide triazine (BT) epoxy-based laminates. Although not among the highest T_g materials, laminates employing bismaleimide triazine resins which have been reacted with epoxy are in wide use. The most common BT epoxy blends are reinforced with woven E-glass to form a laminate that is similar in processability to FR-4 and only slightly more expensive. These laminates are used in applications where T_g values and performance characteristics are slightly greater than those available in conventional FR-4. They fill a need in this "moderately high" T_g requirement area where the more expensive polyimide or cyanate ester resin–based systems are considered overkill. (Table 1.1). T_g values of 170 to 200°C have been achieved with these systems.

Laminates may be tailored to specific applications by varying the epoxy to BT resin ratio and/or reinforcement material. Although the cost is driven down by the addition of more epoxy, T_g suffers in this scenario. The choice of reinforcements must be balanced with the cost impact of the choice. Since one of the key reasons for choosing BT systems is cost, the use of exotic reinforcements, such as quartz or aramid, should be evaluated to determine if the added cost can be offset by the resulting improved performance. In addition, the cost of fabricating multilayer PWBs from laminates employing exotic resins increases significantly.

BT resin–based laminates have displayed brittleness when a lower percentage of epoxy is present in the resin mix. Although the resulting laminate is higher in T_g, the trade-off is, again, the cost of fabrication even when conventional E-glass reinforcement is used. This laminate construction contributes to difficulty in drilling and leads to the formation of microcracks. These small cracks act as sites for chemical and moisture absorption during processing. Electrical performance could also be affected by these small discontinuities in the microstructure. Although the leading suppliers of BT epoxy laminates have worked hard to improve these conditions, multilayer board manufacturers must maintain rigorous process controls in the areas of drilling and preplate hole preparation in order to use these laminates effectively.

TABLE 1.1 High T_g Laminate Systems

Resin	Reinforcement	T_g (°C)
Bismaleimide triazine—Epoxy	Woven E-glass	170–220
Polyimide	Woven E-glass	200 minimum
Polyimide	Nonwoven aramid	220 minimum
Polyimide	Woven quartz	250 minimum
Cyanate ester	Cross-plied aramid	230 minimum
Cyanate ester	Woven S-2 glass	230 minimum
Cyanate ester	Woven E-glass	230 minimum

As mentioned earlier, the choice of BT epoxy laminates for specific applications is usually driven by a need for lower cost, slightly higher performance than FR-4, and ease of manufacture. These attributes can easily translate into greatly reduced multilayer PWB costs and, consequently, lower system costs for a number of applications where leadless chip carriers and low loss is not required. These laminates remain the most effective alternative to conventional FR-4 epoxy for these applications.

1.2.1.2 Polyimide-based laminates. Laminates employing polyimide resin systems have been around since the early 1970s. These laminates went through somewhat of a transformation in the 1980s when one of the constituents of the resin formulation [methylene dianiline (MDA)] was removed because of its carcinogenic properties. Non-MDA polyimides are the only ones available today. Although this has lead to a much safer work environment, it has also contributed to the higher cost of polyimide-based laminates.

These laminates have historically been the high temperature laminate of choice because of their reputation for high performance, reliability, and generally unencumbered availability. Manufacturers of military and other systems where harsh environments are encountered have typically relied heavily on these laminates. Although polyimide-based laminates are more water absorptive than other high performance laminates, system designers are still likely to choose them because of their excellent toughness, processability, and essentially the highest T_g available.

Polyimide-based materials are ideally suited for applications where rework is performed routinely because of their superior adhesion to copper at soldering temperatures. In addition, because these laminates display such a high tolerance for heat, problems such as drill smear seen in lower T_g materials is virtually nonexistent. This is extremely attractive to fabricators because it allows them to reduce the steps required for preplate hole conditioning, thus reducing the overall manufacturing costs of multilayer boards. One of the major impediments to a more widespread use of polyimide laminates is their relatively high cost (Table 1.2). This is especially true in commercial markets where cost has historically been more of a determining factor than in military applications. This is significant because commercial market segments, such as wireless communications and computer applications, are growing at an extremely rapid rate while military segments are experiencing slow to no growth. Work continues at resin manufacturers to reduce the cost of polyimide systems without compromising the properties that make them attractive to designers and fabricators of multilayer PWBs. It is also important to note that the

TABLE 1.2   Laminate Cost Comparison*

Laminate	Relative cost
Epoxy/E-glass	1X
Polyimide/E-glass	2X
Epoxy/Nonwoven aramid	1.5X
Epoxy/woven aramid	6X
Cyanate ester/S-glass	6X
PTFE/glass	3X
Polyimide/quartz	9X

*These costs are relative and may vary to some extent based on supplier and quantities ordered.

choice of reinforcing material has a large impact on the overall cost of any laminate. This is especially true in the case of polyimide-based materials.

As mentioned previously, a number of polyimide-based laminates have also been criticized for their high (3%) water absorption when compared to BT and other epoxy-based materials (<1%). This could be an important consideration when assessing viability for humid and wet environments. Although conformal coatings are available that limit surface exposure after fabrication, it is important to recognize that exposure to liquids during the fabrication cycle is a concern and additional bake/drying cycles may be appropriate.

One of the key contributors to water absorption and other characteristics of laminates is the choice of reinforcements. Polyimide laminates in use today employ a wide variety of reinforcing materials, as can be seen in Table 1.1. The choice of reinforcement is driven not only by mechanical characteristics such as T_g, but by electrical properties which will be discussed later in this chapter. The most common and widely used reinforcement in polyimide laminates is woven E-glass. This borosilicate-type glass fabric is used in single-sided, double-sided, and multilayer applications. Its availability in different thickness and deniers as well as low cost make it an attractive choice for many applications. However, E-glass is not usually considered as a high performance reinforcement because of its shortcomings in CTE and dielectric constant. What is important to note is that, for high T_g applications which may require other key laminate attributes, polyimide-based laminates offer a wide variety of choices because of their proven track record and polyimide resin's compatibility with many advanced reinforcing materials.

1.2.1.3   Cyanate ester–based laminates.   Laminates based on cyanate ester resin systems are characterized by two properties that are extremely attractive to PWB designers and fabricators as well as system designers: high T_g (about 230°C) and low dielectric constant (3.6). We will confine the discussion in this section to high T_g. This resin system was first introduced as part of a PWB laminate more than 20 years ago and was called *triazine*. It displayed severe moisture absorption problems and a dielectric constant not much better than epoxies. Much work has been done since that early introduction, and a fairly robust product has emerged. Cyanate ester laminates are in use today in a variety of high performance applications. These laminates employ the entire range of reinforcing materials and have emerged as strong candidates for military and commercial applications.

Although cyanate ester PWBs are more in demand for high speed applications because of their attractive loss tangent and low dielectric constant, they have not penetrated the high T_g market very strongly, primarily because of cost (Table 1.2). When the cost issue is coupled with some of the multilayer PWB fabrication issues, such as additional bakes and the difficulty in achieving good hole quality due to brittleness, it becomes difficult to choose a cyanate ester solution over polyimide or BT for applications not requiring rigorous electrical performance.

Although other moderately high T_g materials (>150°C), such as modified epoxies and PPO epoxies, are in wide use today, we have not included them in this section on high T_g laminates. These laminates have other desirable mechanical and/or electrical properties and will be discussed later in this chapter.

1.2.2   Low loss PWB materials

We should define here what is meant by low loss in PWB applications. In describing this electrical attribute of PWB resins, reinforcements, and laminates the terms *dissipation factor, tan δ,* and *loss tangent* are often used interchangeably. They all describe what is essentially the power loss through the laminate material or the ratio of total power loss in a dielectric material to the product of the voltage and current in a capacitor in which the material is a dielectric. Dissipation factor is numerically equivalent to the tangent of the dielectric loss angle.

In the case of loss tangent, lower is better. The lower the laminate loss tangent, the more efficient signal propagation becomes. This is a key requirement in high speed, high frequency applications where signal loss is critical. Length of signal propagation becomes critical to PWB and system performance as frequency increases. Low power, high density digital applications such as digital cellular, personal communication systems (PCS), and global positioning systems (GPS) are all applications where low loss PWBs are needed.

TABLE 1.3   **Laminate Loss Properties**

Laminate	Loss (tan δ at 1 MHz)
E-glass/epoxy	0.035*
E-glass/polyimide	0.035*
Aramid fiber (cross-plied)/cyanate ester	0.025*
E-glass/cyanate ester	0.015*
S-2 glass/cyanate ester	0.015*
Quartz/polyimide	0.010*
E-glass/PTFE	0.001

*From IPC—4101, "Specification for Base Materials for Rigid and Multilayer Printed Boards."

Laminates which display loss tangents of 0.025 or less are of much interest to designers, PWB fabricators, and system integrators who must design and produce systems which operate using faint, low level signals that cannot tolerate loss of fidelity. We should note here that the choice of reinforcing material for these applications is very important. Electrical properties of laminate fabrics play a major role in defining the loss tangent of the composite laminate material. Table 1.3 shows some typical tan δ values of high performance, low loss resin/reinforcement combinations. It is clear from these data that the choice of reinforcement materials is equally as important as the resin system. This is especially true as system frequencies increase.

1.2.2.1   PTFE laminates. Materials constructed with polytetrafluoroethelyne (PTFE) resins, commonly referred to as Teflon, generally display extremely low loss, as well as low dielectric constant. This resin system is composed of a high molecular weight molecule with extremely strong bonds attaching fluorine atoms to a carbon center. This configuration makes for a resin that is nonpolar and hydrophobic. It is also thermally stable over a wide temperature range and chemically inert.

Many laminates employing PTFE resins are available in today's materials marketplace (Table 1.4). These laminates are available in varying thicknesses, with many reinforcing fabrics and materials. As mentioned earlier, the choice of reinforcement is critical in developing a laminate with the desired low loss characteristics. Both woven and nonwoven fabrics have been successfully used with PTFE resins. Although E-glass is widely used, a large microwave market exists for randomly dispersed microfiber glass as the reinforcing/filler material. These laminates have been around since the 1960s and are still in wide use because of their ability to provide extremely low loss and extremely low dielectric constant. Other PTFE-based laminates employ ceramic as a reinforcing/filler material to overcome some of the inherent CTE problems that will be discussed later in the section.

TABLE 1.4   PTFE Laminate Constructions

Laminate	Dielectric constant ε_r
Woven glass/PTFE	2.4–2.6
Random microfiber glass/PTFE	2.2–2.33
Ceramic/PTFE	6.15–10.2
Woven glass/ceramic/PTFE	3.02–10.2
Woven quartz/PTFE	2.1–2.5

Much work is under way to develop a high quality, high performing, low cost quartz/PTFE laminate. This is a very attractive material because of the very low dielectric constant (3.5) and very low loss (0.0002) of quartz, coupled with that of PTFE, would yield a composite with properties ideal for very high (K-band and above) frequency applications. Even though this combination would represent a huge performance advantage, it also presents a unique set of problems to the PWB fabricator. These problems, among others, are described in the following paragraphs.

Although these laminates provide unparalleled electrical performance, they do have drawbacks. One of the key problems with PTFE laminates is their very low modulus. These laminates are not very rigid because of the elevated lamination temperatures required to produce the composite laminate. This shortcoming is key in applications where some structural integrity is required of the PWB. PTFE-based laminates also exhibit relatively high CTE when compared to other laminate materials (Table 1.5). This attribute renders them unacceptable for many surface-mount applications where laminate and package CTEs must come as close to matching as possible. In addition, high z-axis CTE may cause plated-through hole failures when PWBs are subjected to temperature cycling or exposed to environments where temperature excursions are common.

TABLE 1.5   CTE Values of PWB Laminates

Laminate	CTE (ppm/°C)	
	x–y	z
E-glass/epoxy	16–18	50–70
E-glass/polyimide	13–15	45–70
Woven aramid/epoxy	6–8	100–150
Epoxy-PPO/E-glass	12–18	150–170
S-glass/cyanate ester	8–10	40–60
Nonwoven aramid/epoxy	7–9	90–110
Nonwoven aramid/polyimide	7–9	75–95

This problem can be minimized with careful selection of reinforcement materials. While some PTFE/glass laminates display z-axis CTE values approaching 280 ppm/°C, other variations, such as silica-filled PTFE, display much lower values which are similar to those of conventional epoxy and polyimide resin–based laminates (Table 1.5).

Another significant problem with PTFE-based laminates is difficulty producing multilayer structures. The most obvious issue here is achieving adequate lamination between layers. Conventional epoxy and polyimide resins are thermoset polymers that permanently cross-link during heating to form a rigid material at room temperature. This can be easily accomplished using lamination temperatures of 150 to 170°C at lamination pressures of 400 to 450 lb/in². PTFE resins soften at the application of heat and eventually melt so that a resin can be fused to itself and resolidify upon cool down. The problem with this approach is that it requires temperatures in excess of 350°C and laminating pressures that exceed 1500 lb/in². Equipment to accomplish this is not readily available in today's PWB industry. This problem may be overcome by using copolymer PTFE bonding films that laminate at much lower temperatures (200°C or below) and pressures (500 to 600 lb/in²). Much work has been done in multilayer PWB fabrication to overcome the inherent difficulties in lamination. This work must continue in order to exploit the performance advantages of a multilayer PTFE PWB. A variety of bonding materials and laminating approaches, including autoclaves, are under development. Autoclaves are widely used in fabricating complex structural composites but could be used in this multilayer application.

In addition to difficulty in lamination, forming plated through holes is also a challenge with PTFE-based laminates. This is especially true in multilayer PWBs. The nature of the resin is nonstick, making plating adherence very difficult. Drilled holes are typically preconditioned with some kind of sodium etch or other treatment designed to modify the surface molecular structure to promote metal adhesion. These are typically costly and difficult to control processes.

Drill smear is also a problem if drilling conditions are not narrowly optimized. Once smeared PTFE resin is deposited on internal copper traces, it is virtually impossible to remove. Plasmas employing carbon tetrafluoride (CF_4) and oxygen have proven to be one of the few effective desmearing agents for PTFE-based laminates. Other drilling, hole conditioning, and plating problems with PTFE-based laminates are related directly to the choice of reinforcement/filler material. In the case of quartz, it is well known that drilling is compromised because of the severe abrasiveness of the quartz fillers or woven fabrics. When this fact is coupled with the high cost of quartz, the difficulty in choosing this laminate approach for low cost commercial applications becomes obvious.

PTFE-based mixed dielectric PWBs. One of the most novel PWB design approaches involves the use of PTFE-based laminates in a technique called *mixed dielectric*. Its premise is based on the integration of FR-4 layers and glass-filled PTFE layers in the same multilayer structure. This approach allows the designer to place high frequency circuits requiring low loss on the higher cost PTFE layer, while concurrently running less critical digital signals on a low cost FR-4 layer. An extremely high level of integration is possible in this construction. In addition, the cost/performance comparison can be performed with much greater fidelity.

Of course this approach is not without problems. One of the most difficult to overcome is preplate hole conditioning. The dilemma lies in the fact that the very aggressive chemical or plasma processes required to condition PTFE surfaces and remove smear are, in fact, too aggressive for FR-4 laminates. What results from this condition is a hole wall whose topography is very rough, causing difficulty during plating or hole walls with loose fibers which also cause difficulty in the plating process as well as potential reliability problems if these fibers break off. More work is needed to define what combination of chemical and plasma processing is adequate to remove smear and render the hole surface acceptable for adhesion of plating. The sequence of these operations may also be important in achieving this end.

Another difficulty with the mixed dielectric approach is the control of flatness in the final PWB. The inherent differences in elastic properties and thermal expansion rates cause PWB warpage after the lamination process. Constrained flattening at elevated temperatures is a possible fix for this difficult problem. This process must be well controlled with respect to heat-up and cool-down conditions. It is also critical to understand the mechanics of the constraining method. Even though a measurable improvement in flatness may be achieved using this method, it is important to note that "spring back," to some extent, will always occur when the constraining mechanism is released from the board. This method also adds considerable cost to the process and resulting parts.

Of course the preferred remedy to this problem is to eliminate it at the lamination step in order to avoid costly additional steps at the end of the fabrication process, as well as eliminate the need for processing a warped board through the drilling, preplate conditioning, and plating operations. As mentioned earlier, one alternative is autoclave or some form of "isostatic" lamination. This type of lamination, or a variation on it, is commonly used in multilayer low temperature cofired ceramics (LTCC) fabrication where complex three-dimensional substrates are required and traditional uniaxial or flat bed lamination is inadequate. The part is placed in some kind of flexible bag with tooling that is evacuated and placed in a sealed chamber which is subsequently pressurized. Selection of tooling plate material for this process is critical in that some attention must be paid to thermal expansion matching in order to minimize stress buildup due to differing CTEs. This process configuration provides for an environment where the forces exerted on the part are exclusively normal to all surfaces exposed to the pressurized liquid or gas. The resulting parts have been shown to display flatness improvement of 5 to 10× over conventional uniaxial lamination.

Although the unpopulated board may be flat at room temperature, reflow temperatures required during the assembly operation may cause warpage due to residual stresses built up during processing. This is almost always expected because of the inherent differences in materials. The amount of warpage expected may be predicted, to some degree of accuracy, by using three-dimensional finite element analysis. If the predicted level of warpage is confirmed by actual results, a trade-off study may be appropriate to determine if the benefits derived from using this type of construction exceed the cost of fabricating the PWBs. This evaluation should include, but not be limited to, equipment purchases, cost of materials, cost of labor, yields, and reliability.

1.2.2.2 Enhanced epoxy laminates. Recent developments in the modification of conventional epoxies has led to laminates which display the processability of conventional FR-4 materials while achieving very good loss tangent. These laminates are based on what is being labeled "high speed, low loss" resins. Loss measurements have been made at frequencies up to 10 GHz, a significant step up from a typical measurement upper limit of 1 MHz. This is a key requirement for the designers making decisions about the viability of organic substrates for high frequency applications. Although the loss tangent of the new materials is an order of magnitude higher than that of PTFE (0.001 for PTFE versus 0.010), this value far exceeds most of the other rigid organic alternatives.

A series of these low loss laminates is being developed to compete with PPO epoxy–based laminates which will be discussed later in this chapter. Two products from this series still under test will reportedly display loss tangents of 0.003. If this proves to be accurate, this class of laminates would be very attractive—not only because of electrical performance, but also because of processability and lower cost when compared to cyanate ester–based laminates. Although these laminates will not fill the requirements of applications where losses in the range of PTFE materials are required, they could make inroads into commercial markets where cost is on par with performance in the material decision. This is especially true in high volume applications where cyanate ester–based solutions may prove too costly. Applications such as wireless modems, wireless local area network (LAN) equipment, and file servers may benefit from a lower cost alternative with good loss and dielectric characteristics.

1.2.2.3 Other low loss laminates. Other laminates that display low loss properties include cyanate ester–based laminates which have been discussed in the high T_g section of this chapter. These laminates have been very popular because they are able to satisfy multiple mechanical and electrical performance requirements. Cost and difficulty processing these materials are driving designers to seek other alternatives. PPO epoxy–based laminates, which will be discussed in the next section (low dielectric constant materials), are also becoming popular laminates for applications where low loss and moderately high T_g are required. These laminates, much like the enhanced epoxy materials, offer low loss characteristics at a lower cost and are easier to fabricate than cyanate esters and PTFE PWBs.

1.2.3   Low dielectric constant materials

Dielectric constant, permittivity, specific inductive capacity, and D_k are all used to describe the property of an insulating material which defines the ratio of capacitance, using the material as the dielectric, to the capacitance when the material is replaced by air. This material characteristic is key to the design and fabrication of PWBs whose function is driven by increased signal speed because signal propagation speed is inversely proportional to dielectric constant ε_r. As dielectric constant decreases, signals experience less delay. For many applications, such as advanced workstations and other high speed digital processing, low ε_r is a must. Many of the materials discussed in the previous section also display low dielectric constants. Of course the PWB materials possessing the lowest dielectric constants are PTFE based. Since we have described these materials in some detail, as well as the issues surrounding the fabrication of PWBs made from them, we will exclude these laminates from this discussion. Again, it is important to note that the use of PTFE-based laminates presents a unique set of fabrication issues to the multilayer PWB manufacturer, and as such, other material alternatives should be explored before the ultimate laminate/PWB choice is made.

1.2.3.1 PPO (polyphenylene oxide)–based laminates. Polyphenylene-epoxy laminates have gained widespread acceptance in many commercial markets because of their superior electrical properties (when compared to FR-4), low cost (when compared to cyanate ester), and very good manufacturability (when compared to PTFE-based laminates). These laminates display a fairly consistent dielectric constant over a very wide range of frequencies. Laminates have been tested at frequencies ranging from 100 MHz to 20 Ghz with very little change in observed ε_r (3.7 to 3.9).

These laminates are very similar in processability to FR-4 and BT laminates with the important added advantage of lower dielectric constant. In addition, since these laminates are in relatively wide use, much data regarding performance and fabrication has been gathered. Laminates are available in a large variety of thicknesses and woven glass styles: from as thin as 0.003 in using 1080 style woven glass fabric to 0.060 in thick with 7628 fabric. The lowest value of dielectric constant is achieved with thinner materials. This is an attractive characteristic for multilayer PWB fabrication.

PPO laminates are often compared directly to cyanate ester materials because of similar electrical characteristics. However, a more appropriate comparison is the delineation of similarities between these laminates and the enhanced epoxy laminates discussed in Sec. 1.2.2.2. One of the more obvious advantages of PPO laminates over enhanced epoxy materials is availability and the commensurate track record they have established over the last few years. However, in terms of performance, cost, and processability, the differences are few.

1.2.3.2 Cyanate ester/quartz laminates.   Although we have discussed cyanate ester–based laminates in previous sections of this chapter, when combined with quartz reinforcement the resulting laminate also displays an attractive dielectric constant. This is worth noting because cyanate ester–based laminates, although more expensive, are fairly available and much fabrication work has been done. While issues associated with this laminate construction are primarily related to fabrication, problems with resin content and laminate consistency still persist.

The cost of any quartz-based laminate is three to four times greater than other laminates using the same resin system. This problem is exacerbated by the commensurately high cost of cyanate ester resins. This choice of laminate makes sense only when niche applications, such as high performance military systems, require the enhanced loss, dielectric constant, and CTE available in this construction.

1.2.3.3 Other low dielectric constant laminates.   As previously mentioned, PTFE-based laminates provide loss and dielectric constant properties which are difficult to match in other PWB laminate configurations. The very large z-axis CTE is, however, a disadvantage in the fabrication of multilayer boards requiring plated through holes. Laminates with cyanate ester–based resins and quartz reinforcements provide a good alternative to PTFE laminates for certain applications, and should be considered when high performance and high reliability are required. Some laminates employing aramid reinforcements have also been used in applications where a slightly higher (3.7 to 4.1) dielectric constant will meet system needs. A more detailed discussion of these laminates will be made later in this chapter.

1.2.4   Controlled CTE materials

One of the key mechanical properties of high performance laminates is CTE. The coefficient of thermal expansion, usually expressed in parts per million per degree Celsius (ppm/°C), represents the amount of change in dimension arising from a change in temperature. Laminate CTEs are driven by a combination of the resin and reinforcement properties. In-plane or x-y CTE is affected most by the reinforcement material and its construction. Out of plane or z-axis CTE is affected more by resin properties. The choice of resin/reinforcement combinations and ratios is key in defining the ultimate CTE of a laminate.

Today's high density designs with ever increasing numbers of I/O's require PWB's not only with enhanced electrical performance, but with the capability to mechanically support different device packaging materials and techniques. Both x-y and z-axis expansion are key to designed performance specifications as well as reliability in environments where extreme temperature excursions occur. Plated-through hole failures are a typical manifestation of excessive z-axis CTE, while damaged and broken solder joints result from differences in PWB and device package material CTE; both represent catastrophic failure modes.

"Controlled" CTE materials represent an attempt by laminate manufacturers to recognize that surface-mounted devices, chip-on-board packaging, and high density interconnections (HDI), among others, will require laminates with CTEs in the range of 6 to 10 ppm/°C instead of 12 to 18 ppm/°C offered by conventional epoxy and polyimide glass constructions. These laminates offer an alternative to the use of metal-clad constraining cores made from copper-invar-copper (CIC) or copper-molybdenum-copper (CMC). These very heavy, expensive, and difficult to fabricate materials have been used for many years in applications where CTE must be constrained in the x-y plane in order to accommodate surface-mount leadless chip carrier (LCC) devices.

1.2.4.1   Aramid-based laminates.   First produced in the late 1960s, aramid generically refers to organic fibers in the aromatic polyamide

family. Known first to most in the PWB industry as Kevlar, the woven version of this material was touted in the 1980s as a possible replacement for glass reinforcements in military PWB applications because of its low in-plane $(x\text{-}y)$ CTE. This approach was fraught with fabrication issues resulting from the limited fabric constructions available, high moisture absorption of the fabric, and the inherent radial expansion properties of the fibers. These problems proved significant enough to cause aramid laminates to flounder in niche military applications for most of the 1980s.

Most of the early fabrication issues encountered in the production of multilayer PWBs from aramid-based laminates resulted from the fact that only woven fabric with fairly large diameter bundles was available from fabric weavers. Today's laminates are made from fabrics of many different deniers and constructions. Multilayer fabricators have overcome most of the issues related to moisture absorption.

One of the most promising developments related to aramid-based laminates is a nonwoven version of the material that can be combined with different high performance resin systems to create laminates with not only good CTE characteristics but enhanced electrical performance as well. Inherent problems with the use of woven fabrics in applications where controlled CTE laminates are required are related to the dimensional instability of finer-weave fabrics and the need to increase resin percentages to avoid resin-starved areas in the laminate. The use of nonwoven laminates overcomes this problem. The nonwoven version also allows the multilayer fabricator to produce a PWB with a much smoother surface finish needed for fine-line etching and laser ablation for the formation of microvias.

This laminate option is also lower in weight than the constraining core approaches discussed earlier, making it a good choice for avionics and satellite applications. The choice of resin system will depend on the application, but reliability of the plated-through hole and thus resin z-axis expansion should be weighed heavily in applications that require a high level of system reliability.

Bibliography

Buffington, Mike: "Revisiting GETEK," *Printed Circuit Fabrication,* August, 1996.

Farquar, Donald S., Andrew M. Seman, and Mike Poliks: "Manufacturing Experience with High Performance Mixed Dielectric Circuit Boards," *IEEE Transactions,* May, 1999.

Harper, Charles A., and Ronald Sampson: *Electronics Materials and Processes Handbook,* McGraw-Hill, New York, 1994.

Hickman, Fred: "High T_g Resin Systems," *Printed Circuit Fabrication,* August, 1996.

Jawitz, Martin W.: *Printed Circuit Board Materials Handbook,* McGraw-Hill, New York, 1997.

Weinhold, Michael: "PWB Laminates for High Performance Applications," *Electronic Packaging and Production,* August, 1998.

2

Microvias, Built-Up Multilayers, and High Density Circuit Boards

Rolf E. Funer

Funer Associates
Greenville, South Carolina

2.1 Introduction

The forces driving electronic products today are to provide more functions at lower cost and often in as small a package as possible. Laptop computers, cellular phones, and handheld video cameras bear witness to this trend. Microprocessors, the "brains" of electronic devices, increase according to the well-known Moore's law (Fig. 2.1), doubling their functions every 18 months. Advances in processing technology have made it possible for microprocessors to add more functions without increasing in size. Putting more functions in the same space necessarily means that there are more input/output (I/O) pads that are closer together. These microprocessors and their pads need to be connected to supporting circuits. Thus, the circuit boards themselves also need more connecting pads spaced closer together. Table 2.1 predicts a rapid increase in connecting pads due to the emergence of small high I/O devices such as chip scale and ball grid array packages. Over the years circuit boards have accommodated this complexity by gradually reducing the sizes of circuit lines and connecting pads and increasing the number of circuit layers, that is, without fundamental changes in the design or the standard process for manufacturing circuits.

There were two breakthroughs which enabled dramatic increases in circuit density. The first was the invention of the *plated-through hole,* which made it possible to connect together circuits on both sides of a board by means of holes that were drilled and then plated to provide conductivity. This greatly increased design flexibility. The second invention, *multilayering,* dramatically increased density. Separate two-layer circuits could be laminated together and then connected to each other with plated-through holes. Over the years circuit boards have grown from two layers to sometimes over 40 layers. While, theoretically, it should be possible to increase layer count without yield penalty, in practice this is not the case. Just as yields fall dramatically as circuit lines below 3 to 4 mils are attempted, so too do yields suffer when layer count gets too high.

All technologies have limitations, and we are reaching the limit for these two tools, now both over 30 years old. Over time, the plated-through hole has gotten smaller and smaller and has reached the practical limit of about 8 mils in diameter, below which it becomes prohibitively expensive (Fig. 2.2). In addition, the plated-through hole is an inefficient feature because it goes all the way through the board

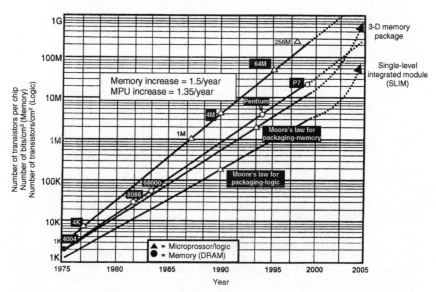

Figure 2.1 Microelectronic density trends. (*Courtesy of R. R. Tummala, Microelectronics Packaging Handbook, 2d ed., Chapman and Hall, New York, 1997.*)

TABLE 2.1   Packaging Trends

	1996	2000	2004
Packaging Trends in Consumer Products			
I/O per chip	100	200	375
I/O density/in^2	100	200	375
Area array pitch (mm)	0.4	0.25	0.20
Board lines/spaces (mil)	5	4	2
Power/chip (W/chip)	1	1	1
Board cost ($/in^2)	1.0	0.6	0.4
Packaging Trend in Automotive Electronics			
I/O per chip	150	200	500
I/O density/in^2	30	100	150
Area array (mm)	0.4	0.25	0.20
Board lines/spaces (mil)	6	3	2
Power/chip (W/chip)	6	12	15
Max. board temp. (°C)	125	155	170
Board cost ($/in.^{2}) (six layers)	1.0	0.70	0.40
Packaging Trend in High-Performance Systems			
I/O per chip	600	1,500	3,000
I/O per MCM	3,000	6,000	10,000
I/O density/in^2	150	300	400
Board lines/spaces	3	2	1
Power/chip (W/chip)	40	60	80
Board cost ($/in^2) (30 layers)	10	5	2.5

SOURCE: Courtesy of National Electrical Manufacturers Institute.

while usually connecting only two layers. This interferes with routing and, as layer counts have increased, the situation has worsened.

Blind and buried vias were developed to partially overcome these problems. A *buried via* is made by drilling and plating a two-sided circuit board and then laminating it into a multilayer (Fig. 2.3). This process helps routing because only the two circuit layers that need to be connected have a plated hole; the rest of the circuit layers are undisturbed. In a similar manner, the *blind via* is made by drilling only part way through the multilayer. Thus, it connects the outer layer circuit to an interior layer without disturbing the layers below the interior layer.

While helpful, blind and buried via boards are expensive because they require several more and difficult process steps and they still have the limitation of a mechanically drilled hole. While significant quantities of blind and buried via boards are used, they are interim solutions. They do not fully meet the need for high I/O density surface mounting or facilitate optimum layer-to-layer wiring.

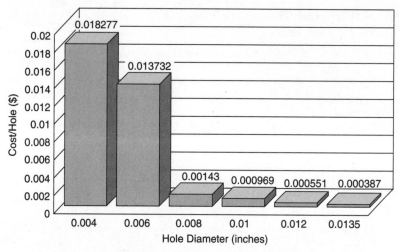

Figure 2.2 Mechanical drilled-hole cost.

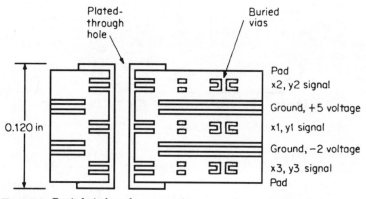

Figure 2.3 Buried via board cross section.

2.2   The Microvia Concept

The *microvia* concept involves making a very small hole, or via, generally by a nonmechanical means, to connect just two layers of circuitry. In comparison to mechanical drilling which, as seen previously, is impractical below 8 mils, microvias can be as small as 2 mils or even smaller. Thus, the density of pads can increase fourfold or more (Fig. 2.4). Since microvias are small, they can also be placed directly in pads, creating even more space savings (Fig. 2.5).

It is possible to build circuits with several layers of microvias (Fig. 2.6). Each microvia connects just the two layers as in an integrated circuit process. Circuit routing can be extremely efficient with optimally short signal lengths. This process consists of building up the circuit one layer at a time; hence the term *built-up multilayer.*

Microvias are electrically superior structures to plated-through holes. Modeling work predicts that the microvia will provide less distortion than a plated-through hole. In actual tests, microvias exhibit no signal distortion for signal pulses with rise times of 100 ps, which is much better than plated-through holes. The level of noise generated by microvias was also less. Cross-talk was <0.1% on equal signal layers and <0.3% on different sequential layers. This is much lower than typical cross-talk from plated-through holes, which is about 2%.[1]

Thus, microvias are becoming the third breakthrough to enable further dramatic increase in circuit density. In addition, they allow for improvement in actual circuit performance, particularly at high speeds and frequencies. Thus, they are essential for built-up multilayers

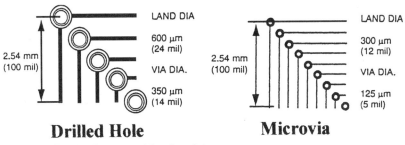

Figure 2.4 Density increase with microvias.

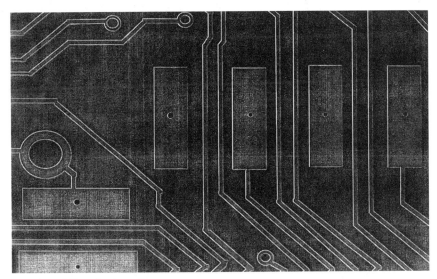

Figure 2.5 Microvias in pad.

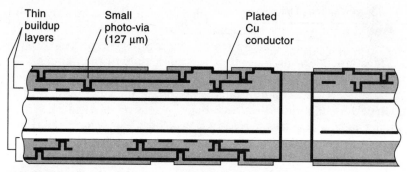

Figure 2.6 Microvia board: two microvia layers on each side.

which will find application in many future products. *Built-up multilayers* can be defined as high density printed circuit boards that employ microvias, generally 6 mils in diameter or less, and that require nontraditional printed circuit fabrication techniques.

This chapter will provide details on how microvia built-up boards are made, the various microvia processes and how they compare, equipment requirements, the imaging technology for high density boards, design considerations, and reliability. Finally, an extensive section on current applications of microvia-based boards illustrates the breadth of the technology.

2.3 General Microvia Methods

There are three approaches currently in use to create microvias: photoimaging, laser ablation, and plasma etching. While all can produce a functional and reliable product, each process has somewhat different capabilities, equipment and process requirements, and costs. It is important to understand these parameters in order to select the best process for a specific application. A summary of the capabilities is shown in Table 2.2. As will be seen later, the processing requirements, equipment, cleanliness, etc., are quite different for each, and this must be understood by the prospective manufacturer when making the selection. This section will cover the major processes and their relative strengths and limitations.

2.3.1 Photoimaging

In this process, a photosensitive dielectric layer, usually an epoxy or acrylic epoxy–based resin, is applied to a substrate and then a microvia pattern is imaged with light energy and unexposed material is developed or removed to form the microvias.[2] Typically, the microvias are located directly above the contact pads. Then the

TABLE 2.2   Comparison of Microvia Processes

Process	Advantages	Disadvantages
Plasma formation of blind vias	Very small features possible High quantity of holes per batch Highly reliable and clean vias Use conventional lamination press Compatible with most resin systems	"Isotropic" profile High material cost (to date) Thin dielectrics only High investment cost No glass reinforcement Needs special coated foils Nonvalue-added steps: (Require photoprocess to open Cu windows, etch back of overhangs) Via size limits to 75 μm Lamination press time Low productivity
Photoimageable dielectric formation of blind vias	Very small features possible High quantity of holes per batch Highly reliable and clean vias Established equipments: (coat/lamination, print/develop) Compatible with all circuitization techniques High productivity 1-mil via capability	Low copper adhesion Nonlevel top surface High investment cost (to date) Thin dielectrics only No glass reinforcement Limited approved materials Clean room handling (potential yield loss from printing defects)
Laser formation of blind vias	Very small features possible Via depth up to 0.4 mm (0.016 in) Vias possible in reinforced material (thermount and glass-reinforced FR-4) Lowest equipment investment cost Materials: Bare resin: liquid or film Coated foils Glass laminate (Nd-YAG only) Faster, cheaper, finer than mechanical drilling Compatible with many resin systems *Good* alternative for few vias and small substrates Potential for finer vias (<1 mil w/YAG or excimer)	Via quality not yet "100%" with FR-4 glass Initial capital investment Low productivity for large area or high via density Bare-resin: Limited material choice, need process development Inconsistent via sidewalls with glass reinforcement Lamination press time

microvias are plated to electrically connect the exposed pad to the next layer (Fig. 2.7). Circuitry can then be formed on the next layer. Additional layers can be built by repeating this sequence. There are both positive and negative working photodielectrics (the previous example illustrates a negative dielectric) and they are available in both films, which are laminated, and liquid coatings, which are sprayed, curtain or flood screen coated, or dipped. The photoimage processes have several advantages: they form vias *en masse* and can produce very small vias. Thus, photoimaging is often a good choice for products that require many small vias such as chip carriers for high I/O count devices or for direct flip-chip attachment to a circuit board.

Photoimaging does require a high level of clean room control since dust or other particulate contamination in the imaging steps can cause improper via development. The initial capital cost for a photoimaging microvia line can be high to provide adequate clean room control. In addition, sophisticated coating equipment is needed if the liquid coating route is selected.

There are currently a limited number of photoimageable dielectrics available. Dielectric thickness is also limited (to typically 2.5 mils or less). This may preclude using this process for certain applications, for example, those that require high temperature resistance or high speed (that might require a low loss dielectric). This is changing as suppliers continue to develop new and better materials. Low copper adhesion has been a problem with many of the dielectrics.

2.3.2 Laser imaging

Here a laser beam is used to remove dielectric material that is coated or laminated over a substrate to form a microvia.[3,4] The via is then plated, as in the previous process, to connect the exposed pad to a circuit layer which is etched on the top surface of the dielectric. Laser

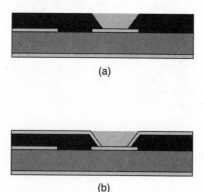

(a)

(b)

Figure 2.7 (*a*) Microvia formed by imaging and development and (*b*) microvia plated to connect pad.

processing is now the most often used method to form microvias. This is the easiest process to put into production; it is relatively simple, and laser equipment has improved greatly in speed and reliability.

Several types of lasers are used. Currently, the most common is a CO_2 laser which has the advantage of being the fastest system. While the CO_2 laser can form vias in organic and glass-reinforced dielectrics, it cannot penetrate copper foil. The UV-YAG laser, while slower, can penetrate copper, so it is the laser of choice when using resin-coated copper as a dielectric layer. Excimer lasers offer the highest via resolution but have the slowest ablation rate. These laser systems are discussed in more detail in a later section. Laser processing can be easier than photoprocessing. Laser processes have less stringent cleanliness requirements and no photomask is required. However, unlike photoimaging, laser via formation is not a mass formation process. Vias are made one at a time, so this can be a much slower and more expensive process. Thus, laser via formation finds its best applications where relatively few microvias are needed. For example, a large multilayer board that has several high I/O ball grid array (BGA) packages requiring microvias may well be an excellent candidate for laser via formation. As a rule-of-thumb, if the microvia content of the circuit board is relatively small and conventional circuit features dominate, then laser processing should be considered first.

Laser processing equipment is expensive, but significant advances in speed and reliability of lasers are resulting in overall lower laser-based microvia costs. For a small-scale process, a laser-based microvia process is usually less expensive to set up than a photovia-based process. This is due to the initial cost of a high quality clean room required for the photoprocess. For large-scale manufacturing, the costs tend to cross over. There is a much wider choice of dielectrics available for laser processing since it is possible to form a via through virtually all types of materials. Typical dielectrics include epoxy coatings and films, with copper foil and dielectrics with glass, aramid, or Teflon reinforcements. High temperature–resistant dielectrics can also be lased. Thus, using laser processing, it is possible to make microvia boards with a broad range of thermal and electrical capability. Very small vias are possible with laser processing and they can be processed to a deeper level than is possible with photoprocessing.

2.3.3 Plasma

This process utilizes a gas plasma, usually a combination of oxygen and carbon tetrafluoride to attack the polymeric dielectric and remove it to form the microvia. The system uses two different frequencies to ignite the plasma—a microwave and a kilohertz frequency. The microwave creates the high etch rate and the kilohertz frequency improves the process uniformity and enhances the etching speed in the z direction.[5] This process, while a mass formation process like photoimaging, has several limitations. Plasma cannot form vias as small as the two previously described processes. This is because it etches isotropically, which tends to underetch the metal beneath the copper surface. This copper overhang may cause problems when the metallization stage is reached. While some printed circuit shops are familiar with plasma because it is used to clean mechanically drilled holes, it is not a generally used process and the equipment is expensive. Plasma is materials-limited in that vias cannot be formed in glass-reinforced dielectrics. For these reasons plasma processing is losing ground to laser and photoprocessing.

2.4   Specific Microvia Processes

2.4.1   Liquid dielectric photoprocess

The original photoimageable liquid dielectric process, surface laminar circuitry (SLC), was developed by IBM at its laboratories in Yasu, Japan,[6] and was the first microvia process to go into mass production. The process flow is shown in Fig. 2.8. A photosensitive epoxy, developed by Ciba Corporation, is the key new material; all other construction materials are standard printed circuit materials. This liquid dielectric is applied by curtain coating to an etched core which is usually made up of several layers of circuitry processed conventionally. After the dielectric is dried, it is exposed using a photomask and developed, removing unexposed dielectric to form the microvias. Glass phototools are used because of their stability and fine feature capability compared to plastic phototools. In the original process, the dielectric had to be planarized after bake to produce a uniform surface. With development of some of the more recent dielectric materials, this step can be eliminated.

After the vias are formed, through holes, if needed, are mechanically drilled and the surface and vias are plated with a flash of copper. Finally, the outer layer pattern is made by applying and developing a resist, then plating up to full copper thickness in typical semi-additive processing. The resist is removed and final finishing (solder mask, nickel/gold plating, etc.) completes the SLC circuit. Additional microvia layers are built by repeating this sequence. Figure 2.9 illustrates a cross section of the completed vias. This process requires stringent cleanliness. The dielectric coating, drying, and imaging are the critical operations. Dust or other contamination on the substrate or photomask can result in a pinhole or undeveloped via.

- Subcomposite Without
  Plated Through Holes

- Apply Photo Dielectric
  (Liquid or Film)
  and Develop Photovia

- Drill Through Holes

- Plate and Etch Circuitry
  or
  Pattern Plate Circuitry

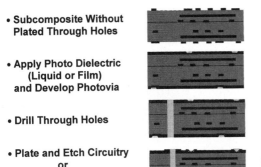

Figure 2.8 SLC process flow.

SLC is in commercial production at IBM in Yasu, Japan, and several sites in the United States. The major applications are portable electronics, such as notebook computers, cell phones, and camcorders, and several other applications that use SLC as a multichip module and flip-chip substrate. SLC with up to three layers of microvias on each side are in production as are other similar photovia processes.[7,8] Built-up boards with the highest density have been made with the liquid photoimageable dielectric processes.

2.4.2 Dry-film photoprocess

This process, pioneered by IBM, at its Endicott, N.Y., laboratories, and DuPont, overcomes some of the process difficulties of the original SLC process.[9] Processing is very similar to that for dry-film resist. The process flow is the same as in Fig. 2.8, except that a film dielectric is laminated, rather than liquid coated, to the substrate. Subsequent steps of imaging, developing, plating, etc., are the same. Usually no planarization is required.

This process is somewhat easier to implement than the liquid dielectric photoprocess. Since the film is preformed, it can be inspected

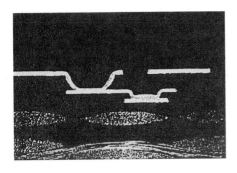

Figure 2.9 Cross section of a two-layer photovia board.

before lamination to ensure there are no pinholes. When laminated, it will "tent" over any surface contamination. Thus, clean room requirements can be less stringent than for the liquid SLC process. However, mask contamination is still an issue, so a good level of control is still needed.

The dry-film process is in production at IBM in the United States and in early production at several other manufacturers in the United States and Japan. Applications are BGA substrates and portable electronics similar to the liquid dielectric process. This process is better for large substrates since it is easier to apply a film, rather than a liquid, uniformly to a large substrate. Several applications, such as network multilayer boards, are under evaluation. However, only single-layer microvias structures are in production.

2.4.3 Resin-coated foil laser process

This process was developed through the collaboration of several electronic materials suppliers, laser equipment manufacturers, and printed circuit firms.[10] It begins with the lamination of a copper foil sheet which has an epoxy film coating. This resin-coated copper foil (RCC) generally has two distinct epoxy layers, a C-staged (or fully cured) layer to provide a controlled dielectric thickness and a B-staged layer to provide fill and adhesion. The process flow is shown in Fig. 2.10. The foil is laminated to the etched substrate in a conventional laminating press. Microvias are formed by several variations. In one approach, small "windows" are etched in the copper and vias are lased using a CO_2-TEA laser. Once the vias are in place, processing is quite similar to the photoprocesses. Vias are plated to join the inner pad and outer conductor, a resist is applied, and the outerlayer circuit is etched.

If a UV-YAG laser is employed, the copper etching to form the windows can be eliminated and holes can be punched through the copper and dielectric at the same time. This is a slower process but does eliminate the need to make a mask and the window-making steps. This option is finding application for prototypes and small-volume production.

In another variation,[11] copper foil is etched down to about 0.2 mil and the surface is oxidized. Then the CO_2 laser can be used to drill through both this tin copper and the resin underneath the copper. This process combines the speed of the CO_2 laser, eliminates the window etching on copper foil, and minimizes registration problems by compensating hole locations in digital form.

The RCC process is gaining wide acceptance because it is a "drop-in" process for most printed circuit firms, requiring no special equipment other than a laser. Clean room control is not as stringent as it is with

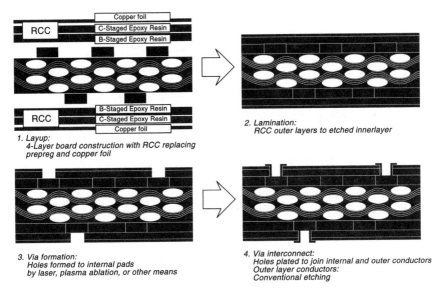

1. Layup:
4-Layer board construction with RCC replacing prepreg and copper foil

2. Lamination:
RCC outer layers to etched innerlayer

3. Via formation:
Holes formed to internal pads by laser, plasma ablation, or other means

4. Via interconnect:
Holes plated to join internal and outer conductors
Outer layer conductors:
Conventional etching

Figure 2.10   Resin-coated copper (RCC) foil process.

the photoprocesses since surface contamination does not interfere with laser drilling. In addition, surface copper adhesion, which is sometimes difficult to achieve, is excellent since this is predetermined at the RCC manufacturing stage.

2.4.4   Any layer inner via hole process (ALIVH)

Developed by the Matsushita Corporation in Japan, the ALIVH process is a laser process that starts with an aramid/epoxy prepreg covered with a removable film. A laser is used to drill via holes through the film and substrate. The film is removed and the holes filled with a copper-based conductive paste. Copper foil is then laminated onto the prepreg and etch patterned to make two-sided circuits. These are inspected and then laminated together to make a multilayer circuit with microvias at each level. Outerlayer circuitization is done conventionally. The process flow is shown in Fig. 2.11.

ALIVH differs from all the previously mentioned processes in that it is a parallel process rather than a sequential process. In a sequential process the circuit layers are built up one at a time, while in a parallel process, individual layers are separately made, and then all are laminated at once. This can have yield advantages in high layer count boards, as shown in a later section. The aramid prepreg is used for ease and speed of laser processing. It also lends dimensional stability

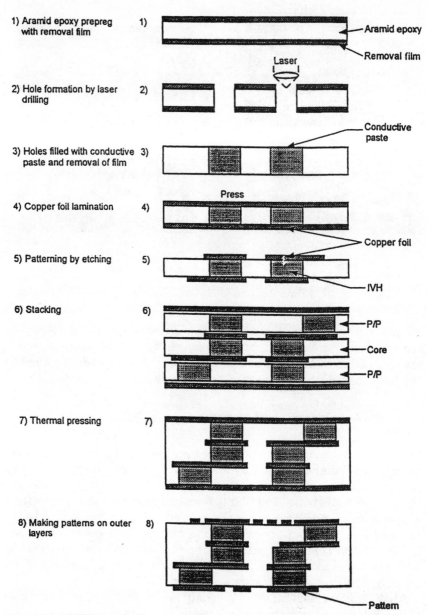

1) Aramid epoxy prepreg with removal film

2) Hole formation by laser drilling

3) Holes filled with conductive paste and removal of film

4) Copper foil lamination

5) Patterning by etching

6) Stacking

7) Thermal pressing

8) Making patterns on outer layers

Figure 2.11 ALIVH process flow.

and good electrical characteristics to the circuit. However, aramid is somewhat expensive (compared to glass reinforcement) and may not be suitable for certain applications because of aramid's tendency to absorb moisture.

While a relatively recent entry, ALIVH has grown rapidly and has been licensed to at least one other firm. The largest application is for cellular phones (picture), followed by subnotebook computers and video cameras. Newer applications are for packaging substrates for semiconductor devices.

2.4.5 Other laser processes

There are a number of variations of laser microvia processes.[12,13] The laser can be used to form vias in film or coated dielectrics through a reinforced dielectric such as a glass prepreg or a Teflon-supported resin (Gore-Ply-trademark Gore Company). The choice will depend on the product requirements. For example, a Teflon-supported dielectric with its excellent electrical characteristics may be very suitable for a high speed circuit. Films, particularly polyimides which can be made very thin and have excellent electrical and thermal characteristics, are used in direct chip attachment applications where very small vias and a thin substrate are desired.

Hitavia is a process developed by Hitachi that uses both mechanically drilled as well as laser vias. It starts with a nonflow epoxy coating on copper foil (resin-coated copper or RCC). Several sheets are drilled mechanically simultaneously with a drill diameter down to 200 μm (8 mils). Then this predrilled RCC is press-laminated onto core material. A small amount of resin that flows into the hole area during lamination is removed by a desmearing process. Holes of 150 μm (6 mil) or smaller are formed by CO_2 laser drilling after the lamination of RCC.

2.4.6 Plasma process

The Dycostrate process, developed by Dyconex in Switzerland, employs a mixture of gases, excited into a plasma state, to etch through dielectrics and form microvias.[14] In the original approach, polyimide-coated copper foil was used as the substrate, holes were etched through the copper, and vias were plasma formed and plated. Additional microvia layers can be made by laminating layers of film and forming vias by the same process. This results in a thin, flexible microvia circuit.

Another variation starts with an epoxy-coated foil which is laminated to a substrate; windows are etched in the copper, and plasma is employed to etch away the epoxy dielectric.[15] Subsequent steps are similar to these. This process is conceptually similar to the RCC laser process, substituting plasma for the laser to make the vias. Plasma processes were among the first microvia processes but have lost ground to laser and photovia processes. Plasma etching equipment is expensive and not conducive to high speed production. In addition, plasma etching is isotropic, resulting in large bowl-shaped vias. Figure 2.12 compares laser and photovias to plasma counterparts.

Applications for plasma-based products have been mainly in Europe and consist of military circuits (where the very light weight and thinness of film-based microvia product is beneficial) and chip carriers.

2.4.7 Buried bump interconnection technology (B²IT)

The B²IT process, developed by Toshiba Corporation, is unique in that it does not use photo, laser, or plasma to form microvias. In the process,[16] conductive bumps are formed on copper foil by screen printing silver paste using a metal mask. In a second step, prepregs, acting as an insulation layer, are pierced by these bumps during lamination under heat and pressure. This results in a double-sided copper-clad sheet with internal silver-filled vias. The copper is etched to form a double-sided circuit. This is inspected and is further processed into a multilayer as shown in Fig. 2.13.

B²IT, like ALIVH, is a parallel process. The formation of the silver bumps is critical; the height and shape of these bumps must be carefully controlled in order to properly pierce the dielectric and connect the conductors. Toshiba reports B²IT can be used to make a broad range of microvia circuits. Many laminate materials, including epoxies, bismaleimides, triazines, and phenolics, can be used and the silver bumps can be used at all levels. The use of silver as the interconnection may retard interest due to concern over silver migration.

Photovia   **Plasmavia**   **Laservia**

Figure 2.12   Microvia shapes.

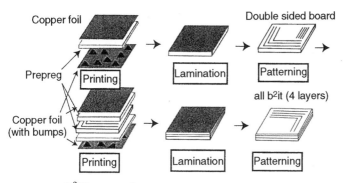

Figure 2.13 B^{2}IT process flow.

However, B^{2}IT has found commercial application for multichip modules including flip-chip assemblies, single-chip carriers, and small portable electronic devices such as pagers and subnotebook computers.

2.4.8 Sand blasting

Another method of forming microvias uses sand blasting. Pioneered by Tokyo Ohka, this system is in development. It requires special film which is several times more expensive than typical patterning films. It does not appear to have any special attributes compared to more common processes and, therefore, its future is unclear.

2.5 Comparative Analysis of the Microvia Processes

In Table 2.2 the key strengths and limitations of the major via formation processes are given. It is valuable to compare directly the two major processes: photovia and laser. The most critical considerations with the photoprocesses are the dielectric itself and the coating method. The cost of dielectric material is high, primarily because the technology is complicated. It is difficult to make a dielectric photoreactive and, at the same time, retain all required thermal, electrical, and mechanical properties, so sometimes these materials are less capable than conventional dielectrics. While there should be some cost reduction as volume production increases, these materials will always be expensive. Several coating methods are in use for the liquid dielectrics: curtain coating, screen coating, extrusion coating, spray coating, and doctor blading. Curtain and screen coating are the two most popular methods. Thickness control and elimination of pinholes are the key difficulties in liquid coating processing. Any variations in thickness or cure can result in difficulty in developing the microvias, particularly if the aspect ratio (the ratio of the diameter of the via to the depth) is greater than 0.5. Clean room control must be excellent to prevent pinholes during the liquid coating processes. These difficulties are lessened by using a film dielectric whose thickness can be precisely controlled and can be manufactured in a very clean environment to reduce contamination. However, one can get pinholes from dust or other contamination on the photomask or the part itself.

These problems are largely overcome with the laser processes. Virtually all organic dielectric materials can be laser processed as is. This includes the materials commonly used for printed circuits: epoxies, polyimides, cyanate esters, even reinforced dielectrics such as FR-4. Thickness control is not as much a problem as the laser can be tuned to "burn" through the dielectric until it hits the pad. Cost of materials currently is high but this is probably due to their newness (and low volume). Cleanliness is less of an issue because any surface contamination is generally burned through. Also, the photomask, a source of contamination, is eliminated. The process issues are relatively minor in comparison to the photoprocesses. When the dielectric or RCC is laminated, it is important that it conforms well over the underlying circuitry so that it is planar. Particles of abraded dielectric must be removed from the vias.

A fundamental difference between the two processes is that the photoprocess is a mass process, forming all the vias at once while the laser process forms vias one at a time. However, with the advent of fast lasers, this is becoming less of an advantage. One laser manufacturer[13] has improved the laser drilling speed of its equipment from 4000 vias/min in 1996 to 20,000 vias/min in 1997 and 34,000 vias/min in 1998. Thus, a 50,000-via panel can be drilled in less than 2 min, which is not that much slower than photoimaging. In summary, the more robust process window for the laser process seems to more than compensate for the one-at-a-time via formation disadvantage. Nonetheless both will continue to be used and each will have value for specific products, which are described in a later section.

The plasma process has not developed into a major process like photo and laser. Plasma produces cup-shaped vias which tend to be somewhat variable. These vias cannot be brought down to as small a size as with the other processes. The plasma equipment is expensive and slow, and, unlike laser processing which advances by leaps and bounds, it does not seem that any significant improvement is likely. This results in a limited and expensive process with no discernible benefits over the other two.

The via size reported in Table 2.3 is the minimum that has been demonstrated in any volume, but it is not necessarily current production capable, which is shown in Table 2.4. All have the capability of fine features (circuit lines and spaces) to complement the microvias.

One of the major differences between processes is the ability to make multiple microvia layers. The metal post (ALIVH) process has demonstrated up to eight layers, a consequence of the benefit of parallel processing where individual layers can be inspected prior to multilayer lamination (see Sec. 2.6). The liquid photoimageable process (SLC) has demonstrated six layers (three on each side) with thin (40-μm/1.6-mil) dielectric which is an excellent production demonstration of high circuit density. Generally, one-layer product has been made with the RCC process. This is probably more a result of newness of the RCC process than an inherent limitation.

TABLE 2.3   Microvia Processes Strengths and Weaknesses

	Photo		Laser		Plasma
	Liquid	Film	RCC	Metal post	Plasma
Multiple-layer demo	Y	N	N	Y	Y
Dielectric thickness control	Difficult	Good	Good	Good	Good
Dielectric cost	Medium to high	High	High	Medium to high	High
Cleanliness required	High	Medium	Medium	Medium	Medium
Via size demo (mμ)	50	100	75	100	75–100
Process issues	Pinholes, thickness control, adhesion	Conformability adhesion	Conformability hole cleaning	Via filling	Via shape, nonuniform via, slow process

TABLE 2.4   Microvia Process Production Capability

	Photovia		Laser			
	Liquid	Film	RCC	Metal post	Plasma	B^{2}IT
Line width	75	75	100	100	75	75
Line spacing	75	75	100	100	85	75
Via diameter	125	125	100–125	175	100	200
Via pad	250	250	250	350	300–350	250
Maximum microvia layers	6 (3/side)	2 (1/side)	2 (1/side)	6–8	4	4
Dielectric thickness	40–80	60–80	60	100–250	50	—
Dielectric constant	3.8	3.8	3.8	3.5	3.5	—
Dielectric T_g (°C)	130	170	170	170	170–>200	—

2.6 Sequential Versus Parallel Processing

The common way printed circuit boards are built is called *parallel processing*. Two-sided circuits are individually made, inspected, and then are all laminated together to form a multilayer printed circuit board. In theory, a 16- or more layer board should have a similar yield as a four-layer board, as long as all the individual circuits are carefully inspected to ensure that any imperfect ones are discarded and that the registration control during lamination is excellent. In contrast, microvia boards are generally made by a sequential process where each individual layer is built up one on top of another. Thus, yield is directly dependent on the number of layers, and the investment in a microvia board becomes quite large if the board has many layers.

The comparison is shown in Fig. 2.14. There are fewer total steps in a sequential process compared to a parallel process if the design has just one microvia layer. But if the design has two or more layers, the number of steps will be greater for a sequential process. This has strong yield implications since yield is usually directly related to the number of layers. Note that the ALIVH process is a parallel process in contrast to the more widely used photovia and RCC laser processes (Fig. 2.15).

A typical microvia board might have two microvia layers on each side and a four-layer inner core. If a microvia layer yield of 92% and a core yield of 95% is assumed, then the yield is calculated as follows:

$$\text{Sequential yield} = 0.95 \times (0.92)\,4 = 0.68\ (68\%\ \text{final yield})$$

$$\text{Parallel yield} = 0.95 \times (0.92)\,2 = 0.86\ (86\%\ \text{final yield})$$

	Sequential	Parallel
Process Strategy	Layers built up one at a time.	Individual 2-sided circuits made, then laminated together
Yield	Directly related to layer count.	Relatively independent of layer count.
Number of Process Steps		
➤ Single Layers	Fewer	More
➤ Multiple Layers	More	Fewer
Processes	All current PHOTOVIA RCC and Liquid Film, Laser and Plasma Processes.	ALIVH B²IT

Figure 2.14 Sequential versus parallel processing.

NOTE: Common Steps (Soldermask, Finish Plate) are omitted.

PHOTOVIA	RCC LASER	ALIVH
1. Coat Dielectric	1. Laminate Foil	1. Drill Holes
2. Expose and Develop	2. Etch Vias	2. Fill Holes
3. Adhesion Promotion	3. Lase	3. Laminate Copper
4. Plate	4. Plate	4. Etch
5. Etch	5. Etch	5. Laminate Foil
		6. Outerlayer Etch

For single layers, Photo and RCC Laser processes have fewer steps.

For 2 or more layers, ALIVH has fewer steps.

Figure 2.15 Sequential versus parallel process steps comparison.

There are other yield factors, such as the individual difficulty of the steps, so this is only a guide. In practice, sequential processing is probably acceptable for up to two-layer designs. However, for three layers and beyond, a parallel process should result in substantially better yields.

2.7 Equipment Requirements

This section will cover just those requirements that are specific to the microvia process itself and not equipment that is required for standard processing. Manufacturing equipment will clearly depend on the chosen process and the size of the operation. Table 2.5 provides a comparative view of the equipment needed to set up an initial capability for the key processes.

2.7.1 Photovia equipment

A benefit of the photovia processes is that the equipment is quite similar to other generally available printed circuit equipment. The key elements are the coating equipment, the imaging and development equipment, and the clean room.

2.7.1.1 Coating equipment. The most common method to apply the liquid dielectrics is by roller coating. Equipment has been developed or modified specifically for dielectric coating. Two prominent manufacturers are Buerkle Pressen and Nubal Electronics. Roller coating can attain speeds of 3 to 10 ft/min and achieve an accuracy of film coating

TABLE 2.5 Process Comparison

| | Photo | | Laser | | |
	Liquid	Film	CO_2	UV-YAG	Plasma
Equipment	Coater Printer Developer	Laminator Printer Developer	Laser	Laser	Plasma
	High level	Medium level	Medium level	Medium level	Medium to high
	Clean room	Clean room	Clean room	Clean room	Clean room
Process Challenges	Cleanliness Registration Coating Uniformity	Cleanliness Registration	Registration Productivity		Registration Productivity Via uniformity

between 0.2 to 0.6 mil from side to side. Both sides can be coated simultaneously. One disadvantage is that different panel sizes require a changeover or adjustments in the coating equipment. A second option is curtain coating. The equipment used is standard equipment used for solder mask application but modified to ensure minimal contamination. Unlike roller coating, curtain coating can accommodate different size panels without a changeover. However, it is often the most expensive method and only one side can be coated at a time. Spray coating is another option. Again, modified solder mask equipment is used. This method seems to have somewhat better results in pinhole minimization. Other methods include screening, dipping, and doctor blading.

Film photodielectric processing usually requires either a standard vacuum laminator or a hot-roll laminator, both of which are often available in printed circuit shops. This is generally faster and simpler compared to the liquid coating methods.

2.7.1.2 Imaging and development. A high powered light source is employed for imaging because the photodielectrics are usually not nearly as sensitive and require more light energy compared to typical resists. It is not necessary to employ highly collimated sources for via formation. Photomasks are a major issue with photoprocessing. Often the microvia circuits are highly precise with fine features, so care must be taken to ensure mask stability. For this reason, glass masks are sometimes employed. Also, stringent control of the surface of the mask is necessary (no contamination or scratches) to minimize any unwanted vias. The development of the microvia must be done carefully to ensure complete development of all vias but not overdevelopment.

Equipment is often modified to ensure greater precision in operating parameters and tighter filtration to remove particulates.

2.7.1.3 Clean room requirements. The highest investment in photovia processing is usually the clean room and associated modifications to minimize contamination and prevent defects due to dust and dirt on the mask and dielectric. As a minimum, a class 10,000 room is required for coating and imaging photomicrovias. Usually, modifications are made to the coating equipment to attain a higher level of cleanliness (down to 1000 to 100) in this critical part of the process. The film photovia process has less stringent clean room requirements, but still needs careful mask control for acceptable yields.

2.7.2 Laser equipment

There are three types of laser equipment in use today for microvia production: CO_2 lasers, which are the most common; UV-YAG lasers; and excimer lasers. All have a place, depending on the application. The UV-YAG laser has the fewest new process steps, so it may be the best choice for a prototyping or small-volume operation. The major cost to initially implement a laser microvia process, based on a UV-YAG laser, is for the laser itself. A CO_2 laser may be best for volume production because of the high power capability, but equipment to image and etch via holes in the copper foil may be required in addition to the laser. The cost to implement is basically the laser itself. Clean room requirements for laser processing, while less severe than for photoprocessing, are still required because the microvia structure will have fine circuit features.

Key laser considerations include

- *Depth of focus.* In order to maintain a consistent penetration depth of the process, the laser beam should remain the same size over the process area. Any change in spot size would lead to a change in the power density (or fluence) which could affect material removal rates.

- *Focused spot size and working distance.* In any machining operation, it is desirable to locate the focusing lens as far away as possible from the machining area.

2.7.2.1 CO_2 lasers. CO_2 lasers are the most common industrial laser. They are available in high power, up to 500 W, and emit light in the infrared where absorption of radiation of dielectrics is >90% and much deeper than ultraviolet (UV) sources. This leads to etching rates of tens of millions of micrometers of material removed per pulse.[13,17] Typical dielectric thickness from 3 to 4 mils can be completely etched in one to three pulses. However, there are drawbacks. CO_2 lasers do not produce a high enough peak power to efficiently machine highly reflective metals, so it is necessary to first etch holes in the copper foil. It is also more difficult to etch dielectrics with reinforcements, particularly glass cloth, and achieve a smooth hole wall surface. CO_2 lasers also remove material thermally so melting or carbonization can occur. Thus, a postcleaning step is often required. However, because of the speed, cost, and reliability, CO_2 lasers are commonly used. Speeds of laser formation in excess of 500/s are reported[18] and this speed is expected to increase.

Recently, new CO_2 lasers have been introduced which combine high power with pulse width and a pulse repetition rate that can be varied over a wide range so that the average power output is highly variable and has the capability of being distributed in different ways to provide flexibility in setting parameters for the manufacturing process.

2.7.2.2 UV-YAG lasers. These lasers are often used with frequency tripling to deliver spectral outputs in the ultraviolet region. The radiation is highly reactive to typical microvia dielectrics as well as the common materials used in printed circuit boards (copper, organic resin, and glass). YAG lasers operate at high pulse rates and can drill a hole in a fraction of the time an excimer laser would take, but slower than a CO_2 laser because it has a smaller spot size (on the order of 1 to 2 mils). However, it can produce energy densities high enough to drill through copper, eliminating the need for a mask and etching of the copper. Running at a pulse rate of several kilohertz, YAG lasers can drill 20 to 30 holes/s in 0.7-mil copper with a 2-mil thick dielectric.[12]

There is speculation as to whether the ablation mechanism is photochemical or photothermal but the results are that, unlike with the CO_2 laser, no thermal damage (melting or carbonization) takes place. Figure 2.16 shows a side wall of a 6-mil-dia via formed in 10-mil thick FR-4. The ablation shows cleanly severed glass bundles. This is very important in order to get a good quality surface that can be subsequently plated. Stepped-down vias can also be formed by drilling through copper on the first level to access the second level (Fig. 2.17).

2.7.2.3 Excimer lasers. Emitting radiation from 0.15 to 0.25 μm, excimer lasers offer the greatest resolution compared to other sources. They are excellent etching sources due to the extremely high absorption rate with common dielectrics like epoxies, polyimides, and aramids. However, because the coupling is limited to 1 μm of material, etching rates are very slow. The cost is also high and corrosive gases are used. Thus, they are limited to applications where only a thin

Figure 2.16   Sidewall of UV-YAG laser-drilled hole.

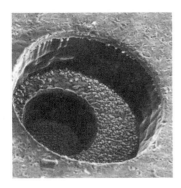

Figure 2.17 Two-level stepped-down via.

dielectric is removed and microvias are very small (<2 mils in diameter) and densely packed.

2.7.2.4 Laser mask technologies. There are a number of different masking technologies that have been developed for laser processing.[19] These masking approaches can be classified into one of three general types: projection imaging, contact masking, and conformal masking. The particular masking scheme selected depends on minimum feature size, throughput required, laser power, wavelength, etc. Projection masking technologies have been further subclassified into either point-to-point or multiple via projection technologies. Point-to-point projection imaging has been the most common masking technology used to date as it provides the most flexibility. Projection imaging involves placing a metal mask in the laser beam's optical path. The image of the mask is then focused through a lens onto the substrate surface (Fig. 2.18). This process has a feature size limit of 1.4 mils. Since computer numeric control (CNC) is possible with point-to-point projection, production can be driven directly from computer-aided design/computer-aided manufacturing (CAD/CAM) data. The use of galvanometric mirrors to move the laser beam allows a small area of the board surface to be processed on a "hole-by-hole" basis before the board is indexed to another position beneath the beam to allow another area to be processed (Fig. 2.19). Although this technique is generally throughput limited for most laser technologies, the relatively low output power and high repetition rate (5000 pulses/s) of ND-YAG lasers makes point-to-point imaging the only viable option. Contact masks involve the patterning of a metal layer applied directly onto the surface of the laminate or dielectric material that is to be ablated. This is illustrated in the RCC foil process where openings in the foil layer are etched to define where the vias are to be placed. Then these openings are flood exposed with the laser beam (Fig. 2.20). The beam passes through the hole in the metal foil and ablates away the dielectric material beneath the hole. Feature sizes of 1 mil are possible with this approach. Most CO_2 laser processes used today are based on contact masks patterned in copper foils. Although a conformal mask provides the highest via resolution capability, it is also the most process-intensive option.

Another contact masking approach involves the use of a metal mask or stencil which is fabricated in thin (50- to 125-μm) stainless steel or brass. Although this approach is limited to via sizes of approximately 3 mils, it allows the use of a reusable robust mask and can significantly reduce via generation cost.

2.7.2.5 Laser production and costs. The production capability for the laser processes is lower than for photoprocessing, which implies that high volume processing should favor photoprocessing.[20] Figure 2.21 illustrates the effect the number of vias per panel has on the cost of the two processes. There is a crossover point at which photoprocessing may become the less costly option. As laser speeds increase, the cost is lessened. In reality, laser processing often has higher yields and this may more than make up for any difference in the microvia formation costs.

Laser drilling costs to drill a 4-mil via in a 14-mil substrate[21] were found to be in the range of $1.25/1000 holes. This compares favorably with mechanical drilling a 6- or 8-mil hole (the limits of mechanical drilling). This was for a much deeper hole, so cost should be even less for a microvia. Another study[22] indicated a cost of under $0.20/1000 holes for drilling a 60-μm hole if drilling speeds are 100 holes/s.

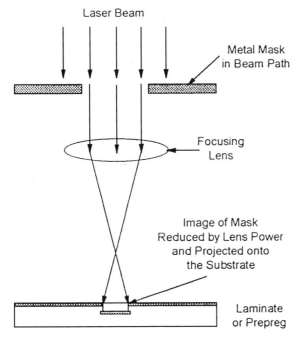

Figure 2.18   Projection imaging system.

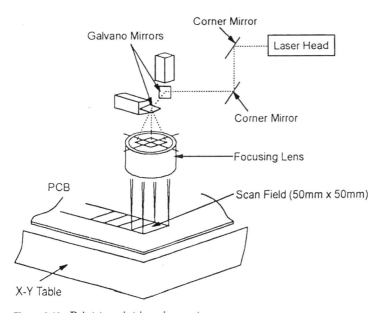

Figure 2.19   Point-to-point imaging system.

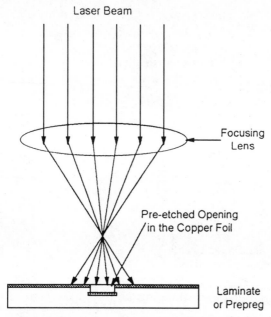

Figure 2.20   Conformal mask imaging system.

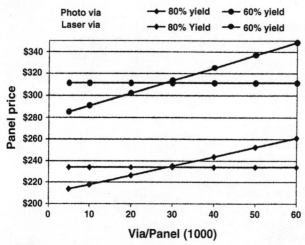

Figure 2.21   Cost versus number of vias for laser and photo processes.

2.7.3 Plasma equipment

Plasma etching equipment consists of a chamber which contains the panels in the desired etching atmosphere. It is generally large and bulky equipment. After the panels are loaded, the chamber is evacuated, the reactive gases are introduced, and the gases are ignited to form the plasma. The total etching time is typically 20 to 23 min, with a total cycle time of 40 min to 1 h. Usually, six panels can be etched simultaneously.

Plasma etching has been used for mechanically drilled hole cleaning, but it is not a process most printed circuit companies employ. Thus, for most, this process represents a new investment.

2.8 Imaging Technology

Very little is accomplished if a manufacturer has a process to make microvias and does not have a way to make fine lines—finer than typically found on conventional circuit boards. Therefore, processes for fine signal lines must also be part of the manufacturing process. Conventional phototool and mask technology is currently being used for line dimensions down to 4 mils and is marginal at 3 mils. This section discusses the newest techniques to make lines that are <3 mils. It will cover

- Imaging equipment including direct imaging and step-and-repeat approaches.
- Resists including precision fine-line resists and liquid resists.

2.8.1 Laser direct imaging

In direct imaging, a laser is used to expose the circuit image directly from CAD data onto the printed circuit substrate. This completely eliminates the phototool and associated material. The problems of phototool dimensional changes, defects, and contamination are eliminated. Further, the laser has a very small spot size that images directly rather than through a mask. This can result not only in a very fine feature but also less variation. Direct imaging is slower than contact mask imaging because the laser must "write" the circuit traces. However, as with laser imaging of microvias, there has been substantial improvement.[23] High powered UV argon-ion lasers have shown significant improvement in price, tube lifetime, and reliability. Resolutions have improved from 500 to 12,000 dpi (dots per inch). Some systems are now fast and powerful enough to image up to 120 22- × 28-in prints/h, making direct imaging rates close to conventional rates. High resolution optical systems for UV wavelengths have improved in energy transmission efficiency, price/performance, and capability to produce very large and accurate lens systems.

A key requirement for good economics is the development of high speed resists. The measure of speed is the amount of light energy required to fully image the resist as expressed in millijoules per square centimeter. Conventional resists have values in the 50 to 200 mJ/cm^2

and are too slow. In order to image a 18- $\times$ 24-in board in 10 to 15 s with current laser systems, a resist should have an exposure sensitivity of 8 to 12 mJ/cm^2.[24] Increasing the speed is not easy. Addition of sensitizers and initiators in the film cause the latitude to diminish and cause problems in achieving thorough polymerization of the resist. Another critical element is the choice of a cover sheet. Cover sheet clarity is important to avoid imaging matte particles, which are usually present in the polyester for easy roll winding and handling. Several suppliers are developing ultrafast resists specifically for laser imaging, and they are beginning to appear on the market.

One system that integrates equipment and resist together[25] uses a drum imaging system and a liquid resist. The imaging unit is based on external drum architecture (Fig. 2.22) similar to drum-based photoplotters. The engine consists of a metal drum cylinder revolving about its axis and an exposure head adjacent to the drum perimeter. Flexible inner layers are held down to the drum perimeter by vacuum and edge clamps. The laser exposure head, moving parallel to the drum axis on very precise and smooth tracks, writes hundreds of individually switched laser beams directly on the resist-coated innerlayer. The combination of the drum rotation (fast axis) and movement of the exposure head along the tracks (slow movement) generates a two-dimensional image on the inner layer area. The resist is liquid and coated with a dip coater to typically 0.1 to 0.2 mil, which is much thinner than conventional resists and images faster. It is positive working which means only the line area is exposed to light which enables a faster exposure than a typical negative working resist where the entire nonline area must be exposed. Throughputs of up to 60 panels/h are claimed. Positive resists are also less prone to damage or contamination, both of which benefit fine-line yield.

It appears that laser direct imaging has the potential to produce submil features. Typical microvia boards will require 1- to 2-mil fea-

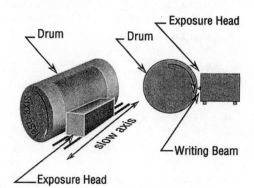

Figure 2.22 Laser imaging drum system.

Figure 2.23 Mil lines by laser imaging. (*Courtesy of Creo.*)

tures and this appears well within its capability (Fig. 2.23). This approach should grow as faster laser equipment and more sensitive resist become more widely available.

2.8.2 Step-and-repeat imaging

An alternative approach to imaging fine features is to use a step-and-repeat and "seamless stitching" technique. This was first developed for flat panel printing which requires fine features over a large area. In seamless stitching, smaller images from one or more masks form one large aggregate image by stepping and repeating,[26] as shown in Fig. 2.24. This method offers very high resolution and is capable of 0.5-mil features. A further benefit is mask/substrate contact is eliminated. However, tool cost is high because of the need to fabricate multiple phototools. Currently, up to a 20-in^2 circuit can be imaged from four master phototools. Thus, it is best used for BGA, chip scale, and other small substrates; large area substrates would require too many photomasks.

2.8.3 New resists

There are several thrusts to develop photoresists for fine features in addition to the laser resist discussed previously: high resolution dry films that work with conventional imaging equipment and phototools, liquid resists applied by various coating methods, and electrodepositable resists.

2.8.3.1 High resolution dry-film resists. High resolution resists are modifications of conventional resists. They are often made thinner to

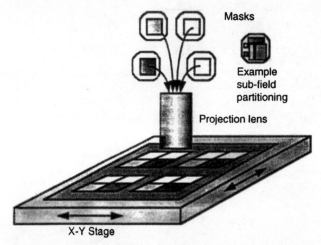

Figure 2.24 Stitching stepper imaging. (*Courtesy of MRS Technology.*) Basic method: seamless "stitching" of smaller images from one or more masks to form one aggregate image through a step-and-repeat process.

reduce light scattering and are formulated so that the resist can be imaged in intimate contact with the phototool.[27] This results in faster exposure and less light distortion and scatter. This requires a high degree of vacuum in the printing frame to ensure that the substrate is flat and coated with a uniform thickness of resist. The exposure units used with precision resists usually have a higher degree of collimation. To reproduce line geometries down to 3 mils will typically require an exposure unit with a collimation half angle of 1.5°. Exposure units are available with close to 0° decimation, which can reproduce 1- and 2-mil geometry in a production mode. Regarding environmental control, it is important to remember that an exposure unit capable of this resolution is capable of replicating dust, lint, fingerprints, hair, and resist chips. Therefore, special care must be used to ensure a high level of cleanliness or an actual lower yield could result with a precision resist.

Light source intensity and spectral output are as important as light collimation. A high intensity, short-term exposure produces better line-edge acuity than low intensity exposures over a longer period of time. A light source with an exposure intensity of 5 mW/cm^2 and greater is recommended for fine-line imaging. The spectral output of most high pressure mercury vapor lamps is compatible with photoresist formulations, but the spectral output shifts as the lamps age, causing a change in the rate of polymerization. An increase in the number of steps between a glossy resist step and the clear copper step is often seen, which can indicate not only bulb age but also a shift in the spectral output. The real effect of collimation and intensity can be seen when the resist is printed off-contact.[28] Figures 2.25, 2.26, and 2.27 show resist images shot with the coversheet on. Figure 2.25 is noncollimated, Fig. 2.26 is low intensity collimated, and Fig. 2.27 is high intensity collimated. Note that the first is bridged while the last two are good. Maintaining contact on the noncollimated source is more critical. The panels in Figs. 2.28 and 2.29 shot with the artwork upside down show the effect of intensity. Figure 2.28 is shot with collimated low intensity light and is bridged, Fig. 2.29 is shot with collimated high intensity light and is good.

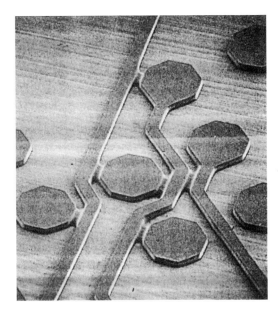

Figure 2.25 Developed resist image exposed with noncollimated light.

Figure 2.26 Developed resist image exposed with low intensity collimated light.

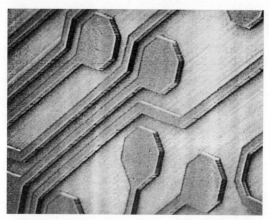

Figure 2.27 Developed resist image exposed with high intensity collimated light.

Figure 2.28 Developed resist image exposed with artwork upside down—collimated, low intensity.

Figure 2.29 Developed resist image exposed with artwork upside down—collimate, high intensity.

The phototool, or artwork, is important. First-generation laser-plotted silver halide artwork should be used for best yields. For very fine-line structures, glass phototools are being used, though this is very expensive.

Figure 2.30 shows an example of 37-μm lines and spaces imaged with a high resolution photoresist. Yield improvements are possible with high resolution resist versus a standard resist for fine lines and defects can be significantly lower (Fig. 2.31).

2.8.3.2 Liquid photoresists. There are two types of liquid resists: conventionally coated resists and electrodepositable resists. The first have basically the same chemistry as dry-film resists but are in liquid form so that they can be coated onto the substrate rather than laminated. Generally, they are applied by curtain or roller coating. They have the potential for fine feature definition primarily because they can be coated thinner than a dry film, usually one-half to one-third as thick, eliminating the distortion and scattering caused by imaging through a thick film. In a comparison of several liquid and a dry-film photoresist,[29] higher levels of performance were possible with the liquid resists. The improvement was especially significant

Figure 2.30 Micrometer lines with high resolution film resist. (*Courtesy of Morton.*)

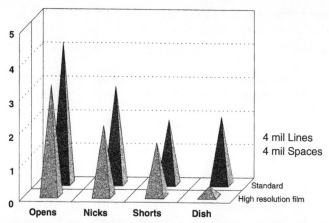

Figure 2.31 Yield comparison: standard versus high resolution film resist.

as feature size decreased (Fig. 2.32). Note that DF refers to the dry-film control and NW to the best liquid resist that was evaluated. Performance improvement is enhanced by the potential for liquid resists to be less expensive than dry film. However, there was much variation in the results due to the processing challenges. Sophisticated methods are required to uniformly apply the resists without defects or contamination. For this reason, liquid resists have been used mainly for standard circuits.

Electrodepositable (ED) photoresists are chemically modified and made into an emulsion, which is stabilized by electrical charges on the emulsion particles. Electrodeposition takes place in the following way: a copper-clad panel is immersed in the ED resist emulsion and an electric current is applied to the panel which disturbs the charges on the emulsion, causing the resist to precipitate, or deposit, onto the panel. This is illustrated in Fig. 2.33. The result is a thin, very uniform coating which is also pinhole-free and conforms very well to the substrate. It is a near-ideal resist for imaging fine features. Positive- and negative-working resists are available. If the circuit contains plated-through holes, the positive resists are preferred because they do not require imaging into the holes. ED resists are in use for microvia boards that have very small features such as 2-mil lines with 3-mil microvias. However, the ED process is a difficult one that requires very careful balancing and replenishment of the components. Equipment is also expensive, though not as expensive as roller or curtain coating. Suppliers, particularly those in Japan, continue to develop more robust formulations. In the future, ED resist should grow into greater use as microvia boards become denser.

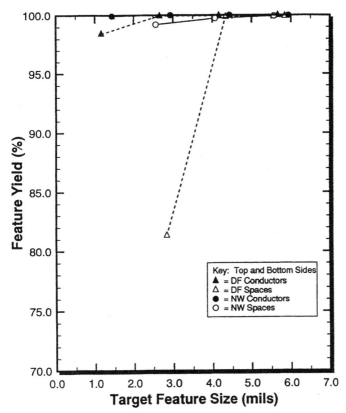

Figure 2.32 Yield comparison: film versus liquid resist.

2.9 Laminate Substrates

For the most part, laminate substrate materials based on FR-4, or conventional circuit board substrates, are utilized. It is important that these substrates are quite uniform and smooth with minimum warpage or distortion. Dimensional stability is another important characteristic in order to ensure that there is minimal shrinkage during processing. Research is in progress to develop improved substrates. Some of these are based on nonwoven laminates to eliminate the distortions caused by the weaving process. Others are based on high modulus reinforcements such as aramids. This is expected to be an important research focus as microvia structures become denser with smaller features.

2.10 Dielectric Materials

The key material in a microvia-based circuit is the dielectric in which the microvia is formed. The needs of the application may dictate the

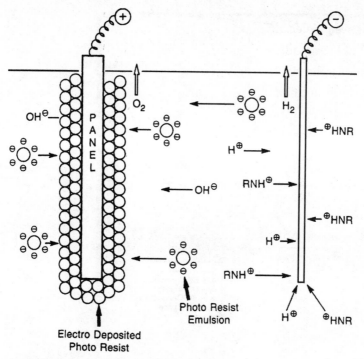

Figure 2.33 Electrodeposited photoresist process.

choice of the dielectric and, in turn, the process. Photoprocessing, which requires a photoimageable dielectric, has the most limited range of dielectrics which are usually epoxy resins modified by a photosensitive group. The key criteria for a photoimageable dielectric are shown in Fig. 2.34. In contrast, many types of dielectric materials are laser processable, so the range of materials available—and the possible applications—is much broader. These include epoxies, polyimides, and bistriazine (BT) modified epoxies (used for chip carriers and BGA substrates). Dielectrics are also available with reinforcements like glass and aramid. A representation of the dielectrics is shown in Table 2.6 which illustrates the range of dielectrics and highlights their strengths.

The photodielectrics tend to have somewhat lower electrical and thermal properties, likely due to the inclusion of the photoimaging chemistry. However, they are capable of typical assembly operations such as solder attachment and wire bonding. The polyimides have the highest thermal resistance. The aramid laminates have the best x-y dimensional strength, which may be critical for a circuit with very fine features or one being built on a large panel. Copper adhesion is highest for the RCC materials. This is because the foil is prebonded to the

- Pin hole free dielectric material
- Controllable and adjustable PID thickness
- Self-leveling characteristics leading to good planarity
- High resolution to handle 25-μm vias
- Copper peel strength over 10 Newtons/cm
- "Drop in" processability and equipment
- Electrical and physical properties compatible with conventional laminate resins

Figure 2.34 Photoimageable dielectric criteria.

TABLE 2.6   **Properties of Typical Microvia Dielectrics**

| | Photoimageable | | | | |
	Liquid*	Film†	RCC	Polyimide	Aramid/epoxy
Flammability	V-O	V-O	V-O	V-O	V-O
Water absorption (%)	—	<2	1.04	2–3	0.5
Dielectric constant					
@ 1 MHz	4.1	4.1	3.43	3.4	3.9
@ 1 GHz	3.4	3.4	3.1	—	3.5
Dissolution factor					
@ 1 MHz	0.022	0.025	0.0251	0.002	0.024
@ 1 GHz	0.017	0.020	0.024	—	0.022
Electrical Strength (V/mil)	>2000	2700	1760	>2000	>750
Insulation Resistance	E13-E14	1E+14 Ω	1.6×10^6	—	—
Volume Resitivity (Ω)	—		1.8×10^8	$>10^{16}$	$>10^8$
T_g (DMA)(°C)	125–130	130–150	160	>400	205
CTE (ppm)(x-y)	60–70	60–70	57	20	8–12
Solder float (s)	—	>60	116	—	—
Copper Adhesion (lb/in)	>6	5–7	6–12	—	7

*Probelec (Ciba-Geigy).

†Vialux 81 (DuPont).

epoxy resin. The others must be plated which can also be a difficult process to control.

Microvia circuits will find application in high speed circuits. Laminate materials based on a fluorocarbon membrane from W. L. Gore offer very low dielectric constant and loss. These may find application in microprocessor chip carriers, for example. Their properties are shown in Table 2.7. Speedboard N is epoxy-based and C is cyanate ester–based.

TABLE 2.7   **Fluorocarbon-Based Laminate Properties**

Material properties	Speedboard N	Speedboard C
Dielectric Constant ASTM D-2520	@ 1 MHz: 3.0	@ 1 MHz-10 GHz: 2.6–2.7
Loss tangent	@ 1 MHz: 0.02	@ 1 MHz-10 GHz: 0.004
Glass transition temperature (°C)	140	220
Resin (%)	66–68	66–68
Flow (%)	<4	<4
Thickness (pressed) (mils)	1.5, 2.0, 3.2, 3.4, 4.5	1.5, 2.0, 3.2, 3.4, 4.5
Color	Opaque/yellow	Opaque/tan
Flammability (UL)	94V0	94V0

SOURCE: W. L. Gore.

2.11   Design Considerations

This section is not intended to be a comprehensive design guide, but rather to offer some direction on when to consider a microvia rather than a conventional circuit. This section does not attempt to cover the electrical parameters but primarily the geometrical considerations. For more detailed information on microvia design, consult the Institute for Interconnecting and Packaging Electronic Circuits' "Design Guide for High Density Interconnect Structures and Microvias," IPC-2315. Another excellent reference (Evan Davidson, et al., "Package Electrical Design," in *Microelectronics Packaging Handbook,* R. R. Tummala, E. J. Rymaszewski, and A. G. Klopfenstein, eds., Chapman and Hall, New York, 1997, pp. I-199 to I-313), provides an overview of high performance electronic package design.

Basically, a microvia substrate contains smaller geometries than conventional substrates. The microvia is used to reduce size and weight and to enhance electrical performance. Its nature allows innovation in three-dimensional packaging as well. All printed circuits share a common attribute in that they must route signals through conductors and there are limits to how much routing each can accomplish. The factors that define the limits of their wireability as a substrate are

- The pitch or distance between the vias or holes in the substrate.

- The number of wires that can be routed between those vias.

- The number of signal layers required.

The method of producing the via will be important for routing efficiency. The microvia will typically occupy only the signal layers that it connects to while plated-through holes are routed completely through the printed circuit board, precluding any use of that space on any of the conductor layers. These factors can be combined to create an equation which defines the wire routing ability of a technology.

Historically, most components have had their leads on the periphery either on two or four sides; however, newer technologies, BGAs, chip scale packages, etc., based on area array interconnection of component I/O, are much more space conservative while allowing coarser I/O pitches to be used. However, these area array devices will require very dense routing to interconnect the devices because they contain so many more I/Os. This is where microvia substrates with their much greater routing density potential come into play. To calculate the wiring density required for a specific design, the components are assembled and the area usage calculated using the maximum space that each part occupies. Wiring capacity (W_c) is the most common definition of printed circuit board density. The microvia board has a much higher wiring capacity (Fig. 2.35) which can enable more compact and higher functionality products. The wiring capacity will depend on the specific microvia process that is chosen.

CAD packages that can accommodate and utilize microvia design rules are available. Table 2.8 lists several of them. These tools can follow the rules set by the designer and automatically design all the chip, package, and board-signal nets. Typical design rules are given in Table 2.9.

2.12   Electrical Performance

Microvia-based circuits would be expected to perform differently than conventional circuits. The main reasons are that the microvia is much smaller and has a different geometry compared to a plated-through hole and it is possible to pack much more functionality in a smaller space through smaller features and more efficient circuit routing. The microvia, which can be viewed as virtually a continuation of the signal line, is a more efficient and less noisy "connector" for signal lines than

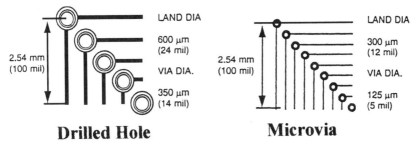

Figure 2.35 Wiring capacity comparison.

TABLE 2.8   CAD Vendors

Company	Product
Ansoft (412-261-3200) Pittsburgh, Pennsylvania	Maxwell-3D
Cadence Design Systems (408-943-1234) San Jose, California	DF/Sig/Ansoft
Contec Microelectronics (408-434-6767) San Jose, California	SI/Plane
IBM Microelectronics (800-IBM-2784) Hopewell Junction, New York	ASX, COSMIC
Hewlett-Packard (800-343-3763) Westlake Village, California	HFSS
MacNeal-Schwendler (800-624-6442) Milwaukee, Wisconsin	MSC/EMAS
Mentor Graphics (800-547-3000) Wilsonville, Oregon	BS500/XFX3D
Meta-Software (408-369-5400) Campbell, California	HSPICE
Pacific Numerix (602-483-6800) Scottsdale, Arizona	Explorer
Quad Design (805-988-8250) Camarillo, California	XTK/XFX-3D
Quantic Labs (800-665-0235) Winnipeg, MB, Canada	Greenfield-3D/Phidias/Boardscan
Swanson Analysis Systems (412-746-3304) Houston, Pennsylvania	ANSYS

a plated-through hole which is a larger circular hollow cylinder. In addition, thinner dielectric layers are also possible. These benefits should lead to improved performance.

In one study, the performance of a microvia and a plated-through hole was modeled using a lumped element equivalent circuit of both types.[30] These circuit models are shown in Figs. 2.36 and 2.37. The results indicate the microvia parameters are much smaller than for the plated-through hole. This implies that, for an equivalent rise time pulse, the microvia will provide less distortion to the rise time. Alternatively, the microvia will be able to pass a faster rise time pulse at an equivalent level of distortion. This was validated by building and testing both types of circuits.[31] At 25-ps rates, distortion was much less. The data indicated that the microvia circuits should be acceptable in applications approaching a 10-GHz signal speed, which is much beyond the capability of the plated-through hole circuit.

In another study, the signal line resistance of a conventional multi-chip module was found to be 17× greater than a microvia-based design.[32] The results are shown in Fig. 2.38 in a study of SLC microvia board performance. This is attributed to the design improvements possible with the thin dielectrics and efficiency of the microvia board. The microvia structure also had less signal delay at 100 MHz and above.

TABLE 2.9   Typical Design Rules For Microvia Boards

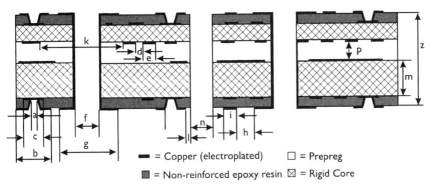

■ = Copper (electroplated)   □ = Prepreg

■ = Non-reinforced epoxy resin   ⊠ = Rigid Core

Illustrations are not to scale and are not representative of actual layups

Symbol	Feature	Conventional size (μm/mil)	Current size (μm/mil)	Proto minimum size (μm/mil)
a	Minimum microvia hole diameter (as imaged)	100/4	75/3	75/3
b	Surface via pad diameter	350/14	300/12	250/10
b-1	Via in SMT pad width	300/12	250/10	250/10
c	Landing pad diameter	350/14	300/12	250/10
d	Conductor/pad spacing on rigid innerlayer	125/5	75/3	75/3
e	Conductor width on innerlayers (etched)	125/5	75/3	75/3
f	Minimum finished hole size—fhs (plated through hole)	250/10	200/8	150/6
g	Surface via pad (fhs + annular ring × 2)*	fhs + 350/14	fhs + 300/12	fhs + 250/10
g	Minimum through hole pad diameter	600/24	500/20	400/16
h	Trace spacing on rigid outerlayer	125/5	100/4	87/3.5
i	Trace width on rigid outerlayer	125/5	100/4	87/3.5
k	Drill hole plane clearance on innerlayer (fhs + annular ring × 2)*	fhs + 600/24	fhs + 500/20	fhs + 700/28
l	Surface conductor to unplated hole	250/10	200/8	200/8
n	Minimum unplated hole diameter	350/14	300/12	250/10
z	Minimum board thickness	625/25	500/20	200/16
m	Minimum core thickness	100/4	62.5/2.5	50/2
p	Minimum prepreg thickness	100/4	62.5/2.5	50/2
	Minimum plated thickness	25/1	25/1	30/1.2

*Annular ring (AR) is pad allowance per side of hole. Pad diameter equals finished hole size (fhs) + 2 × AR. Annular ring in these instances is the same as manufacturing allowance.

NOTE: All copper thicknesses are ½ oz (17 μm = 0.007 in). Plating thicknesses for 2 to *n*-1 layer to be nominal 17 μm (700 μin), minimum 12.5 μm (minimum 500 μm).

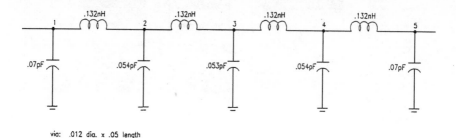

via: .012 dia. x .05 length

Figure 2.36   Subtractive via model.

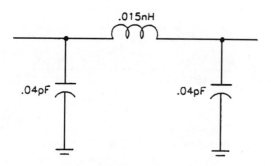

Figure 2.37   Sequential via model.

via: .006 dia. x .001 length

For applications above 200 MHz, the authors concluded that only the microvia design would be suitable. Thus, as designs require higher speed capability, microvia technology will be beneficial due to its inherent ability to transmit signals in a cleaner way and to reduce line lengths through efficient routing.

2.13   Reliability

Reliability testing of microvia boards has been carried out by several of the major original equipment manufacturers in the electronics industry. IBM, Motorola, SONY, and others have carried out extensive in-house evaluations to demonstrate the reliability of microvia boards for their applications. In the public sector additional information has been made available. In Europe IVF, the Swedish Institute of Production Engineering Research published a microvia report in 1997.[38] This report analyzed electrical properties of microvias as well as the reliability of microvias. In the United States, The Interconnection Technology Research Institute (ITRI), located in

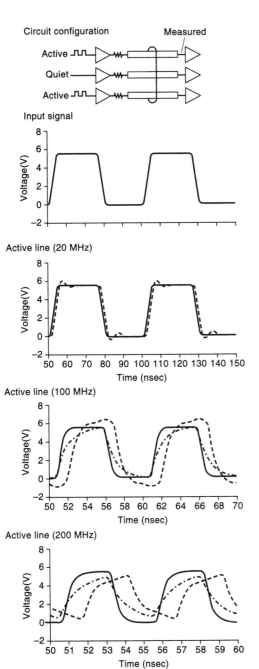

Figure 2.38 Microvia signal propagation simulation.

Austin, Texas, has published several reports on microvia reliability.[39,40] This represents the most extensive set of reliability data thus far made public on microvia substrates. Results of this work is detailed in the following section. Further work is in progress on assembled microvia structures.

The ITRI reliability studies utilized a test vehicle (Fig. 2.39) that was initially designed by Motorola and updated by ITRI. The design assesses the general capabilities of microvia boards with respect to key features such as minimum via size, conductor width and spacing, and interfacial adhesion, while also providing a metric for assessment of interconnect reliability. In Fig. 2.39, *A–F* represent blind via chains from 1 to 5 mils, *G* provides front to back connections, *H–L* are surface insulation resistance patterns, *M* is a dielectric breakdown area, *N* is a RF structure, *P–S* are packaged assembly sites, *T* is a flip-chip assembly site, *W* is a wire bond assembly site, *V* is a BGA assembly site, *W–X* are serpentine test areas, and *Y* is a *z*-axis comb pattern.

2.13.1   Liquid-liquid thermal shock

Testing included liquid-to-liquid thermal shock to test the reliability of via chains, assembled components on the boards, and laminate integrity. High temperature storage testing was carried out to evaluate the effect of long, high temperature storage on the different microvia technologies. Adhesion testing was done to access the copper-to-dielectric peel strengths. Time to delamination and dielectric-withstanding voltage tests were also performed.

The liquid-to-liquid thermal shock test is used to test the reliability of the via chains, assembled components, and integrity of the board construction. Initial resistance measurements are taken before the boards are pretreated prior to testing. The boards are pretreated with five flip-chip on-board (FCOB) solder reflow cycles to simulate assembly and rework processes. The reflow cycle uses a ramp-up stage and a cool-down stage (approximately 2.5 min each), with the board being subjected to a temperature greater than 183°C for 1 min with a maximum of 225°C. The liquid-to-liquid thermal shock test has a 5-min soak at −55°C and +125°C with a 4-s transition time. It is carried out to 2000 cycles. The resistance of the via chains is measured as received and after simulated FCOB solder reflows, with 0, 100, 250, 500, 1000, and 2000 cycles, using an automated four-point resistance measurement system.

Figure 2.40 summarizes the results of the liquid-to-liquid thermal shock data for various PWB suppliers. The chart shows the minimum reliable via diameter initially and after 250, 500, 1000, and 2000 thermal shock cycles for each supplier. The via diameters given are the actual measured via sizes obtained from cross-sectioning as given by

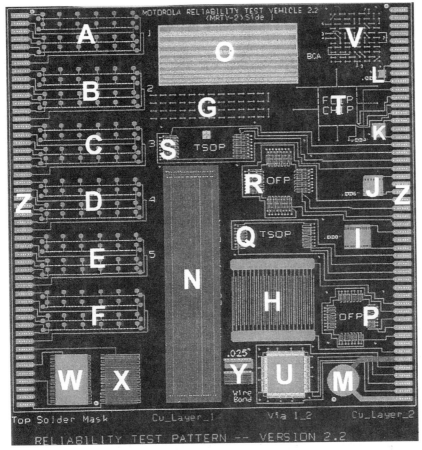

Figure 2.39 Motorola/ITRI test vehicle.

the largest opening in the dielectric before plating. The data show that all three of the tested technologies, at least from some suppliers, were able to survive 2000 cycles. Thus, all three technologies are capable of producing reliable vias. The smallest vias that maintained reliability to 2000 cycles were made by the laser process.

Some of the typical types of failures are shown in Fig. 2.41. As can be seen in these cross sections, the failures were cracking along the bottom of the via and corner-cracking due to thin copper plating at the via bottom.

2.13.2   High temperature storage

High temperature storage is a test used to examine the effect of long-term high temperatures on the different technologies. Boards are

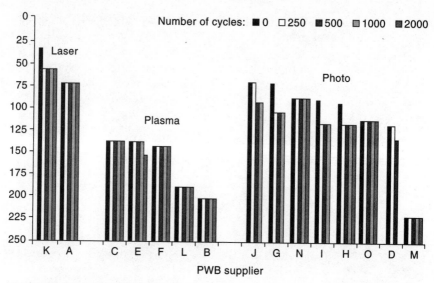

Figure 2.40 Liquid-liquid thermal shock results.

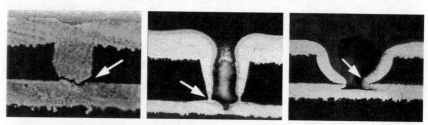

Figure 2.41 Thermal shock failures.

initially electrically tested and then stored at 125°C for 1000 h. The resistance of the via chains is checked during testing and after 1000 h, and none of the technologies shows any opens with all having increases in resistance of less than 10%.

2.13.3 Adhesion testing

Peel testing is a high-density interconnect (HDI) copper-to-dielectric adhesion test. Peel test sites with widths of 0.4, 0.8, and 1.6 mm and a length of 3.8 cm are used to characterize the copper-to-dielectric adhesion of the different technologies. Low adhesion values are a cause for concern, especially during rework and also for products required to pass vibration and drop tests. Boards for each condition, initial and after high temperature storage, are tested with a minimum of three peel strips pulled per side.

Peel adhesion testing showed that all technologies had a copper-to-dielectric adhesion greater than 5 lb/in, except for one supplier (Fig. 2.42). The foil technologies, in general, had higher adhesion than the photovia ones, although one photovia suppliers' process produced samples with the highest adhesion at 14 lb/in.

2.13.4 Time to delamination

A thermomechanical analyzer (TMA) capable of determining dimensional changes to within 0.0025 mm over the specified temperature range is used to determine the "time to delamination" of the different board technologies. IPC spec TM-650 no. 2.4.24.1 is the reference document for this test. Test specimens approximately 0.25 × 0.25 in in size are used. These specimens should be taken from the same location of the test vehicle for all the different PWB technologies tested. Test specimens are preconditioned by baking for 1 h at 110°C and then cooled to room temperature. This baking process allows moisture in the board to be removed. Specimens are mounted on the stage of the TMA and a force of 0.005 N is applied. The samples are equilibrated at 35°C and then the temperature is increased at 10°C/min up to 260°C. The time to delamination is taken as the time at 260°C when an irreversible dimensional change in the sample first occurs.

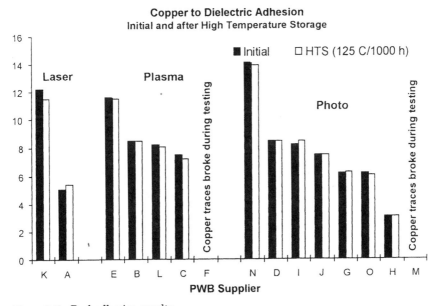

Figure 2.42 Peel adhesion results.

As shown in Fig. 2.43, all of the technologies survived at least 10 min at temperature, with most over 20 min. All failed within 45 min at temperature, except for two suppliers' samples which did not show any indication of delamination. The majority of samples showed delamination within the core substrate as the main locus of failure. Typical values for FR-4 multilayer printed wiring boards are generally less than 10 min at temperature.

2.13.5 Dielectric withstanding voltage

Dielectric withstanding voltage of the microvia technologies is checked following IPC TM-650, Section number 2.5.7. A special test structure is used for this measurement and the samples are tested in an as-received condition from the fabricators. The samples have a DC voltage applied which is increased from 0 to 5000 V (the limit of testing equipment) at a rate of 100 V/s. All samples survived voltages in excess of 1500 V/mil, with most exceeding the limit of the testing equipment (>2500 V/mil).

2.13.6 Conclusions

Conclusions that can be drawn from a reliability standpoint are

1. Many suppliers are capable of manufacturing HDI printed wiring boards using sequential build-up technologies achieving line/space

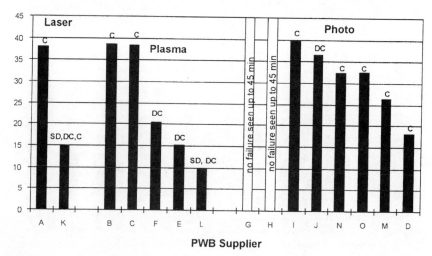

Figure 2.43 Time to delamination results. Where failure was observed: SD = soldermask to HDI dielectric; DC = HDI dielectric to core; C = core.

and via/pad dimensions smaller than conventional blind and buried PWB technology.

2. Photovia suppliers are capable of resolving reliable 75-μm vias. Laser suppliers are capable of resolving smaller reliable vias, with one supplier achieving 34-μm vias. Plasma via suppliers are not able to successfully resolve microvias as small as photo or laser technology. The ability to form and plate vias smaller than 75 μm, providing a reliable interconnect, needs improvement for most of the suppliers tested. Fifty-micrometer lines and spaces was the best resolution achieved, with 75-μm lines and spaces being the most typical.

3. All three types of via forming technologies showed failures during liquid-to-liquid thermal shock testing:

   - Laser via boards having 50-μm (actual 58-μm) vias survive 2000 cycles of liquid-liquid thermal shock. Laser process thus demonstrated the smallest vias of any technology that could survive this thermal shock test. The 34-μm vias failed after 250 cycles.

   - The plasma via boards all survived 2000 cycles of thermal shock with no opens or large increases in resistance seen (<10%), although the plasma vias were all larger than the other technology vias.

   - The photovia boards showed failures during thermal shock but some suppliers lasted through 2000 cycles.

4. Peel adhesion testing showed all technologies had copper-to-dielectric adhesion greater than 5 lb/in. The foil technologies, in general, had higher adhesion than the photovia ones, although one supplier's photovia process had the highest adhesion with 14 lb/in.

5. All technologies survived delamination testing longer than 10 min at 260°C, which is comparable to most FR-4 multilayer printed wiring boards.

6. Most of the suppliers showed the need for improvements in via and layer registration. Via breakout was not an issue here, however, due to the large capture pad size of 250 μm.

2.14 Applications

Microvia applications are growing rapidly. It is possible to segment them into two general groups:

- *High density substrates.* For high value applications such as computers and certain telecommunications and lower value products typified by cellular phones.

- *Chip carrier substrates.* For area array semiconductor packages such as ball grid arrays (BGAs) and miniature designs such as chip scale packages (CSPs).

Any assembly that has over 120 attachment pads per inch and array components of less than 40-mil pitch will likely benefit from microvia technology. Following is a review of several current applications. The breadth of the applications is apparent.

2.14.1 IBM Thinkpad board

This product was manufactured, with several design modifications, over a period of several years and is probably the largest volume microvia product made to date at an annual production of several million per year. Thus, it has extensive and very reliable field experience. The construction (Fig. 2.44) utilizes the SLC process (photoformed vias) with an eight-layer conventional core and two layers of microvias (one on each side). The SLC design allows the use of IBM's flip-chip technology. With these thinner, lighter unpackaged chips, IBM was able to reduce size and weight, important considerations for a portable computer.

2.14.2 Sony digital camcorder

This microvia board (Fig. 2.45) enabled Sony to produce its ultrasmall camcorder which is about the size of a paperback book. It simply could not be built with a conventional circuit board. The microvia board itself is 5.1 × 4.6 in, has 606 pads/in^2 (compared to a maximum of 100 to 120 for a conventional circuit board), and accommodates 20 high density chip scale packages. This dramatically illustrates the potential for microvia substrates to provide high levels of I/O connections. There are two microvia layers on each side and a four-layer conventional

Figure 2.44 One IBM Thinkpad board.

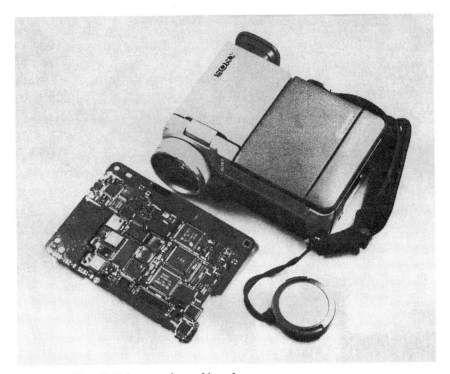

Figure 2.45 Sony digital camcorder and board.

core, and it is built using the SLC process. A cutaway view of the assembly with the chip scale package is shown in Fig. 2.46. The CSP is flip-chip mounted on a 20-mil pitch with 20-mil solder balls.

2.14.3 Casio radio pager

This is a typical high volume, low cost application where microvia enables high functionality in a small package. The board has a two-layer core and one microvia layer on each side with 3.5-mil lines and 6-mil vias (Fig. 2.48). Pager boards are built with both laser and photovia technology.

2.14.4 NTT P201 hyper digital phone board

This board is built by the Matsushita ALIVH process. It has six laser-formed microvia layers with 6-mil lines and 6-mil vias (Fig. 2.47). Attachment pad density is 261/in^2. A closeup of the board reveals how effective Matsushita's ALIVH technology is in reducing the overall size of the board by avoiding fan-out to via holes and pads (Fig. 2.49). In the earlier phone two boards were used (RF and processor). These

Figure 2.46 Cross section of a camcorder board. (*Courtesy of Prismark.*)

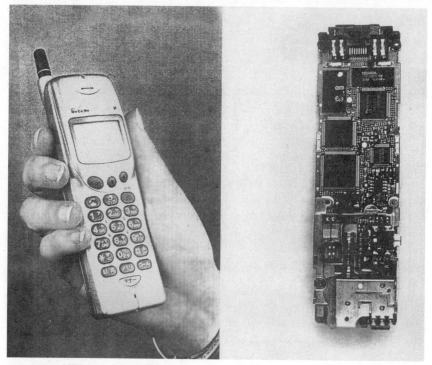

Figure 2.47 NTT hyper digital phone.

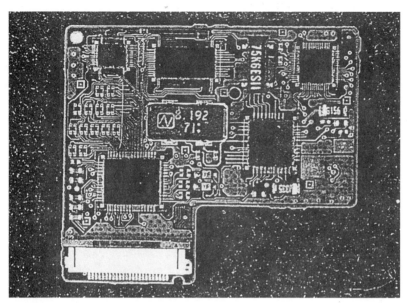

Figure 2.48 Casio radio pager board. (*Courtesy of Prismark.*)

Figure 2.49 Close-up of ALIVH board.

have been reduced to one board weighing 6.2 g versus 21 g. The resulting phone is thin and lightweight weighing a total of 3.3 oz (its nearest competitor weighs over 5 oz). Production of cellular handsets will exceed 100 million units in 1999. Many of them will utilize microvia technology.

2.14.5 Single-chip carrier substrate

Chip carriers demand the highest level of circuit and attachment density and are excellent candidates for microvia technology. Often via-in-pad designs are used and the low inductance connections of the microvia are important for the high frequency signals from the chip. Other requirements for this application include flip-chip and/or wire bonding capability. For flip chip the substrate must be compatible with solder bumping and underfill materials and processes. With the emergence of ball grid array and other area array packages, the number of microvia-based products is growing rapidly. The single-chip module (Fig. 2.50) shown here is used for high end application-specific integrated circuits (ASICs) in workstations and other computers. It has four microvia layers (two on each side) and a two-layer conventional core. The features are 2-mil lines and spaces and 4-mil vias. Total I/O count is over 500.

2.14.6 Multichip module chip carrier substrate

The chip carrier substrate shown here (Fig. 2.51) has 2-mil line and space and six layers of 3-mil microvias (three on each side). This

Figure 2.50 Single-chip carrier substrate.

Figure 2.51 Multichip module substrate.

extremely dense substrate supports four chips with a very high utilization of the surface area. Again, this is used in sophisticated computer applications. In the past this level of substrate functionality
could only be had with ceramic substrates at a cost of 5 to 10 times
that of the microvia option.

2.14.7   Multilayer board with high I/O BGAs

Assembling the high I/O count BGAs onto the motherboard represents a
challenge. By using a microvia solution, these dense area assemblies can
be accommodated, layer count can be reduced, and the costly alternative,
mechanically drilled blind vias, can be avoided. The example shown here
(Fig. 2.52) utilizes a single layer of microvias over a conventional 12-lay-
er multilayer. Generally, laser processing is used for these large area
boards and RCC foil is the preferred dielectric material.

2.14.8   High density backplane

Backplanes are used extensively in communications applications to
route information. They are growing in complexity with the rise of the
Internet and other needs to rapidly send, sort, and correlate data. The
assembly generally consists of daughter cards which are attached with
connectors perpendicularly to the backplane. These typically have

Figure 2.52 Multilayer board.

through-hole connectors, but to accommodate the increased density, manufacturers are developing surface-mount connectors with much higher and denser I/Os. Instead of through holes on 100-mil centers, these newer connectors have attachment pads on 40-mil centers. These backplanes will preferably use a microvia layer to accept the high number of attachment pads. A prototype example made with the RCC laser forming process is shown in Fig. 2.53 along with the surface-mount connector. This will impose several additional requirements on the microvia process. It will have to be quite rugged with excellent pad bond strength to support the daughter cards and very uniform and coplanar across the surface to ensure reliable solder attachment for the long connectors. Therefore, processes which use a resin-coated foil or prepreg-coated foil, which have excellent adhesion, are preferred.

2.15 Economics

Microvia technology is in early stages. Costs inevitably are higher than conventional circuitry now, but will decrease over time as manufacturing efficiencies are found, production volumes increase, material costs decrease, and competition increases (Fig. 2.54). Nonetheless it is useful to look at microvia costs. This section analyzes several aspects of cost. Then the relative cost contributions of materials and

Figure 2.53 Microvia backpanel and surface-mount connector.

the various process steps for the three major processes are analyzed. The cost of a specific design built by the major processes is modeled. Microvias allow substantial design improvement and the cost benefit of this redesign from a conventional circuit is shown. Finally, examples of current prices of the major processes are given.

Three processes, a plasma formation process, DYCOstrate, a photovia process, SLC (both described previously), and HDI, which utilizes a polyimide film dielectric and laser vias with sputtered metallization, were compared.[33] The method used was a technical cost model (TCM), a tool specifically designed for multichip module fabrication cost analysis. The first objective was to determine the relative costs of the process steps required for fabrication. The second was to evaluate the effect of increasing production.

This study demonstrates that different processes have different relative costs of process and materials. For the DYCOstrate part, electrical testing/visual verification (13.3%) and laminate materials (12.8%) were the most expensive factors. Plating (12.2%), photoresist operations (12.1%), and gold plating (9%) are cost significant operations due to equipment and materials costs. The remaining nine operations each contributes less than 8% of the circuit cost.

In the SLC model, the materials, yield, and equipment for coating and curing the photosensitive dielectric is the driving cost factor

CIRCUIT COSTS ($)

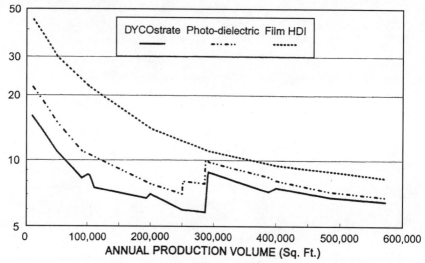

Figure 2.54 Microvia process circuit costs versus volume.

(28.8%) and inspection/electrical testing is second (11.3%). The photoresist operations (10.2%), laminate materials (8.4%), and gold plating (7.6%) are cost significant operations.

For the HDI process the most expensive factors are sputtering (33.3%) and materials (including the purchased adhesiveless polyimide flex microcircuitry). Other factors are the initial via cleaning and metallization (17.3%), laser drilling (6.9%), and electrical testing/verification (6.3%).

Secondly, the effect of increasing production was determined. Figure 2.54 shows these reductions to be significant in all cases. A steady-state pattern reflecting an economy of scale appears for DYCOstrate at about 300,000 ft²/y, indicating limited savings with higher production volumes because of the need for additional capital investment. At the baseline assumption of 300,000 ft²/y, a 17% cost savings can be realized through yield improvements from 83 to 95%. For the baseline production volume of 300,000 ft²/y and 95% yield, the cost per square foot of circuit is about $82/ft². The SLC cost curve shows similar dependencies. The HDI's cost curve is dominated by the capacity and cost of the sputtering and metallization line.

A second study compared the costs of a specific circuit fabricated by photo, plasma, and laser processes.[34] The modeling was done with an 11- × 3-in board built on an 18- × 24-in panel (10 pieces/panel). Depending on the technology used, construction varied from six to eight layers.

Six process variations were studied:

- *laser 1.* A laser ablates the copper outer layer, then changes frequency to create vias through the underlying dielectric.

- *Laser 2.* Conventional etching is used to pattern the vias on the outerlayer copper and then laser drilled dielectric to form the vias.

- *Laser 3.* The dielectric is deposited, laser vias are formed, and the outer layer copper is plated.

- *Photo 1.* A two-core construction is laminated, then the outer layer copper foil is patterned, photosensitive dielectric deposited, and photo defined followed by copper plating and patterning.

- *Photo 2.* A three-core construction is used followed by dielectric deposition and photodefining. Copper is plated and patterned. This option, while less standard, eliminates a series of outer layer processing steps but may incur yield loss in lamination.

- *Plasma.* Uses a resin-coated copper which is etched to remove the copper and then vias are formed with oxygen plasma chambers followed by copper plating and patterning.

This study assumed an equal yield of 95% and a production volume of 500,000 boards. Costs per board ranged from $13.83 to $16.91, a variance of ±10% from the median cost. These projections are shown in Fig. 2.55. There does not appear to be a clear winner or loser.

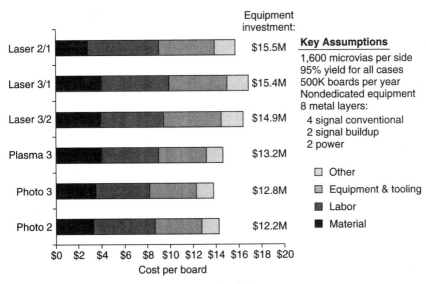

Figure 2.55 Cost comparison of microvia technologies.

This highlights the importance of yield. In general, the laser processes appear to cost more. But if practitioners can make process improvements and, for example, drive yields higher than 63% while other processes are below 50%, then laser is the most competitive process. Considering the relative complexity of the processes, this may indeed be the case.

In order to achieve an optimally low cost, it is critical to take advantage of the potential design benefits of microvia technology. Singer and Bhatkal[35] analyzed the impact of design for several different types of circuit applications: a laptop motherboard, a hard disk drive controller, and a CPU multichip module. Eleven different microvia processes were modeled.

Two basic benefits of microvia are layer count reduction and area reduction. In case 1 (Fig. 2.56), layer count is reduced from 12 in the conventional construction to 8 in the microvia version. Cost reduction is actually small and the microvia version can even be more expensive. This is because some of the microvia processes have a significantly higher per layer cost than the conventional printed circuit process. Case 2 allows for a 33% area reduction and layer count reduced from 8 to 6. As shown in Fig. 2.57, cost reduction can be as much as 30%. This is because 60% more parts can be made on the same size panel. In case 3, the most aggressive, design area is reduced by 75% and layer count is reduced from 10 to 6. The panel can now yield 300% more parts and a resulting cost savings of up to 70% (Fig. 2.58). While this illustrates the benefit of microvia technology, it must be understood that, in order to achieve this small size, the manufacturing process will be more difficult. Nonetheless, this study dramatically demonstrates that size reduction is significantly more effective in reducing cost than eliminating a few layers. However, if layer count reduction is large, then many steps are eliminated and cost savings could be significant.

In any discussion of cost and price, it is important to note that, with a new technology, change is inevitable. Thus, this section should be viewed as a starting point and so prices will be changing as volumes and competition increase and as processes improve. There are three application categories for microvias: chip packaging, high end (computers, workstations, some telecommunications), and low end (cellular phones, pagers). The consulting firm BPA developed probable pricing for these three categories as shown in Fig. 2.59. Note that these prices are in dollars per square inch per layer. Sarnowski[36] surveyed several Japanese suppliers and found that a typical four-layer construction with two microvia layers was priced (all prices shown are per square inch total board) at $0.75 to $1.00 for low density product and $1.25 to

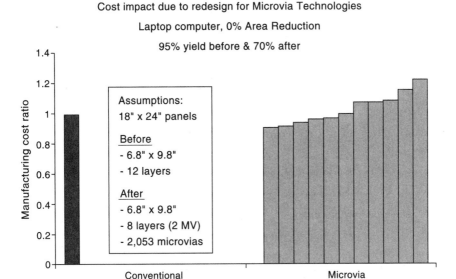

Figure 2.56 Cost impact of layer count reduction.

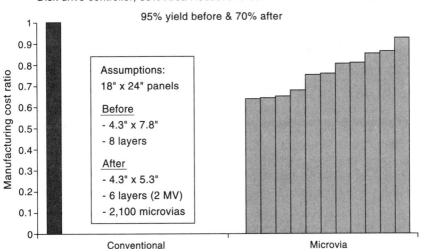

Figure 2.57 Cost impact of 33% area reduction.

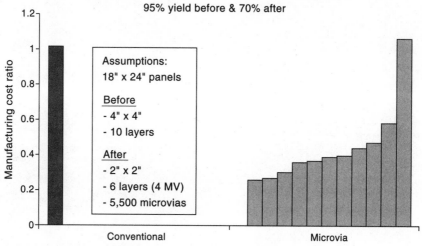

Figure 2.58   Cost impact of 75% area reduction.

$1.50 for high density product. The year 2000 projection is that these prices will come down aggressively by 50 to 60%. Prismark[37] projects that microvia-based motherboards have fallen in price by 18%/y and will continue to do so over the next few years. This will not make microvia boards less expensive than conventional boards on an area basis because they are also decreasing in price. However, as shown here, many products will be less expensive when redesigned to incorporate the benefits of microvias.

2.16   Future Opportunities

Microvia technology is in early stages. As has been true with other technologies, there will be opportunities to improve both process and product. Material suppliers will develop a greater range of materials with improved properties, particularly for higher speed circuitry. This will extend printed circuits into applications currently dominated by ceramics and TeflonR-type boards.

Laser processing will continue to improve. It should be possible to approach a 1000-via/s drill speed, which will make it possible to use laser processing for very high count via products and put further pressure on photovia processes.

Applications currently are mostly for portable electronics but microvias will find their way into many other applications such as

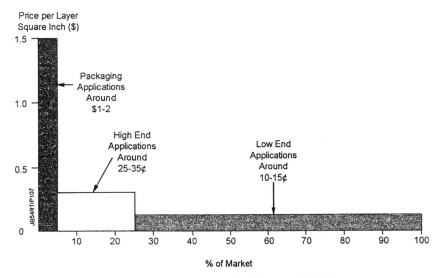

Figure 2.59 Probable pricing of microvia boards. (*Courtesy of BPA.*)

high layer count multilayers. Ultimately, microvias should find application in virtually all printed circuit segments, except for the very low cost products.

There are opportunities for new manufacturers, particularly in the United States and Europe. The capital to properly outfit a microvia plant is high, and small- to medium-size firms may find it advantageous to partner or enter into joint ventures to share the burden.

Costs will inevitably come down. Even today, as seen here, microvia boards can be competitive on a functionality basis. Further cost reductions will occur as manufacturing volumes increase and resultant efficiencies develop, yields improve, and materials costs decrease. This will mean microvia processes will be the most cost effective approach for many applications.

Acknowledgment

I owe a strong debt of gratitude to Jack Fisher, chief technical officer of the Interconnection Technology Research Institute who has authored the section on reliability and allowed the use of the ITRI reliability study data. This is the most extensive reliability data available which compares the various microvia processes and is a major contribution to this work. Jack has also contributed other data on microvia processes which I have incorporated into the main body of the text.

References

1. T. Flottmann, private communication.
2. R. Carpenter et al., "SLC: An Organic Packaging Technology for the Year 2000," *IPC Expo Proceedings*, March 3–7, 1996, p. S14-2-1.
3. A. Cable, "Laser Micro-Via Drilling in Circuit Boards," *Circuitree*, December 1998, pp. 30–34.
4. R. D. Schaeffer, "An Overview of Lasers for Microvia Drilling," *Future Circuits International*, vol. II, 1998, pp. 159–164.
5. D. A. Doane, "Is There a Plasma in Your Future?" *Circuitree*, February 1998, pp. 62–76.
6. S. Tsukada et al., "Surface Laminar Circuits and Flip Chip Packaging," *Proceedings 42d ECTC (Electronic Components Technical Conference)*, May 1992, pp. 222–227.
7. M. Nakamura et al., "High Reliability High Density Build Up Printed Circuit Board for MCM-L," *ICEMCM (International Conference for Electronics and Multichip Modules) Proceedings*, 1995, pp. 36–41.
8. B. McDermott, "The Practical Application of Microvia Technology," *Circuitree*, March 1998, pp. 48–60.
9. C. Gonzoles et al., "Epoxy-based Aqueous Processable Photodielectric Dry Film," *ECTC Proceedings*, 1998, pp. 132–143.
10. "RCC and Multifoil," Allied Signal Application Bulletin T13303.01b, Allied signal, La Crosse, WI, 1996.
11. S. Nakahara, "Laser Technology in Japan," *Printed Circuit Fabrication*, September 1998, p. 64.
12. R. D. Schaeffer, "An Overview of Lasers for Microvia Drilling," *Future Circuits International*, vol. 2, 1998, p. 159.
13. J. A. Morrison, "Lasers and the Fabrication of PWB Vias," *Industrial Laser Review*, November 1995, p. 9.
14. H. Schmid et al., "Low Pressure Plasma Etching in Microvia Formation," *Proceedings*, IPC Technical Conference, April 26–30, 1998, p. S14-2-1.
15. G. Brist et al., "Plasma Microvia Etching," *Printed Circuit Fabrication*, January 1997, p. 34.
16. Y. Sato et al., "Buried Bump Interconnect Technology," *Circuitree*, February 1997, p. 10.
17. A. Kitai, Sr., et al., "A Flexible Production System for Blind Via Drilling," *Circuitree*, September 1998, p. 10.
18. H. Nakahara, "1998 JPCA Show, Part 1," *Printed Circuit Fabrication*, October 1998, pp. 64–68.
19. T. Tessier et al., "Casting Light on Recent Developments in Laser-based MCM-L Processing," *ICEMCM Proceedings*, 1996, p. 006.
20. Mason Hu, "HDIA Using Laser Ablated Microvias," *Circuitree*, March 1998, p. 86.
21. Gary Nightingale, ESI, private communication.
22. J. Tourne, "Microvias, A New Cost Effective Interconnection Technology," *ICEMCM Proceedings*, 1995, pp. 71–76.
23. Barry Ben-Ezra, "Direct Imaging Comes of Age," *Circuitree*, March 1997, pp. 80–86.
24. M. Ehlin, "High Speed Photoresists for Laser Direct Imaging," *Printed Circuit Fabrication*, vol. 21, no. 7, July 1998, pp. 40–47.
25. E. Halevi and Y. Atiya, "A Cost Effective Laser Direct Imaging System," *IPC Proceedings*, April 26–30, 1998, pp. 12-1-1.
26. D. Naugler, private communication and MRS Technology Literature P/N 93-5200-FAM-1-03, September 1997.
27. E. Hayes and J. Caraway, *Circuits Manufacturing*, April 1989, pp. 44–48.
28. Samuel C. Miller, "Collimated Light Form a User's Point Of View," *Printed Circuit Fabrication*, vol. 12, no. 6, p. 22, June 1989.
29. C. Sullivan, R. Funer, and R. Rust, "Yield and Uniformity Comparisons of Liquid Photoresists With Dry Film as a Control," *IPC EXPO Proceedings*, 1996, S8-6-6-S8-6-1.
30. T. Flottmann, private communication.

31. J. Schroeder, private communication.
32. Y. Tsukada et al., "A Novel Solution for MCM-L Utilizing Surface Laminar Circuit and Flip Chip Attach Technology," *Proceedings ICEMM,* 1993, p. 252.
33. H. Holden, "Comparing Costs for Various Build Up Technologies," *ICEMCM Proceedings,* 1996, pp. 15–21.
34. A. Singer, et al., "A Cost Analysis of Microvia Technologies," *IPC Proceedings,* 1997, p. 11.
35. A. Singer and R. Bhatkal, "Quantifying the Impact of Design on Microvia Printed Circuit Board Cost," *Future Circuits International,* no. 3, 1998, pp. 151–154.
36. T. Sarnowski, private communication.
37. C. Lassen, private communication.
38. M. Lindgren, Per Carlson, and Rolf Sihlfon, "Sequential Build-Up Boards, Low Cost Multilayer Carriers," Swedish Institute of Production Engineering Research, Molndal, Sweden, 1997.
39. "High Density Printed Wiring Board Microvia Evaluation Phase 1 Round 2," Interconnection Technology Research Institute Report 97971501-G, Austin, Tex., July 15, 1997.
40. "October Project High Density Printed Wiring Board Evaluation Phase 1, Round 1," Interconnection Technology Research Institute Report 9606002-G, Austin, Tex., June 1996.

Substrates for RF and Microwave Systems

Stephen G. Konsowski

Consultant, Glen Burnie, Maryland

John W. Gipprich

Northrop Grumman
Baltimore, Maryland

3.1 Introduction

Substrates are used in all types of electronic applications but they may not always be called substrates. Instead they can be called *printed circuits* or *printed wiring boards* or *circuit boards.* The basic functions are, nevertheless, essentially the same in all cases. A printed wiring board is composed of an insulating material that carries conductors on its outside surfaces and very often within itself as buried conductors, usually in layers. The board may be rigid or flexible. The simplest substrates are used for watches and children's watchlike games and toys. These may have only one layer of circuitry or two at most. Slightly more complicated games, such as those used in function-dedicated toys, are minicomputers that use cartridges containing electronic devices on printed wiring boards made up of alternate layers of conductors and insulators. The conductors have connections made from conductors on one layer to the next or to lower conductor layers that pass through the insulating layers in the form of holes in the insulation. These are plated with a conductor material and allow the circuit to become three dimensional. At the higher end of the scale, some equipment such as VCRs, video cameras (camcorders) and the like, personal desktop computers, and laptops or notebooks and similar equipment use slightly more complex printed wiring boards that may be composed of as many as 20 layers of circuitry. Some of these conductors are very narrow and are formed with great precision and may represent one form or other of transmission line technology. The requirement to use transmission line technology, which is a means of preserving the integrity of very high frequency signals, arises from the high switching speed at which many of these devices operate. As products become more versatile and include more functions, the trend to higher speed circuitry will continue, and this will be reflected in more rigorous performance demands on components and the substrates carrying their circuitry. The types of circuits just described are generally digital in nature and function, but the high speed signals they carry have very similar substrate and conductor requirements as do microwave and radio frequency (RF) circuits. Examples of high performance digital equipment may provide an appreciation of the types of substrates that have been utilized in microwave and RF circuits for some time.

3.2 Functions of Substrates

Microwave and RF circuits are physically constructed from conductors in the form of wires, or, more commonly, conductor tracks on a dielectric or substrate material, generally in the form of a PWB. In the case of RF and microwave circuitry, the tracks are called *conductors* and the insulator plate is called a *substrate*. Substrates (substrata), so called because they lie beneath circuit conductors, serve as mechanical supports for conductor circuit tracks, providing insulation between conductors, but, at high frequencies, the substrate becomes part of the circuit as well. A typical RF/microwave circuit topology is shown in Fig. 3.1.

Conductors lie atop the insulator, which in this case is a ceramic material. Conductors can be attached to the insulator (dielectric) by various methods. Commonly used combinations of conductors, dielectrics, and attachment methods are listed in Table 3.1.

Conductors are the carriers of electrical current and substrates are the carriers of electromagnetic waves that are related to the current being carried in the conductors. The mechanical function of substrates can be broken down into two categories:

- Physical or mechanical support for conductors, plated-through holes, connectors, and components

- Thermal paths to remove unwanted heat

Figure 3.1 View of microstrip circuitry.

TABLE 3.1 Typical RF and Microwave Circuit Substrate Materials

Substrate material	Conductor	Conductor attachment method
Aluminum oxide (Al_2O_3)	Cr-Au	Vacuum deposition
Aluminum oxide	Ta-Ni-Au	Vacuum deposition
Aluminum oxide	Thick-film Cu or Au	Printing/firing
Epoxy/glass	Cu/Ni/Sn/Pb	Lamination
Polyimide/glass	Cu/Ni/Sn/Pb	Lamination
Cyanate ester/glass	Cu/Ni/Sn/Pb	Lamination
PTFE/glass	Cu/Ni/Sn/Pb	Lamination
PTFE/aramid	Cu/Ni/Sn/Pb	Lamination

In most electronic circuits, printed wiring boards are generally made of plastic. The plastics can be epoxy/glass or polyimide/glass, but in RF and microwave applications, higher performance materials are often required. These can be polytetrafluoroethylene (PTFE) or Polysulfone™ or other plastic material generally reinforced by a woven or randomly oriented fabric contained within the dielectric matrix. It is this reinforcement that gives the dielectric its dominant mechanical properties. Glass or aramid fibers are the principal reinforcement materials. The mechanical properties that determine a

PWB's ability to provide physical support for its own weight and the weight of the components and connectors are shown in Table 3.2.

The mechanical strength of most dielectric materials is sufficient to provide physical mounting support for the components that comprise the active (semiconductors) and passive portions of the circuits (resistors, capacitors, inductors, and connectors). This physical mounting support must act to maintain the mechanical integrity of the circuit board and also the components in the case of rigorous environments such as airborne or automotive applications, which include high levels of vibration and shock. Components are mounted to the substrates by their electrical connections (solder pads if the components are small) and by adhesives or by plastic or metal straps fixed to the board by fasteners such as bolts. Another mechanical feature which is common to substrate circuits is the construction known as a plated-through hole (PTH) or via. In the plated-through hole, the conductor tracks on the top of the substrate are electrically connected to buried tracks within the substrate or laminate. Such through holes and vias are shown in Fig. 3.2.

TABLE 3.2   Mechanical Properties of Commonly Used Substrate Materials

Material	Flexural strength (lb/in^2)	Conductor peel strength (lb/in)	TCE - in-plane (ppm/°C)	TCE out-of-plane (ppm/°C)
FR-4 epoxy/glass	60,000	8	14	44
Polyimide/glass	65,000	8	15	55
Al$_2$O$_3$ (alumina)	24,000	>20	8	8
PTFE/glass	45,000	7	15	50
PTFE/aramid	40,000	8	6	62

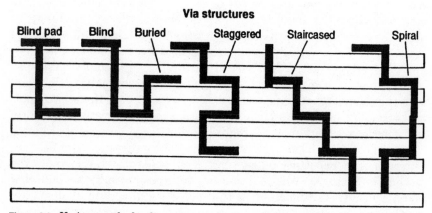

Figure 3.2 Various methods of carrying conduction to internal conductors in multilayer boards and substrates.

Since vibration can be encountered in many applications, one can expect flexure of the substrate between mounting points. The amplitude of the flexure is a function of the stiffness (mechanical property) and mass of the substrate and its components, the applied vibration energy and the frequency of the vibration.

Over time, flexure of the tracks and through holes with respect to the board can cause cracking of the tracks and through holes. These cracks usually are detected by intermittent electrical performance, during either vibration or temperature cycling. Therefore, if the application must endure these environments, it is imperative that care be taken to analyze the mechanical performance, both statically and dynamically. In addition to providing a means of mechanical support for these circuit elements, the substrates also can be used to provide thermal paths for conducting heat away from components so that they can be cooled and maintained at a temperature within the limits specified for their operation. This is an important feature, which cannot be overlooked or diminished. Components will perform predictably if certain conditions are met, and one of them is the temperature at which they are held. This does not necessarily mean that their temperatures must be held within a highly restricted range, but device input and output parameters may exceed desired limits if the device temperatures are not controlled to predictable limits.

3.3 Characteristics of Microwave Circuits

The most significant characteristic of a microwave circuit is that its circuit elements are comparable in size to the wavelength at its operating frequency. *Microwave frequencies* are typically defined as the frequencies in the range from 1 GHz to several hundred gigahertz. At these frequencies, wavelengths are on the order of 30 to about 0.1 cm. Since the wavelengths are small, phase changes occur rapidly along the length of the circuit, and, as a result, the circuit analysis and design techniques used at lower frequencies are no longer adequate. For this reason, distributed circuit techniques are employed. These involve the use of transmission line techniques and analysis that deal with the phenomenon of waves traveling along a line. In a distributed circuit, transmission line properties, such as characteristic impedance, velocity of propagation, attenuation, and phase shift, must be considered in the design. These are influenced directly by the material properties of the substrate such as the dielectric constant and the dissipation factor, as well as the substrate thickness. In the following paragraphs, we will discuss these material properties and review their relationship to the propagation of electromagnetic waves and to the properties of transmission lines.

3.3.1 Permittivity and permeability

Two very important properties of materials that affect electromagnetic waves are permittivity and permeability. Permittivity relates to electric fields and permeability relates to magnetic fields. *Permittivity* is a term that describes how a material under consideration allows itself to become polarized by an electric field imposed upon it from outside itself. *Polarization* refers to the realignment of charged particles within the dielectric along the lines of the imposed electric field. When an external electric field is imposed, it causes a force to be exerted on each particle, and, as a result, the charges in the dielectric are made to realign themselves or move from their resting place with positive charges being moved in one direction and negative charges being moved in the opposite direction.

The permittivity, ε, of a given material is expressed in relation to the permittivity, ε_0, of a vacuum of air. (In practical terms, the value of permittivity of air is the same as for a vacuum or what is commonly referred to as "free space.")

This relationship is given as

$$\varepsilon = \varepsilon_0 \kappa$$

or stated another way

$$\kappa = \frac{\varepsilon}{\varepsilon_0}$$

where κ is known as the dielectric constant of a material. The value of $\kappa = 1$ for air. The value of ε_0 is 8.85×10^{-12} F/m.

When electromagnetic fields vary in time, the representation of permittivity, ε, of a material must be modified to account for the loss mechanisms that become significant at higher frequencies. In general, the permittivity is complex and can be written as

$$\varepsilon = \varepsilon' - j\varepsilon''$$

where ε' is the real part of the permittivity and ε'' is the imaginary part of the permittivity. The ratio of $\varepsilon''/\varepsilon'$ is known as the loss tangent of the material and relates to the dielectric loss of an electrical circuit. The dielectric loss of a circuit is directly proportional to the loss tangent of the material and increases linearly with increasing frequency. The loss tangent is also known as the *dissipation factor.*

In similar fashion, *permeability* is a measure of the magnetic property of a material that allows a magnetic field to be set up within itself by another magnetic field imposed upon it from outside. The relationship of the relative permeability of a material to the permeability of air, μ_0, is

$$\mu = \mu_r \mu_0$$

where μ_r is the relative permeability of a given material.

The relative permeability of air is 1, so the value of μ for air is μ_0 which is 4×10^{-7} H/m. Values of dielectric constant and relative permeability for commonly used circuit materials are given in Tables 3.3 and 3.4. It should be stated here that the dielectric constants of insulating materials can vary a great deal but their relative permeabilities are about the same, namely, 1. The relative permeability of most metals can be assumed to be approximately 1, except for iron, nickel, and cobalt which can be much greater than 1. These three metals are said to be ferromagnetic because they tend to behave magnetically very much like iron.

3.3.2 Propagation of electromagnetic waves

An electromagnetic wave propagating in free space or in a simple dielectric medium, at large distances from the transmitter, may be well represented by a uniform plane wave. A uniform plane wave propagates with the electric field normal to the magnetic field and with no electric or magnetic field in the direction of propagation. Since the

TABLE 3.3 Typical Permittivity of Commonly Used Microwave Materials

Dielectric	Permittivity @ 25°C, 1 kHz, expressed as *relative* dielectric constant
Polytetrafluoroethylene (PTFE)	2.04
Polyethylene	2.25
Glass-reinforced PTFE	2.55
Cyanate ester/glass	3.6
Polysulfone	3.8
Quartz	3.8
Polyimide-glass (woven)	4.2–4.6
Epoxy-glass (woven)	4.3
Low temperature cofired ceramics	5.0–6.0
Ceramic-loaded PTFE	6.0–10.0
Beryllium oxide (beryllia)	6.9
99% Aluminum oxide (alumina)	9.7

TABLE 3.4 Typical Permeability of Commonly Used Microwave Conductors

Conductor	Relative permeability μ_r
Gold	1
Aluminum	1
Tin	1
Copper	1
Silver	1
Nickel	600
Iron-nickel-cobalt	1100
Tantalum	1.01
Chromium	1.05
Titanium	1.01

electric and magnetic fields lie entirely in the transverse plane, these waves are referred to as *transverse electromagnetic* (TEM) *waves*. This mode of propagation is also the principal mode for most two-conductor transmission lines. For electromagnetic waves propagating in the TEM mode, the velocity of propagation, v_p, of the wave in a given medium is related to the permittivity, ε, and the permeability, μ, of the medium by

$$v_p = \frac{1}{(\varepsilon\mu)^{1/2}}$$

In air or vacuum, the speed of light is c, where

$$c = \frac{1}{(\varepsilon_0\,\mu_0)^{1/2}}$$

which is 3×10^{10} cm/s.

To determine the speed of electromagnetic waves in any other medium relative to their speed in air, one simply uses the dielectric constant and relative permeability in the relation

$$v_p = \frac{1}{(\kappa\varepsilon_0\mu_r\mu_0)^{1/2}}$$

which yields

$$v_p = \frac{c}{(\kappa\mu_r)^{1/2}}$$

Since it is convenient and often important to know the speed of electromagnetic waves in various media, this relationship allows a quick and simple computation of that speed in terms of the speed of light in air or in a vacuum. The speed of electromagnetic waves in any medium other than a vacuum is always less than it is in a vacuum. Thus, we often express the speed of electromagnetic waves in a given medium as, for example, 80% of c (where c is the speed in a vacuum). This would yield a speed of $0.8 \times 3 \times 10^{10}$ cm/s or 2.4×10^{10} cm/s.

For dielectric materials (with the exception of ferrite materials) the permeability, $\mu_r = 1$, and the velocity of propagation, v_p, is simply

$$v_p = \frac{c}{(\kappa)^{1/2}}$$

Since this is true for most dielectric materials, for the remainder of this chapter when we speak of dielectric materials, we will assume that $\mu_r = 1$.

As we have just stated, electrical signals travel at different speeds in air and in circuit board dielectrics, and that speed is determined by the dielectric constant of the medium. The higher the dielectric constant, the slower the signal speed. This brings us to the concept of wavelength and electrical length. The wavelength, λ, of an electromagnetic wave, which is the distance the wave travels in one period, is related to the frequency of the wave, f, and the velocity of the wave, v_p, by

$$v_p = f\lambda$$

We can also express v_p in terms of its radian frequency, $\omega = 2\pi f$, as

$$v_p = \left(\frac{\lambda}{2\pi}\right)\omega$$

or

$$v_p = \frac{\omega}{\beta}$$

where $\beta = 2\pi/\lambda$. The term β is often referred to as the *phase constant* and is expressed in radians per unit length. It follows that $\lambda = v_p/f$ may be written as

$$\lambda = \frac{2\pi}{\beta}$$

where λ is expressed in units of length.

The wavelength in a dielectric medium may also be expressed in terms of the velocity of propagation in free space, c, and its dielectric constant, κ, as

$$\lambda = \frac{c/f}{(\kappa)^{1/2}}$$

Also, since the wavelength, λ_0, of a wave propagating in free space is $\lambda_0 = c/f$, the wavelength, λ, in a dielectric medium can be expressed in terms of its free space wavelength as

$$\lambda = \frac{\lambda_0}{(\kappa)^{1/2}}$$

Often we may want to express the length of a circuit in electrical units (radians or degrees) rather than in wavelengths. Remembering that there are 2π radians in a wavelength, the electrical length, θ, of a circuit may be written as

$$\theta = \left(\frac{2\pi}{\lambda}\right)l$$

or

$$\theta = \beta l \quad \text{radians}$$

where l is the physical length of the circuit. The electrical length may also be expressed in degrees as

$$\theta = \left(\frac{360}{\lambda}\right) l \quad \text{degrees}$$

3.3.3 Transmission lines

As mentioned previously, circuit design at microwave frequencies involves circuit techniques that make use of distributed circuits. These distributed circuit elements are often made from short sections of transmission lines. A transmission line is a system of parallel conductors used to guide electromagnetic waves from one point to another. Usually, the conductors are separated by a dielectric material and are uniform in cross section and separation. There are many forms of transmission lines, some of which will be discussed later in this chapter.

3.3.3.1 Transmission line analysis. For the purpose of analysis we will examine the uniform two-conductor line of Fig. 3.3a. We can analyze the transmission line in terms of its distributed circuit parameters, as defined in Fig. 3.3b.

The equivalent circuit of an infinitesimal length of line consists of a series resistance per unit length, R, and a series inductance per unit length, L, followed by a shunt conductance per unit length, G, and a shunt capacitance per unit length, C.

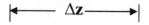

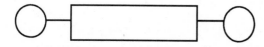

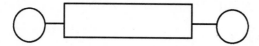

Figure 3.3a Circuit diagram of a transmission line element.

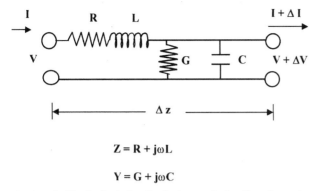

$$Z = R + j\omega L$$

$$Y = G + j\omega C$$

Figure 3.3b Equivalent circuit of a transmission line element.

We can derive from the analysis two important transmission line parameters in terms of the distributed elements. The first is the characteristic impedance, Z_0, and the second is the propagation constant, γ. The solution to the transmission line equations show that, in general, two oppositely traveling waves propagate along the line. This may be described mathematically as

$$V(z) = V_1 e^{-\gamma z} + V_2 e^{+\gamma z}$$

and

$$I(z) = I_1 e^{-\gamma z} + I_2 e^{+\gamma z}$$

where V_1 and V_2 are complex constants that describe the magnitude and phase of the voltage waves traveling in the positive and the negative z directions, respectively; $\gamma = (ZY)^{1/2}$ where Z is the series impedance per unit length; Y is the shunt admittance per unit length, and $I_1 = V_1(Z/Y)^{1/2}$ and $I_2 = -V_2/(Z/Y)^{1/2}$. If we let $Z_0 = (Z/Y)^{1/2}$, then $I_1 = V_1/Z_0$ and $I_2 = V_2/Z_0$. The parameter, Z_0, is known as the *characteristic impedance* of the transmission line and it relates the ratio of the voltage, $V(z)$, and the current, $I(z)$, at any point along the line, z, as

$$Z_0 = \frac{V(z)}{I(z)}$$

The characteristic impedance is defined in terms of its circuit elements as

$$Z_0 = \left(\frac{R + j\omega L}{G + j\omega C}\right)^{1/2}$$

which, in general, is complex. For lines with low loss, where R and G are negligibly small, the characteristic impedance may be approximated by the real term as

$$Z_0 = \left(\frac{L}{C}\right)^{1/2}$$

The second important transmission line parameter is the propagation constant, γ, which relates to the magnitude and phase of a wave traveling along the line. It can be written in terms of its circuit elements as

$$\gamma = [\,(R + j\omega L)\,(G + j\omega C)\,]^{1/2}$$

or

$$\gamma = \alpha + j\beta$$

where α is the attenuation constant and β is the phase constant. For lines with low loss, the resistive components are small and the real and imaginary parts of the propagation constant may be approximated by

$$\alpha = \frac{1}{2}\left(\frac{R}{Z_0} + \frac{G}{Y_0}\right)$$

and

$$\beta = \omega\,(LC)^{1/2}$$

where $Y_0 = 1/Z_0$.

3.3.3.2 Lossless transmission lines. For many practical transmission lines the losses are small enough that, for the purpose of analysis, the line may be considered to be lossless. For lossless lines, where R and G are equal to zero, the characteristic impedance, Z_0, and the propagation constant, γ, become.

$$Z_0 = \left(\frac{L}{C}\right)^{1/2}$$

and

$$\gamma = \alpha + j\beta = j\beta$$

where $\alpha = 0$ and

$$\beta = \omega\,(LC)^{1/2}$$

The characteristic impedance is purely real and the propagation constant is purely imaginary. The expressions for the voltage, $V(z)$, and current, $I(z)$, now become

$$V(z) = V_1 e^{-j\beta z} + V_2 e^{+j\beta z}$$

$$I(z) = (1/Z_0)\, (V_1 e^{-j\beta z} - V_2 e^{+j\beta z})$$

If we consider the terminal conditions shown in Fig. 3.3b, then $V_1 e^{-j\beta z}$ is referred to as the *incident wave* and $V_2 e^{+j\beta z}$ is referred to as the *reflected wave*. We can define a *reflection coefficient*, ρ, as the ratio of the voltages, V_1 and V_2, as

$$\rho = \frac{V_2}{V_1}$$

or

$$\rho = \left|\frac{V_2}{V_1}\right| e^{j(\theta_2 - \theta_1)}$$

where $|V_1|$ and $|V_2|$ are the amplitudes of V_1 and V_2, and θ_1 and θ_2 are the phase angles of V_1 and V_2, respectively. We can now rewrite the expressions for $V(z)$ and $I(z)$ as

$$V(z) = V_1 e^{-j\beta z}\, (1 + \rho e^{+j2\beta z})$$

and

$$I(z) = \left(\frac{I_1}{Z_0}\right) V_1 e^{-j\beta z}\, (1 - \rho e^{+j\beta z})$$

If we define $z = 0$ to be at the load end of the line, then, at $z = 0$, the load impedance, Z_L, becomes

$$Z_L = \frac{v(0)}{I(0)}$$

or

$$Z_L = \frac{Z_0\, (1 + \rho)}{(1 - \rho)}$$

If we solve for ρ, then

$$\rho = \frac{Z_L - Z_0}{Z_L + Z_0}$$

It is interesting to note that if the impedance of the load, Z_L, terminating the line is equal to the characteristic impedance, Z_0, then $\rho = 0$ and there is no reflected wave. In this case we say that the line is terminated in a "matched load." If we terminate the line with an open circuit, then $1/Z_L = 0$ and $\rho = 1$ which represents a total reflection. If we terminate the line with a short circuit, where $Z_L = 0$, then $\rho = -1$ and the wave is totally reflected with a phase shift of $180°$.

We often want to compute the impedance looking into the terminals at the source end of the line at a distance $z = l$ from the load. This impedance is known as the input impedance, Z_{IN}, and is written as

$$Z_{IN} = \frac{Z_0 [Z_L + jZ_0\tan (ßl)]}{Z_0 + jZ_L\tan (ßl)}$$

This equation is widely used in the design of distributed circuit elements that use transmission line sections. For example, if $ßl = \pi/2$ or 90°, then Z_{IN} becomes

$$Z_{IN} = \frac{Z_0^2}{Z_L}$$

and the circuit behaves as an impedance transformer, commonly known as the *quarter wave transformer.* If $Z_L = 0$ (a short circuit), then Z_{IN} becomes infinite (an open circuit). Conversely, if the load is an open circuit, the input looks like a short circuit.

Other examples include short sections of line where $ßl$ is small and the input impedance can be approximated by

$$Z_{IN} = Z_L + j \left(\frac{Z_0^2 - Z_L^2}{Z_0}\right) \tan ßl$$

For $Z_0 > Z_L$

$$Z_{IN} = Z_L + jX \qquad \text{inductive}$$

and for $Z_0 < Z_L$

$$Z_{IN} = Z_L - jX \qquad \text{capacitive}$$

3.3.3.3 Transmission line losses.

In the preceding paragraphs we dealt with lossless transmission lines because it was convenient for the purpose of analyzing low loss circuits. In reality, all transmission lines have losses and we often need to account for them. If we reexamine the expression for $\alpha = \frac{1}{2}(R/Z_0 + G/Y_0)$, we can associate the series resistance per unit length, R, with the conductor losses and the shunt conductance per unit length, G, with the dielectric losses. The attenuation due to the conductor losses may be written as.

$$\alpha_c = \frac{1}{2} \left(\frac{R}{Z_0}\right)$$

where R is a function of frequency. At very low frequencies, the current is evenly distributed over the whole cross section of a conductor, whereas at very high frequencies, currents do not penetrate deeply into the conductor and flow only near the surface. The skin depth, δ, represents the depth at which the magnitude of the penetrating currents falls to $1/e$ (36%) of the magnitude at the surface. The number e is approximately 2.7, the base of the natural logarithm. The skin depth, δ, is related to frequency by

$$\delta = \frac{1}{(\pi f \mu \sigma)^{1/2}}$$

where σ is the conductivity of the conductor. The attenuation, α_c, due to the conductor losses cannot, however, be calculated in general since it is dependent on the geometry of the transmission line. Another factor affecting the conductor loss is the surface roughness. As the frequency increases, the skin depth decreases. Depending upon the surface finish of the conductor, a point is reached where the surface roughness approaches the magnitude of the skin depth. When this condition is reached, the effective surface resistivity increases rapidly relative to the value for a smooth finish. The effective surface resistivity is increased by a factor of k, where k is a function of the geometry and of the frequency, usually between 1 and 2.

The attenuation due to the dielectric losses, $\alpha_d = \frac{1}{2}(G/Y_0)$, is also a function of frequency but is not dependent on the geometry of the circuit. It may be expressed in terms of the dielectric properties as

$$\alpha_d = \left(\frac{1}{2}\right) \omega (\mu \varepsilon)^{1/2} \tan \delta$$

where $\tan \delta$ is the loss tangent of the dielectric. If we compare the two components, we find that the conductor losses are proportional to the square root of the frequency, whereas the dielectric losses are proportional to the frequency.

3.3.3.4 Transmission line types.

As previously mentioned, a transmission line is a system of conductors used to guide electromagnetic waves from one point to another. Generally, a transmission line may be characterized by the mode of the electromagnetic wave that it propagates. There are three basic types. The first type includes the transmission lines that propagate the TEM wave as its principal wave. This wave contains neither electric fields nor magnetic fields in the direction of propagation, and is thus named transverse electromagnetic (TEM). The basic geometry of a TEM line is one consisting of two or more uniform parallel conductors generally separated by a dielectric. Examples of TEM lines are the coaxial cable, the two-wire line (twin lead), and stripline. TEM lines operate over large frequency bandwidths from DC to very high frequencies.

The TEM wave propagates without dispersion, that is, the velocity of the wave is constant with frequency. A pulse traveling down the line will not spread. The characteristic impedance of a TEM line is constant with frequency. The second type of transmission line includes lines that propagate quasi-TEM waves. These waves are basically TEM but have small electric or magnetic field components in the direction of propagation. The basic geometry of quasi-TEM lines are planar circuits with nonuniform or mixed dielectrics. Examples are coplanar waveguide structures and microstrip lines. These lines operate over wide frequency bandwidths from DC to very high frequencies with low to moderate dispersion. Some pulse spreading occurs. The characteristic impedances of these lines vary a small amount with frequency. These lines are also prone to radiation at high frequencies.

The third type of transmission line includes lines that will not propagate the TEM wave but will support TE or TM waves. The TE wave has no electric field in the direction of propagation (transverse electric), whereas the TM wave has no magnetic field in the direction of propagation (transverse magnetic). Examples are the hollow waveguides and fiber optic cables. TE and TM lines do not operate down to DC since these waves do not propagate below a certain cut-off frequency. Single mode propagation is limited to relatively narrow bandwidths. These lines are very dispersive, that is, the velocity of propagation varies with frequency. A pulse traveling along a line would experience large spreading and distortion. The characteristic impedance of these lines changes significantly with frequency.

3.3.3.5 Commonly used transmission lines. Some commonly used transmission lines are shown in Fig. 3.4. Transmission lines (*a*), (*b*), and (*c*) are widely used and are important in many applications but are generally not used in PWBs and ceramic substrates. We will, however, describe them here briefly.

The transmission line in (*a*) is a parallel two-wire line often referred to as *twin lead*. It consists of two closely spaced round wires, usually supported by a dielectric material. Its range of impedances is from 300 to 500 Ω. It is widely used at ultrahigh frequency (UHF) and lower RF frequencies, particularly for television antenna hookup, but it is rarely used at microwave frequencies because it is unshielded and radiates energy. This radiation is undesirable since it often results in excessive losses as well as poor isolation from nearby circuits.

The coaxial line (*b*) overcomes this shortcoming by placing one conductor concentrically around the other such that all of the electromagnetic energy is contained between the two conductors inside the outer conductor. Thus, the coaxial line is completely shielded. Coaxial lines operate with moderately low line losses and with moderate power-handling capability. Signals propagate in the TEM mode without dispersion from low frequencies through microwave frequencies.

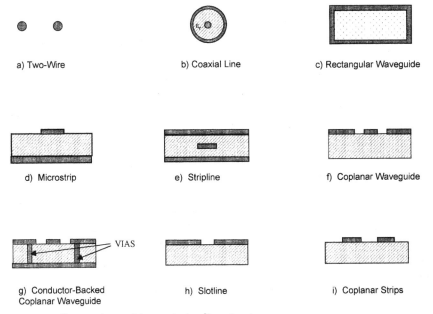

a) Two-Wire

b) Coaxial Line

c) Rectangular Waveguide

d) Microstrip

e) Stripline

f) Coplanar Waveguide

g) Conductor-Backed
Coplanar Waveguide

h) Slotline

i) Coplanar Strips

Figure 3.4 Commonly used transmission line structures.

The upper frequency limit is determined by higher order mode propagation which is a function of the coaxial line conductor diameters.

The rectangular waveguide (c) is a transmission line in which the top and bottom conductors are shorted at the ends with two vertical sidewalls. Signals propagate only at frequencies that are greater than the waveguide cut-off frequency, which is the frequency at which the width of the waveguide is equal to one-half of a wavelength. This propagation is *dispersive,* that is, the velocity of propagation is not constant with frequency and the wavelength is not precisely inversely proportional to frequency. As the frequency is doubled, the wavelength is not exactly halved. The bandwidths are generally limited to about 40% without the propagation of higher order modes. Other waveguide structures include circular waveguides and ridged waveguides. The primary advantages of the waveguide structures are that they are lower in loss and can handle higher power levels than other transmission line structures.

3.3.3.6 Planar transmission lines. The planar transmission line structures of (d) through (i) are the most commonly used for PWBs and ceramic substrate applications because the planar nature of these structures lend themselves more readily to PWB and ceramic substrate processing where conductors are either printed or screened. A comparison of these planar transmission lines is shown in Table 3.5. We will examine these structures in more detail in the following paragraphs.

3.3.3.7 Microstrip lines. The microstrip line of (*d*) is the most widely used planar transmission line and consists of a flat conductor on top of a dielectric substrate with a ground plane on the other side. Propagation is quasi-TEM with some dispersion. This is a result of the wave traveling partly in the air above the line and partly in the dielectric below the line. The velocity of propagation is inversely proportional to the square root of an effective dielectric constant that is between that of the dielectric constant of the substrate material and that of air. The effective dielectric constant is a function of the microstrip conductor width and the substrate height and varies slightly with frequency. The bandwidth of a microstrip circuit is large and may be used at frequencies up to 100 GHz. The realizable range of impedances is typically from 15 to 100 Ω depending on the dielectric constant of the substrate material and the substrate thickness. Microstrip circuits are small in size and have excellent integration capabilities with chip and lumped devices. It is particularly easy to connect series elements. Microstrip structures also lend themselves to multilayer circuits because the ground plane isolates the microstrip from other buried structures. The line losses are somewhat higher than other transmission lines with poorer isolation between circuits. The microstrip line is unshielded and some radiation occurs for thicker substrates depending on the dielectric constant of the substrate and the frequency of operation. For example, a 50-Ω microstrip on a 0.025-in-thick alumina substrate may operate at 10 GHz with only minimal radiation. A 50-Ω line on an organic substrate of a dielectric constant of about 3 would require a substrate thickness of less than 0.015 in to avoid excessive radiation losses at 10 GHz.

3.3.3.8 Coplanar waveguide. Another planar circuit commonly used is the coplanar waveguide (CPW) (*f*). It consists of a center strip

TABLE 3.5 Comparison of Microstrip, Stripline, Coplanar, and Slot Line Waveguide Features

Characteristic	Microstrip	Stripline	Coplanar strips	Slot line
Attenuation loss	Low	Low	Medium	High
Impedance (Ω)	15–110	20–150	25–125	50–300
Radiation loss	Low-medium	Very low	Medium	High
Dispersion	Low	Low	Medium	High
Connect shunt elements	Difficult	Difficult	Easy	Easy
Connect series elements	Easy	Difficult	Easy	Difficult

located between two semi-infinite ground planes located on the same surface of a dielectric substrate. The backside of the substrate is unmetalized. Its advantages are that it can achieve a wide range of impedances from 25 to 125 Ω with various line widths and gaps. The impedance is a function of the width-to-gap ratio. The center conductor width can be made small to reduce radiation. It offers better isolation than microstrip especially when constructed on a higher dielectric constant substrate. It provides very good integration capabilities for chip and lumped elements. Series and shunt connections are relatively easy. Propagation is quasi-TEM and has moderate dispersion and loss, somewhat more than microstrip. CPW circuits tend to take up more space in a layout than in a microstrip circuit. The major disadvantage of CPW is that thick substrates are required to keep the structure away from the chassis or other grounds. Integration into multilayer substrates is poor since the structure is unshielded.

3.3.3.9 Conductor-backed coplanar waveguide. Conductor-backed coplanar waveguide (*g*), sometimes called coplanar waveguide over ground or CPWG, offers some of the advantages of pure CPW with the performance of microstrip. The propagation is primarily that of a microstrip line, that is, most of the energy is in the microstrip mode where the electric field extends mostly between the center conductor and the bottom ground with only fringing fields located between the center conductor and coplanar grounds. The CPWG line (microstrip with coplanar grounds) offers the advantage of less radiation loss than a pure microstrip line because the fringing fields of the wave are terminated at the coplanar grounds. Another advantage is that connections can be made at the I/O's with ground, signal, or ground probes. The structure does require metal vias or plated-through holes to connect the coplanar grounds to the backside ground. Also, the structure is susceptible to unwanted parasitic modes.

3.3.3.10 Slot lines and coplanar strips. Two other planar transmission lines are the slot line (*h*) and coplanar strips (*i*). The slot line consists of two conductors on the same side of the substrate of semi-infinite extent separated by a narrow gap or slot. The coplanar strip structure is a special case of the slot line where the two conductors are of finite width. Propagation is non-TEM with high dispersion and loss. The advantages include easy integration capabilities with chip and lump elements and the realization of higher characteristic impedance values (40 to 300 Ω). Other advantages are when used in combination with microstrip, that is, when located in the microstrip ground plane, certain circuits can be realized that offer constant 180° phase shifts independently of frequency. These circuits are particularly useful in mixer networks. The major disadvantage of these structures is that they suffer from the same disadvantages of CPW and do not integrate well into multilayer substrates.

3.3.3.11 Stripline.

The stripline circuit (*e*) consists of a flat center conductor buried between two coplanar semi-infinite ground planes. The principal mode of propagation for the stripline structure is the TEM mode. This wave propagates when the two ground plans are maintained at the same potential. The propagation of the TEM wave is nondispersive and operates over wide bandwidths. A wide range of characteristic impedances can be realized—from 20 to 150 Ω. The losses in stripline are usually less than the losses in microstrip and higher than the losses in coaxial cables. The loss depends on stripline dimensions. The structure is similar to the coaxial line in that the electromagnetic fields extend only between the center conductor and the ground planes. No fields extend outside the ground planes; therefore, the structure does not radiate. Stripline structures lend themselves well to multilayer circuits because the structures are well shielded and may be isolated from other buried circuits. The stripline circuit, however, is susceptible to unwanted modes. Since the stripline structure is a three-conductor system, modes other than the principal TEM mode may be supported. When a potential difference is maintained across the two stripline ground planes, parallel plate waves may propagate. These modes may be TEM, TE, or TM. The particular mode that propagates depends on how the stripline is excited as well as the boundary conditions of the stripline package. Often the unwanted modes are excited at locations where the signals are launched. Asymmetrical connections to the stripline structure are particularly susceptible to exciting these unwanted modes.

Another source of excitation occurs at circuit discontinuities. Once excited, these parasitic modes may propagate along with the desired stripline TEM wave but the stripline conductor does not guide them. This is undesirable since this energy is lost to the desired transmission and may couple to other circuits in the same package. This coupling is generally referred to as *cross-talk*. Cross-talk between two circuits in the same stripline package may be suppressed by isolating the circuits with the use of conducting vias or plated-through holes that electrically short the two ground planes. Since the coupling between the two circuits is a result of parallel plate mode propagation, shorting the two ground planes with a row of conducting vias or shorting posts at locations between two circuits would not allow that wave to propagate across that fence. In practice, the vias or plated-through holes are not perfect shorts but act as small inductive posts. The amount of isolation, therefore, is not perfect and depends on the via diameter, the via height, and the spacing between vias. Table 3.6 shows the modeled isolation for via fences with vias of several diameters and spacings and for two dielectric materials. The isolation values are for single and multiple rows. Another technique to suppress parallel plate waves is to place a via fence on both sides of a stripline center conductor that follows the conductor along its route. This is shown in Fig. 3.5. The two fences should be close to the stripline conductor and the spacing between the two fences should be less than a half wavelength. This structure provides a waveguide below cutoff around the stripline conductor such that the parallel plate mode may not propagate. The stripline TEM wave, however, is not affected. These fences are particularly important around asymmetric structures such as vertical coaxial connections to the stripline.

3.3.4 Dielectric constant and characteristic impedance

It is usually important to control the impedance of microwave and RF circuits to within a few percent of a nominal value so as to minimize energy reflections at interfaces between the circuit tracks and circuit components. In the RF world, circuit components, as well as input and output impedances, are generally 50 Ω. Therefore, circuit track impedance should be 50 Ω as well. Assuming that circuit tracks are

TABLE 3.6   Isolation (dB) at 13 GHz for Via Wall Structures with Various Diameters, Spacings, and Row Separations

Via diameter (mils)	Via spacing (mils)	ε_r	Number and distance between rows (mils)					
			1	2/20	2/40	2/80	3/20	3/40
6.8	20	6.0	40	73	80	84	110	121
6.8	50	6.0	15	28	32	34	42	52
6.8	100	6.0	5	12	13	13	20	22
17	40	6.0	38	N/A	74	84	N/A	112
20	50	2.2	37	N/A	70	76	N/A	102
20	100	2.2	17	N/A	32	32	N/A	46

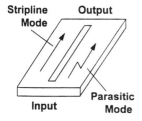
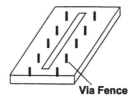

Figure 3.5 Use of via fence to suppress parasitic modes.

configured as transmission lines, a 50-Ω impedance can be easily achieved for the three most common configurations. Table 3.5 lists some of the features of those configurations.

The impedance of each of the transmission line structures is a function of the dielectric constant and thickness and of the geometry of the conductors. These dimensions can be easily calculated using various available software programs that enable designers to select design parameters that will result in a desired impedance. However, since microstrip is the most commonly used transmission line configuration, we will illustrate a simple manual approach that can be used as well.

In an effort to achieve miniaturization, which is desirable in ever-shrinking product size, dielectric materials are often selected that have a higher dielectric constant than typical PWB materials. This is done because higher dielectric constant materials permit lower ratios of track width to track height above a ground plane. This height above ground in a microstrip is the dielectric thickness. Figure 3.6 shows this feature for a microstrip track.

The effect of dielectric constant on conductor track width and characteristic impedance is shown in Fig. 3.7. In this figure, the dielectric thickness has been selected as 0.025 in. It can be seen from the figure that if a certain impedance value is desired, the line width must increase significantly as lower dielectric constant materials are used. This does not aid in reducing the size of the PWB. One way to reduce the size of the PWB is to use a high dielectric constant material. This allows the tracks to become smaller as dielectric thickness is reduced. Aluminum oxide (alumina or Al_2O_3) is a material that has been used for this purpose. This

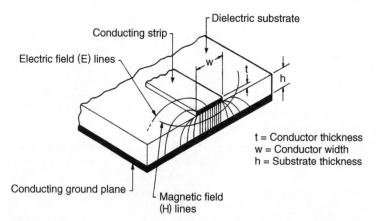

Figure 3.6 Microstrip transmission line. [Copyright © 1989 by ISHM (now IMAPS, International Microelectronics And Packaging Society). Reprinted with permission from the technical monograph entitled "Materials and Processes for Microwave Hybrids" by Richard Brown.]

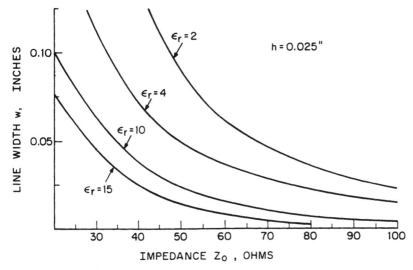

Figure 3.7 Effect of dielectric constant on conductor line width and characteristic impedance. [Copyright © 1989 by ISHM (now IMAPS, International Microelectronics And Packaging Society). Reprinted with permission from the technical monograph entitled "Materials and Processes for Microwave Hybrids" by Richard Brown.]

material is brittle and has a practical limit in size of about 2 in on a side for a thickness of 0.025 in. Several companies have produced a more flexible and less brittle material with essentially the same electrical properties, and marketed it under different trade names. The materials are essentially polytetrafluoroethelyne (PTFE) with the addition of fillers that have higher dielectric constants, resulting in a composite ε_r of approximately 6.0 or 10.0. The value of 10.0 is roughly the value of the dielectric constant of alumina. The PTFE has a very low dissipation factor, and is known for its ability to resist water absorption, making it highly desirable as a microwave dielectric with predictable properties under high humidity conditions. The material's flexibility allows it to be machined to any desired shape, adding to its attractiveness. Etching the copper cladding (foil) into tracks forms conductors. Being clad with copper on top and bottom makes it an ideal candidate for microstrip circuits. Dielectric thickness is usually 0.025 when the ε_r is 10.

To execute a microstrip design it is necessary to find the values of line width w, h, the dielectric thickness, and the ratio w/h for specific dielectric constant materials, since they all interact. The term *effective relative dielectric constant*, or ε_{eff}, is used to account for the inhomogeneity of the total microstrip dielectric, since that configuration has air above the track and a higher dielectric constant material below it. The relative dielectric constant of the dielectric below is ε_r and the dielectric constant of the air above is 1. The effective relative dielectric constant is made up of the combination $\varepsilon_0 = 1$ and $(\varepsilon_r - 1)$ multiplied by a number called the filling fraction, q. Therefore

$$\varepsilon_{\text{eff}} = \varepsilon_0 \text{ air} + q\ (\varepsilon_{r\ \text{substrate}} - 1)$$

or

$$\varepsilon_{\text{eff}} = 1 + q\ (\varepsilon_r - 1)$$

To find the value of ε_{eff} the filling fraction, the value of q, must be found first. Use of a graph (Fig. 3.8) allows this value to be calculated rather easily:

1. First, assume that $\varepsilon_{\text{eff}} = \varepsilon_r$.
2. Next, calculate the air space characteristic impedance, Z_{01}, from the relation $Z_{01} = \varepsilon_r^{1/2}$ times Z_0, the desired characteristic impedance— say 50 Ω.
3. From the value of Z_0, move to the curve for q and find the corresponding value for q.
4. Use that value for q to obtain an approximate ε_{eff} from the relation $\varepsilon_{\text{eff}} = 1 + q(\varepsilon_r - 1)$.
5. Replace ε_r in step 2 above with ε_{eff} and repeat steps 2, 3, and 4. Several iterations will yield the desired ε_{eff}.[*]

3.4 Multilayer circuits

With the drive toward lower costs, lighter weight, and smaller volumes for today's and future RF and microwave systems, the trend for achieving higher packaging densities is moving toward multilayer circuitry. Many advanced applications require the integration of low loss microwave structures, high speed digital circuitry, and DC power circuits into the same package. Figure 3.9 shows an example of a typical mixed signal application using multilayer cofired ceramic printed wiring technology. In this example, signals are routed on several layers buried within the substrate as well as on the top surface. Microwave tracks include buried stripline located between internal ground planes and microstrip tracks at the top surface. Digital and power lines are routed between the buried ground planes. Connections are made between layers and to the top surface with metal vias. Chips or other packaged components are mounted into the circuit on the top surface of the substrate. Other examples include chips placed into substrate cavities, stripline to microstrip feed

---

*Steps 1 through 5 above are courtesy of IMAPS. [Copyright © 1989 by ISHM (now IMAPS, International Microelectronics And Packaging Society). Reprinted with permission from the technical monograph entitled "Materials and Processes for Microwave Hybrids" by Richard Brown.]

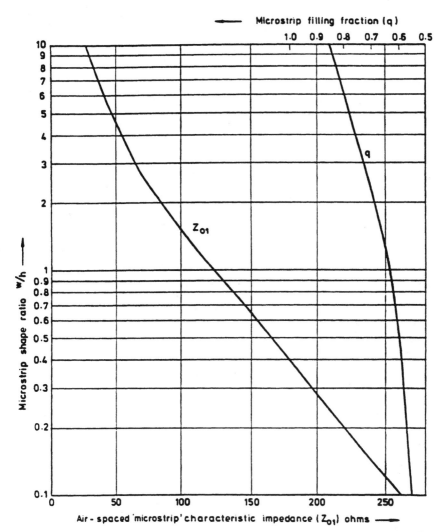

Figure 3.8 Graphic evaluation of w/h and ε_{eff}. [Copyright © 1989 by ISHM (now IMAPS, International Microelectronics And Packaging Society). Reprinted with permission from the technical monograph entitled "Materials and Processes for Microwave Hybrids" by Richard Brown.] (*Courtesy IMAPS.*)

throughs, and coplanar structures. In the following paragraphs the fabrication and processing of these multilayer substrates is described in detail.

Examples of multilayer design. The following illustrates several multilayer design approaches accomplished through various media. The first media group, ceramics, consists of two basic types: high temperature cofired ceramic (HTCC) and low temperature cofired ceramic (LTCC). The words "high" and "low" refer to the firing temperature. HTCC requires about

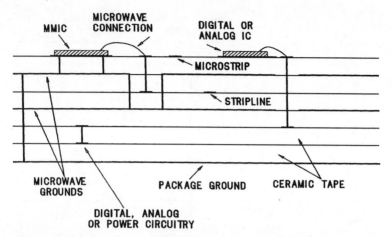

Figure 3.9 Multilayer ceramic structure showing various buried conductors and ground planes.

1600°C while LTCC fires at 850 to 950°C which does not seem like a low temperature, but when compared to 1600°, it is not very high. The distinction is important to processors of ceramic materials because different equipment is required for each type. The second, printed wiring boards, consists of several different materials. The two major groups, ceramics and printed wiring boards, differ in processing, mechanical and electrical properties, and also in design requirements. A brief overview of the processing steps for both ceramics and printed wiring boards is presented here for clarity and understanding of the differences as well as the appropriateness of either for specific applications.

The difference between HTCC and LTCC materials lies in their basic makeup and, consequently, in their processing. It is difficult in most cases to differentiate one from the other by visual properties. However, their finished or fired properties are significantly different. For example, the HTCC group has a dielectric constant of nearly 10 while the LTCC group has a dielectric constant typically from 4 to 8. Another difference is that the loss tangent of HTCC can be much lower than that of LTCC as well. The principal difference for microwave designers is the metal systems that each uses. The HTCC materials use refractory metals, such as tungsten and molybdenum, for conductors which have a resistivity that is several times higher than the silver systems that are used in LTCC substrates. This can be a significant factor in performance because the high resistivity causes electrical losses that can be unacceptable for certain circuits. Both HTCC and LTCC dielectrics begin as specially prepared powders that are ground and milled to a very fine and uniform particle size. However, the HTCC powders are made up essentially of aluminum oxide (alumina) whereas the LTCC powders are mostly glass with some alumina and other minerals added to raise the dielectric constant. During HTCC firing, the alumina particles fuse together to form a hard material. During LTCC firing, the glass melts and encases the alumina particles which themselves do not fuse together because the firing temperature is not high enough for that to happen. This accounts for the basic differences between HTCC and LTCC. What follows is a brief description of the basic process steps for both HTCC and LTCC, which are very similar. Differences are noted where applicable.

Since the finished properties are different, one should expect the design details to be different also. This is true from the standpoint of conductor lines' starting dimensions and desired height of the conductors above ground, because the materials are prepared in tape form similar to a sheet of computer printer paper. The final structure of the substrate or ceramic board is achieved through the stacking of individual layers. The tape is made from a basic ceramic powder mixed with binders and solvents that allow the mixture to take the form of a tape. These solvents and binders are organic in nature. Baking evolves the solvents, and the remaining binders are burned out as the higher temperatures of firing are reached. After this has been accomplished, as higher temperatures are reached, the glass and ceramic tapes shrink. The degree of shrinkage is dependent on the amount of organic materials in the initial tape. The applicable shrinkage factor of a given tape must be known before conductor features and dimensions can be specified correctly. Conductors are printed onto the tape before firing, using a metal slurry or paste that contains the conductor metal in fine particulate form. Consequently, their dimensions change according to the shrinkage that takes place during the firing cycle. Typical shrinkage can be between 8 and 14% depending on the material makeup. The main processing differences between HTCC and LTCC are the firing ovens, the processing temperatures, and ambient atmospheres. HTCC ovens operate at temperatures above 1600°C and employ hydrogen gas to some extent. LTCC ovens can be the same as those used in thick-film processing and they see temperatures up to 1050°C. The firing time is also longer for HTCC substrates. As was noted, despite similar appearances between the finished HTCC and LTCC product, the fired properties are different and must be accounted for in the design.

3.4.1 Basic HTCC and LTCC processes

There are eight basic processes required to fabricate a cofired ceramic substrate device: blanking, via formation, via filling, pattern printing, cavity formation, lamination, firing, and postprocessing.

3.4.1.1 Blanking.
Blanking is the process whereby sections of green (unfired) tape of an appropriate size for processing are cut from a large roll of tape. Blanking can be accomplished with a die that may also provide holes for tooling and registration, or by a laser or an automatic blanker or a razor blade guided by a template.

3.4.1.2 Via formation.
Holes or vias are formed in green unfired tape by a punching process. The blanked green tape is attached to frames

having an opening for a sheet of green tape. These frames are then placed in a punch machine. The punch machine is numerically controlled and can punch several sizes of vias in any required pattern. Punching is accomplished by calling up the appropriate data for the layer of tape and directing specific punch and die sets to operate. Each punch head operates pneumatically to send the punch through the green tape and into the die.

3.4.1.3 Via filling. Vias are filled with specially formulated conductive material applied through a stencil in a printing process. Stencils are typically made from 0.002-in-thick stainless steel. Stainless steel stencils are made by photoimaging the via pattern and chemically etching the holes in the stencil. Quality factors are completeness of fill without overfill and accuracy of placement of the fill material. The *aspect ratio*, defined as the ratio of the tape thickness to the formed via diameter, is an important design factor.

3.4.1.4 Pattern printing. Conductor patterns are printed on the green tape using the process for printing thick-film conductors. In general, printing on green tape is of better quality and higher resolution than that possible on a fired thick-film substrate. One reason is that the printing is always done on a flat surface without the warping and topography characteristic of multilayer thick film; another reason is that the paste is deposited on a porous surface which inhibits the tendency of the paste to spread. Cofired conductors are often now as wirebondable as thick-film conductors. To achieve extremely high bond success rates, it is necessary to use postfired or plated metalization on top of the cofired metalization.

3.4.1.5 Cavity formation. Some designs are enhanced by the use of cavities that hold semiconductors and other components. Routing each layer of tape in the required location, using either a mechanical router or a laser, forms cavities. The preferred method of cavity formation depends on the tape system being used as well as the type of cavity being formed. The laser is normally used for speed and accuracy.

3.4.1.6 Lamination of tape layers. Individual sheets of processed tape used in a substrate are aligned, collated, and pressed together to form a cohesive laminate ready for subsequent firing. Two distinct methods of lamination are used. One is isostatic and the other is uniaxial in nature. Isostatic lamination is a process that employs a flexible plastic or rubber bag that holds the tape layers that are to be laminated. The layers are placed in a fixture and the entire assembly is then placed into the bag. The bag is put into a chamber that is filled with a fluid which is then put under considerable pressure. Since the hydrostatic pressure on the layers is uniform, this is called the *iso-static process*. Isostatic lamination is used to laminate flat substrates and especially substrates that contain cavities where it would be impossible to apply pressure to the surfaces of the cavities (internal walls and floors). Uniaxial lamination is accomplished by placing the layers into a standard hydraulic press that is capable of applying heat to the layers during pressing. This process is used for parts that do not contain cavities.

3.4.1.7 Substrate firing. Laminates are fired in a furnace to burn off organic binders and to sinter ceramic and metallic constituents. *Sintering* is a term applied to the action of the particles of ceramic and glasses which depicts a flowing together and fusing of the constituents. Burn-out and firing can be accomplished in one continuous operation, and, with proper care, the part does not experience thermal gradients across itself as furnace temperatures rise and fall. The thermal mass of the furnace with shelves and hardware should be made substantial so that the mass of the laminates and fixtures does not affect the temperature profile.

3.4.1.8 Postprocessing. Finished parts may be metalized with either a screen-printed thick film or a vacuum-sputtered and etched thin film conductor, depending on the planned use of the part. When post-processed metalization is used, shrinkage tolerances become of prime importance since the postprocessed pattern is based on a certain shrinkage. If that shrinkage is not achieved within an allowable toler-ance, the pattern will not fit the substrate.

3.4.2 Shrinkage of tape ceramic substrates in processing

The most prominent design consideration associated with cofired ceramic structures, either HTCC or LTCC, is their shrinkage during firing. The amount of shrinkage is a function of the composition of the tape, the design of the part (the presence of ground or power planes, or the presence of cavities), and the lamination and firing processes. In general, the nominal shrinkage for a given composition can be charac-terized very accurately, but the other factors mentioned can cause an uncertainty of up to several tenths of a percent to occur. One-tenth of 1% translates to 0.001 in of length for every inch in fired ceramic. That is significant because the diameters of vias that must line up with tracks are only several thousandths (mils) and a serious misalignment could occur between layers because of that uncertainty. A part that is 6 in in length can have a misalignment of as much as 3 mils. This factor is especially important if the part is designed to use postprocessing such as printing of very fine-line patterns that must align closely with vias or with a fine line connector whose dimensions are fixed.

The amount of shrinkage or the shrinkage factor representative of a particular lot of a specific composition of tape processed in a given facility must be known before the artwork for the printing patterns can be generated. In addition, matching of the postfired metalization to the cofired pattern with the attendant uncertainty in the location of features on the cofired part may not be possible unless adequate overlap of the two patterns is allowed. It is helpful to include an alignment mark when printing the top layer metalization to account for the alignment of the postfired layer to be printed on the substrate.

3.4.3 Properties of LTCC tapes and conductors

Since the LTCC process allows the use of multilayer boards with high electrical conductivity buried layers, the properties of some typical commercial available materials are given in Table 3.7. There is no attempt to represent any supplier's datasheet here, but these listed properties have been furnished and verified by measurements. For specific suppliers and the property data desired, consult the IMAPS whose address is given at the end of this chapter.

A well-known and long used tape that has been the mainstay of the hybrid industry is considered very manufacturable. However, its insertion loss is relatively high (electrical conductivity is low compared to a pure gold or pure copper conductor). Typical uses for this system are for frequencies up to X-band. The system offers cofired, wire bondable gold, tin-lead solderable gold-platinum, and silver metal formulations. Postfire copper may be used and gold-tin or gold-germanium solders may be used for brazing to form vacuum-tight seals.

A competitive product family is available that yields considerably lower losses than the previously mentioned system. This family consists of cofired, wire bondable gold, tin-lead solderable refractory metals, silver metal formulations, and cofired resistors as well. A cofired capacitor tape is offered as a development product. This product allows buried capacitors to be included within the cofired structure which can enhance packaging density as well as eliminate some solder or wire bond connections to those capacitors. This second system's gold and silver products must follow a different firing profile from each other which results in different dielectric properties for the ceramic material. This must be kept in mind when designs are under way if changes in dielectric constant and dissipation factor could affect circuit performance.

A third system features a very low dielectric constant tape material. The dielectric constant is 3.9 which, in combination with a spe-

TABLE 3.7   Electrical and Mechanical Properties of Cofired Tape Systems

Property	Vendor A	Vendor B	Vendor C	Comments
Unfired thickness (mils)	4.4, 6.5, 10.0	5.0, 10.0	1.9, 5.5	Fired thickness is less than 80% of unfired
Dielectric constant @ 10 GHz	7.85	5.9	3.9	
Dissipation factor @ 10 GHz	0.006	0.001	0.0006	
Breakdown voltage (V/mil)	>1000	>1000	>800	
Insulation resistance (Ω) @ 100 V DC	>10^{12}	>10^{14}	>10^{12}	
In-plane shrinkage (%)	12.7 ±0.2	15.25 ±0.3	19.0 ±0.4	
Out-of-plane shrinkage (%)	20.0 ±0.5	26.5 ±1	40.0 ±0.5	
Fired density (g/cm^3)	3.1	2.50	2.38	
Surface roughness (μin)	8.7	Not specified	<10.0	
Camber (mils/in)	<3	<3	<4	
Flexural strength (MPa)	320	170	90	
Thermal expansion coefficient (ppm/°C)	5.8	7.0	6.8*	
Thermal conductivity (W/m-K)	3.0	2.0	2.0	

*Custom tailorable from 3.0 to 9.0 ppm/°C.

cially prepared gold conductor material, has the lowest insertion loss of any materials so far. It has been used for applications as high as W-band (94 GHz) where it yielded losses comparable to those of polished quartz. The gold is wire bondable. Postfired resistor paste, solderable, and brazable conductors have been used successfully with the low dielectric constant cofiring tape. Although this commercial system is more difficult to manufacture than the two commercially available materials systems, its superior electrical properties have made it the material of choice in high performance applications.

3.4.4 Design of substrates and packages

3.4.4.1 Panel size. Depending on processing equipment capabilities, LTCC panels can be as large as 8 in on a side. After tooling holes have been punched, the material available for product is under 7 in. After firing, the shrinkage reduces the dimensions to just over 5½ in. Therefore, a finished part dimension of 6 in on one side translates to an unfired panel size of 8 in. The maximum unfired panel size that can be accommodated by the hole punching equipment is 8 in on a side. While these dimensions may seem somewhat small compared to printed wiring boards, it should be noted that the feature sizes of RF and microwave conductors in LTCC are generally smaller than those used on printed wiring boards. Consequently, the smaller physical size of LTCC parts need not be a limiting factor for designs.

3.4.4.2 Circuit features. These guidelines provide helpful information for the designer who intends to use HTCC or LTCC substrates for microwave circuits. The guidelines are concerned with cavities, conductors, ground/power planes, and vias.

3.4.4.3 Cavities. A *cavity* per se is an opening in the surface of a fired or finished part that, in general, exposes another layer of tape ceramic. Cavities may also be used as through openings to allow a chip, such as a power device, to be mounted onto a heat sink metal plane beneath the ceramic. A *ledge or shelf* is an exposed layer of tape ceramic internal to the cavity caused by a change in cavity shape. A *wall* is a section of tape between cavities or along the perimeter of a part. A *base* is the bottom of the cavity. These features are shown in Fig. 3.10 and the recommended dimensions are listed in Table 3.8.

3.4.4.4 Conductor tracks. An *external conductor* is a conductive track on the top surface of a part. It can be printed on unfired tape and then fired with the ceramic (cofired) or printed onto the fired part (postfired). An *exposed conductor* is a conductive line on a cavity shelf. An *internal conductor* is a conductive line which, after lamination, is completely contained within the finished part. It is printed on unfired tape and cofired with the ceramic. The limitations on conductor line widths and spaces are shown in Table 3.9 and the significant features of conductor tracks are illustrated in Fig. 3.11.

3.4.4.5 General recommendations. Minimize the number of conductor layers that require conductor pattern printing so as to minimize fabrication time and cost. Use the widest possible lines and spaces to make fabrication of the part less costly, keeping in mind the design imped-

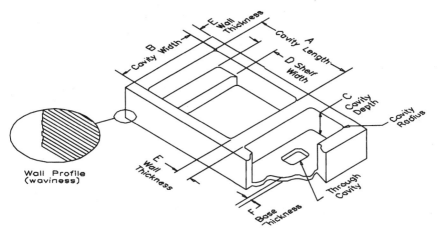

Figure 3.10   View of a cavity in HTCC or LTCC ceramic.

TABLE 3.8   Recommended Fired Dimensions for Cavities in HTCC and LTCC Parts

Feature description	Dimension
C/B—maximum ratio of depth/width	1:2
D—minimum shelf width (in)	0.050
E—minimum wall thickness (in)	0.125
C/E—maximum ratio of depth/wall	1:2
F—minimum base thickness (in)	0.060
Minimum cavity dimension (in)	0.100
Maximum cavity area (percent of finished part dimensions)	50%
Minimum corner radius (unfired) (in)	0.016

ance of tracks where a specific impedance is required. Avoid using tracks that are not at 0 or 90° (angles between 0 and 90° present printing and fired uniformity problems).

3.4.4.6   Ground and power planes.   A *gridded plane* is a cross-hatched printed plane with spaces between lines. A *solid plane* is made up of large sections of continuous metal. It may be between 25 and 90% of the total area of a layer. An *external plane* is a conductive plane on the top surface of a finished part. It can be cofired or postfired. An *exposed plane* is a conductive plane on a cavity shelf that extends into the finished part. It must be printed on unfired tape and cofired with the ceramic. An *internal plane* is a conductive plane contained completely within the finished part. It must be printed on unfired tape and cofired with the ceramic. The features of ground and power planes are illustrated in Fig. 3.12.

TABLE 3.9   Typical Dimensions of Printed Conductor Tracks and Spaces

Feature description	Dimension
A—minimum line width (in)	0.008
B—minimum space between tracks (in)	0.007
C—minimum distance, line edge to via edge (in)	0.010
D—minimum space, track/plane (in)	0.010
E—minimum space, track/part edge (in)	0.015
Maximum conductor tolerance	
Track <0.025	80% of designed width dimension ±0.001
Planes>0.025	80% of designed width dimension ±0.001

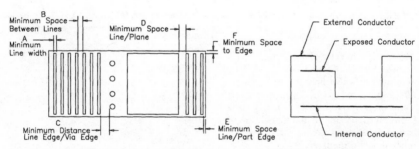

Figure 3.11   Significant features of conductor tracks in LTCC and HTCC.

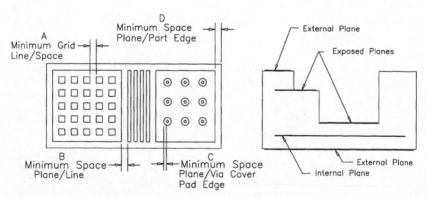

Figure 3.12   Features of metal planes in LTCC and HTCC.

Ground and power planes are used to distribute power with low ohmic loss and low inductance. Ground planes provide the return for transmission lines as well and are often used to provide shielding. They can cause difficulties in the fabrication of a cofired part because metalization shrinks at a different rate from ceramic, resulting in warping, and there is a need for ceramic to ceramic contact within the structure during firing to ensure structural integrity. A metal plane interferes with this requirement. However, experienced fabricators are capable of minimizing these effects. Some design approaches to minimize these effects are given here after the features of planes which are illustrated in Fig. 3.12 and the dimensions of which are given in Table 3.10.

To allow for intimate contact between layers of tape, use a gridded ground plane if possible. Where such an approach is not allowed because of electrical requirements, use partial solid ground planes under stripline and microstrip conductors. For a microstrip configuration the plane is positioned directly under the signal track, and extends three track widths from each edge of the track (seven track widths in total). For stripline the plane is positioned both above and below the signal track. In that way, there is an adequate ground and the ceramic tape layers are bonded together as well. In order to minimize warping, it is a good idea to keep the amount of metal evenly distributed on each level and throughout the part. Do not place a ground plane totally in one quadrant, for example, with any additional ground planes on that level.

3.4.4.7 Vias. The metal-filled holes in tape used to connect to different layers, either electrically or thermally, are known as vias. Since a via requires both a punching step and a filling or printing step, there are many aspects of their use in circuitry that must be addressed. The various features of vias are shown in Fig. 3.13 and dimensions are given in Tables 3.11 and 3.12.

TABLE 3.10   Typical Dimensions for Ground and Power Planes

Feature description	Dimension
A—minimum grid line/space (in)	0.010/0.010
B—minimum space, plane to line (in)	0.015
C—minimum space, plane to via cover pad edge (in)	0.020
Maximum percent coverage	
Internal	<50
External	100
Minimum space, plane to edge of part (in)	0.030

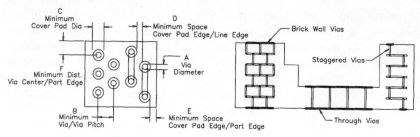

Figure 3.13 Via features in HTCC and LTCC.

TABLE 3.11 Typical Fired Dimensions for Vias

Feature description	Dimensions (in)
A—via diameter	(See Table 3.12)
B—minimum via/via pitch	3 times via diameter
C—minimum cover pad	1.5 times via diameter
D—minimum space cover pad edge/track edge	0.015
E—minimum space cover pad edge/part edge	0.020
F—minimum distance via center/part edge	0.050

TABLE 3.12 Typical Via Diameters for Various Tape Thickness (All Dimensions in Inches)

Tape thickness	Minimum via diameter	Maximum via diameter
0.0040	0.0060	0.0120
0.0050	0.0060	0.0150
0.0060	0.0060	0.0150
0.0100	0.0100	0.0030

A *brick wall* set of vias is a series of overlapping vias without cover pads. Cover pads are used on the top and bottom of each printed layer to ensure that the via makes contact with metalization tracks or ground planes on the surfaces of the layer containing the via. *Staggered vias* are offset vias connected with cover pads and tracks. *Through vias* are vias stacked on top of each other.

3.4.4.8 Discrete components. Many ceramic factories are capable of producing designs with several discrete components. These components include buried and surface resistors and capacitors. Any one part may include one or more of these discrete component styles.

Resistors. One of the advantages of LTCC technology is that, due to the compatibility of the firing temperatures and similarity of materials, discrete, embedded components, such as resistors and capacitors, can be cofired within the multilayer structure to yield an integrated monolithic package. Cofiring allows the discrete components to be embedded in the package or placed on the surface. Resistors may also be thick-film printed onto the surface of fired packages and postfired. Generally, surface resistors can be laser trimmed to provide resistor values with tolerances less than 10%. The expected tolerance for resistors used with LTCC is given in Table 3.13.

The resistance of a resistor is determined by its intrinsic resistivity and its length, width, and thickness. Stated mathematically

$$R = \frac{\rho l}{A} = \frac{\rho l}{tW}$$

where ρ is the intrinsic resistivity (Ω-cm), l is the length, W is the width, and A is the area (section area defined by the thickness and the width).

If a resistor had the form of a cube, its resistance would be dependent on its length, width, and thickness as well as its intrinsic resistivity. The sheet resistivity is spoken of in terms of ohms per square. One square of a resistor is that dimension of length that is equal to the resistor's width. The resistor pastes that are used in thick films and with LTCC are formulated to have a specific resistance when printed and fired in one square. Regardless of the actual dimensions of a square, the resistance will be the same when measured across one square. Therefore a resistor of 100 Ω could be made from paste with a 10-$\Omega/\square$ sheet resistivity, a width of 10 mils, and a length of 100 mils because there are 10 squares in that resistor (length divided by width is 10). The range of paste sheet resistivities available is as follows with all units being in ohms per square: 10, 100, 1000, 10,000, and 100,000. These values apply to both buried and surface resistors. Due to the nature of cofired materials, those pastes that are designed to be cofired are made up of different materials from the surface resistor pastes and each must be used only for the purpose it was created. A typical resistor configuration is shown in Fig. 3.14. Typical dimensions are shown in Table 3.14.

Capacitors. Embedded capacitors can be incorporated into LTCC parts by cofiring capacitor material with the tape and metalization.

TABLE 3.13 **Resistor Tolerances for LTCC**

Type of resistor	Feature	Minimum size (in)	Tolerance
Buried	No more than two sheet resistivities per layer	0.020×0.020	±30%
Surface	N/A	0.010×0.010	±10%

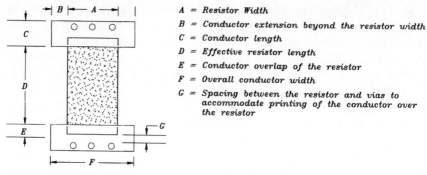

$A = Resistor\ Width$

$B = Conductor\ extension\ beyond\ the\ resistor\ width$

$C = Conductor\ length$

$D = Effective\ resistor\ length$

$E = Conductor\ overlap\ of\ the\ resistor$

$F = Overall\ conductor\ width$

$G = Spacing\ between\ the\ resistor\ and\ vias\ to\ accommodate\ printing\ of\ the\ conductor\ over\ the\ resistor$

Figure 3.14   Cofired resistor configuration.

TABLE 3.14   Cofired Resistor Dimensions

Letter	A	B	C	D	E	F	G
Dimensions (in)	0.030	0.010	0.040	0.050	0.020	0.060	0.010

Changing its dielectric constant or the thickness (distance between the capacitor plates), or the area enclosed by the capacitor plates, or any combination of the three factors can alter the capacitance of a given capacitor. Increasing the dielectric constant increases capacitance, decreasing the separation of the capacitor plates increases capacitance, and increasing the area increases capacitance as well. The mathematical statement is

$$C = \frac{\varepsilon_0 k A}{t}$$

where C is the capacitance, ε_0 is the permittivity of free space (8.85 $\times$ 10^{-12} F/m), k is the material dielectric constant, A is the area of the capacitor, and t is the thickness of the dielectric. For a multilayer configuration of n parallel plates, the value for a single-plate capacitor is multiplied by the number of plates, n.

With LTCC multilayer processing, two design techniques can be used to embed capacitors into LTCC parts. First, a capacitor plate can be formed through the punching of vias in the green tape and then filling them with a high k dielectric paste, followed by printing metal electrodes on both the top and bottom of the tape. Since the capacitance depends on the thickness of the fired tape, to maximize the value, it is important to minimize the thickness. In like fashion, since the capacitor area is determined by the diameter of the via, it is desirable to use the largest via diameter that is compatible with ease of printing. The possible via diameter can range from 15 to 25 mils, based on the tape thickness as well as the aspect ratio of the via diameter to the tape thickness. There are two disadvantages to this technique that limit the capacitance to a low value. First, the capacitor value is limited by tape thickness. Second, large via diameters, although not difficult to punch, are very difficult to fill by standard printing techniques. However, vias can be connected in parallel to increase the value of the capacitor. A capacitor value of 50 to 100 pF can be obtained by using this technique.

The second technique is to print a thin capacitor plate onto the surface of the green tape. For this, a gold paste of desired dimension is first printed on the tape. A thin print of high k capacitor paste is then applied to the metal electrode. This is then followed by another print of metal electrode on top of the capacitor paste to form a capacitor plate. To avoid shorting, the dimension of the printed capacitor paste should be slightly bigger than the dimension of the metal electrodes. The tape layer with printed capacitor is then stacked in the desired sequence, laminated with the other layers of tape, and subjected to cofiring. In this case, the capacitor thickness can be reduced to approximately 1 mil and the area can be increased to meet the design need. Second, the capacitance value may be increased by placing parallel plates on adjacent tape layers. A large capacitance value can be obtained by this technique.

3.4.5 Printed wiring board technology for RF and microwave circuits

The second type of multilayer substrates, namely, printed wiring boards, are made up of a large group of different materials. What they have in common are a few essentials: copper is clad onto an organic dielectric material that is usually reinforced with a fiber of one kind or another. The copper can be applied in thin-sheet form (from 0.125 oz/ft^2 in weight to over 2 oz/ft^2) to the organic dielectric sheet. The common way of referring to 1-oz/ft^2 copper is 1-oz copper and it is 0.00137 in thick. This copper can be either rolled from a copper ingot to the desired thickness or electrodeposited from a chemical bath onto a polished steel mandrel and then peeled away from the mandrel as the desired thickness is achieved. Another method of applying the copper to the dielectric is by direct electrodeposition onto sensitized areas of the dielectric surface. The distinctions will be explained later.

Printed wiring boards can be classified as rigid, flexible, rigid-flex (which combines the attributes of both rigid and flexible boards), or molded. Molded boards are seldom encountered in RF and microwave circuits. The first three can be further divided into single-sided (copper on only one side of the organic), double-sided, or multilayer. The circuit pattern is created by imaging the conductor pattern on copper sheets using a photoresist material and one of two imaging techniques: screen printing and photoimaging. The photoresist material acts as a protective layer that defines conductor tracks while any unwanted copper from the foil sheet is etched away chemically during processing. This is called the *subtractive method* and is the most commonly used technique to create conductor patterns from the copper foil which has been fixed to the dielectric. These techniques are used with a variety of dielectric materials. Some of the most common dielectric materials used for microwave applications are epoxy/E-glass laminates (E-glass is an electronic grade of glass), polyimide film for flexible printed wiring, and polytetrafluoroethelyne (PTFE) with glass fiber reinforcement. Rigid-flex boards are usually a combination of both materials. The rigid PWB is fabricated from copper-clad dielectric materials. Quartz and aramid fibers are commonly used for high circuit speed and for surface-mounted components as well. The organic materials may be selected from a number of formulations and include flame-retardant epoxy, polyfunctional epoxy polyimide, and polytetrafluoroethelyne.

The base laminate is first cut into working panels that are drilled or punched to form registration holes. This is important because these holes are used as reference points on the various layers of a multilayer board, they permit the conductor on various layers to be aligned properly as desired, and they allow holes drilled after the layers have been laminated to line up with their proper pads on the buried layers. Multilayer boards are usually made from stacking together double-sided boards that have been imaged and etched using an adhesive layer or two of unclad dielectric that is in the partially cured state called the "B" stage. These layers act to hold the opposing double-sided panels together and be a dielectric as well that separates conductors on adjacent layers. This is depicted in Fig. 3.15.

Double-sided boards are usually fabricated from laminates with 1-oz copper that has been clad to both sides of a dielectric. RF and microwave printed wiring boards typically can have up to 10 layers or more of circuitry. Some of the buried layers are used for ground and power. The sequence in making a multilayer board is as follows. The individual double-sided boards are brought together in the desired sequence to achieve the desired lay up to form a "book." The book is then laminated under pressure and heated to the appropriate thickness. The outer layers have not been etched up to that point and the laminate appears as a double-sided copper-clad laminate of the

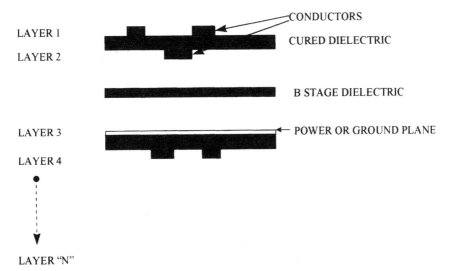

LAYER 1

LAYER 2

LAYER 3

LAYER 4

LAYER "N"

CONDUCTORS

CURED DIELECTRIC

B STAGE DIELECTRIC

POWER OR GROUND PLANE

Figure 3.15 Typical lay up of a multilayer printed wiring board. (Vias are not shown.)

same thickness. After lamination, the book is processed in the same manner as a thick double-sided board. The book is drilled to add the via holes and then processed as if it were a double-sided board using plated-through holes.

When circuit design requires them, or circuit density forces them to be used, techniques such as blind and buried vias can be used. When these techniques are used, the inner layer pairs are fabricated as double-sided boards complete with plated vias, and then assembled into books for processing into multilayer boards. In this way the inner layers may be connected without a through hole that goes through all the layers of the structure.

Flexible multilayer boards may be fabricated in very similar fashion to the rigid multilayer boards. The basic differences here are in the dielectric materials, which are flexible, and in the copper cladding, which is often rolled copper. Rolled copper is used instead of electrodeposited copper for improved ductility and reduced conductor cracking. Dielectric materials can include polyimide, polyester terephthalate, and fluorinated ethylene propylene (FEP) Teflon.* To bond the various layers together, a variety of adhesives may be used. These include acrylics, epoxies, phenolic butyrols, polyester, and PTFE. In some cases processes exist that allow lamination of the conductor directly to the dielectric film without the use of an adhesive.

---

*Teflon is a registered trademark of the E. I. du Pont de Nemours Company.

Rigid-flexible or rigid-flex circuitry consists of single or multiple flexible wiring layers selectively bonded together using a modified acrylic adhesive or an epoxy prepreg ("B" stage) bond film. Cap layers of rigid-core copper-clad laminates may be bonded to the top and bottom surfaces of the circuit for improved stability in the vicinity of connectors where some mechanical stress can be expected as connector mating is accomplished.

The printed wiring boards described, along with some of their materials and construction features, are used extensively for digital and analog circuits while microwave applications tend toward materials that have low dielectric constant and also low loss. For the purpose of this discussion, low dielectric constant can mean a value of 4.0 or lower and low loss or low tangent delta can mean 0.002 or lower. The lowest loss is, of course, ideal. Lower dielectric constant and lower loss mean higher board material costs, sometimes several times higher than if a standard epoxy glass board were used. It is important to know, based on a careful analysis, just how low the dielectric constant and loss must be such that the proper applicable material is specified and unnecessary costs are avoided.

While low loss in the dielectric system is of considerable concern, loss in the conductors is a greater contributor to overall loss in a circuit. For frequencies below those at which RF and microwave circuits operate, the wave considerations are generally not important. However, at RF and above, electromagnetic wave theory is used extensively. The electromagnetic energy is carried in the dielectric as a wave guided by the conductors and their grounds (microstrip, stripline, and coplanar guide). The waves traveling in the dielectric cause currents to flow in the conductors and ground planes. In order for the energy coming out of such configurations to nearly equal the energy put into the configurations, losses due to resistance in the conductors must be controlled. At microwave frequencies, currents in the conductors travel very close to the surfaces of the conductors. In other words, the currents are "bunched" together near the outside surfaces of printed wiring board conductors. The resistivity of the conductors will determine the loss to be expected at microwave frequencies. Therefore, the best way to reduce conductor losses at microwave frequencies is to ensure that the highest electrical conductivity conductors are used.

Since the currents in conductors at microwave frequencies are carried close to the outer surfaces it is necessary that the conductor material at those surfaces be of high conductivity. In the case of copper printed wiring board tracks, unfortunately, the opposite is usually true. This happens because of two conditions. To promote and enhance adhesion of the conductor to the dielectric, the surface of the copper to which the dielectric is attached is prepared with an oxide. This oxide has a lower conductivity than the base copper and also consists of long crystals which provide a rough surface. To passivate and prevent corrosion of the copper surfaces not in intimate contact with the dielectric (the top surface and side walls of the conductor), a coating of plating is used. This plating is usually a tin-lead alloy. If a metal plating is not used, another coating such as solder mask or a conformal organic material is applied. The tin-lead plating has a relatively poor electrical conductivity that is lower than that of the base copper. This increases the resistivity (lowers the conductivity) of that part of the conductor near to the outside surfaces. The phenomenon associated with the bunching of current near the outside surfaces is called the *skin effect* and is directly related to the frequency at which the circuit operates. The relationship discussed earlier is given here in alternate form by the following equation:

$$\delta = \left(\frac{\rho}{f \pi \mu} \right)^{1/2}$$

In the equation, δ is the skin depth, ρ is the resistivity of the conductor, f is the frequency of operation, and μ is the magnetic permeability of the metal. The skin depth is the depth of penetration of the wave's electric field into the metal. Its definition is the distance from the surface of a metal conductor to the depth at which the magnitude of a penetrating magnetic field falls to a value of $1/e$ (about 36%) of the magnitude at the surface. The letter e in this definition is the number e (~ 2.7), the base of the natural logarithm. From the equation, it is seen that the skin depth becomes deeper as the resistivity increases, and becomes shallower as the frequency and permeability increase. Since the skin depth is larger with poorer conductivity metals, the wave can penetrate more deeply into such metals, allowing some power from the wave to be absorbed into the metal. This power is then dissipated or lost as heat. Power is lost in the conductor because of the resistivity of the conductor. The resistivity of a conductor chosen for a microwave circuit should almost always be as low as possible. Referring to the equation for skin depth, it is easy to see that as resistivity increases, the alternating electromagnetic wave propagates more deeply into the conductor. Ideally, since power will be lost if any of the wave penetrates the conductor, the conductor should have no electrical resistance at all, and should be a perfect conductor. However, this ideal case is not achievable, and we are faced with accepting the effects of Maxwell's eddy currents which are generated in a metal in opposition to the imposed changing magnetic field. These currents dissipate a part of the wave's energy in flowing through the conductor and meeting resistance. As the frequency increases, the skin depth is shorter, and the eddy currents are then bunched into a decreasing thickness of a shell of the conductor. This can result in high energy loss if the resistivity of the conductor is high, as is the case with tin-lead solder over copper. This means that a poor conductor (high resistivity) can cause serious problems with power loss at very high frequencies. Table 3.15 lists the electrical conductivity of several typical PWB conductors.

In order to illustrate the importance of keeping the electrical resistance of the outer surfaces of RF and microwave conductor tracks low, the skin depths of aluminum and copper are compared in Table 3.16 for several frequencies. It should be noted that these conductivity and skin depth guidelines apply to ceramic boards as well as organic printed wiring boards. At RF and microwave frequencies, there is a distinct advantage to using LTCC technology over HTCC. In that construction, silver, which has a low resistivity, may be used in buried layers whereas HTCC buried conductors consist (of necessity) of refractory metals that have a very high resistivity compared to silver. While these guidelines are important, not all circuits will fail completely if they are not followed. Therefore, an analysis of the circuit and the design features, such as conductor width and length, should be performed to determine whether the anticipated loss can be tolerated.

Other factors affect the total loss of an RF or microwave circuit. These are radiation of the electromagnetic energy out of the dielectric and coupling to other conductors in the multilayer board such as DC power and digital signal lines. These issues were discussed earlier in Sec. 3.3.3.11 which deals with cross-talk and parallel plate waves. A detailed discussion of these factors is beyond the scope of this chapter, but the reader is referred to the article by James D.

TABLE 3.15 Conductivity of Typical Printed Wiring Board Conductors

Material	Symbol	Conductivity (S/m $\times 10^6$)
Silver	Ag	61
Copper	Cu	58
Gold	Au	41
Tin	Sn	20

TABLE 3.16 Skin Depth of Aluminum and Copper

Frequency (MHz)	Skin depth in copper (thousandths of an inch)	Skin depth in aluminum (thousandths of an inch)
10	0.8	1.0
100	0.2	0.3
1000 (1 GHz)	0.08	0.1

Woermbke and books by C. A. Harper and by S. G. Konsowski and A. R. Helland in the Bibliography of this chapter for a detailed treatment of these factors.

3.4.6   Organic printed wiring boards

3.4.6.1   Isotropy (courtesy of Rogers Corporation). While ceramic materials play a large part in RF and microwave circuits, there is an increasing demand for organic PWB materials. These can be the typical FR-4 type that have been the mainstay of the PWB industry for many years, but more typically, they belong to a class of higher performance materials. One of these high performance materials with a long history is fiberglass-reinforced polytetrafluoroethylene (PTFE). Both woven fiberglass and microfibers (short fibers of glass randomly oriented in the dielectric) have been used for the reinforcement material with PTFE. This material evolved over many years into a stable, widely available commodity with very predictable properties. Two of these properties of interest to RF and microwave circuit designers are dielectric constant and dissipation factor. Laminates of this type usually have different values for dielectric constant and dissipation factor in all three directions. This is due to the physical differences of the weave in plane and out of plane as well as different glass densities in plane. They show the random glass microfiber structure is less anisotropic than a woven fabric structure. This difference may possibly be explained by considering the woven fabric structure, especially at lower fiber content, to be a series of alternating polymer-rich and fiber-rich layers. The z direction field, in effect, sees a series capacitor network while the x-y fields see a parallel capacitor network. The fine scale uniformity of glass microfibers in the PTFE matrix of the microwave laminates may account for a significant advantage in isotropy of dielectric constant compared with PTFE materials depending on a woven fabric for reinforcement.

Anisotropy, the condition in which the dielectric constant varies in value depending on the direction of the electric field with respect to the axis of the material, can be expected to lead to extra fringing capacitance, most evident in resonator elements, in narrow lines and in edgewise couplers. Localized variation in anisotropy, especially characteristic of woven fabric-reinforced composites, may degrade performance. This is particularly true at higher frequencies in situations where a narrow coupling gap or very narrow conductor lines are used. To simplify microwave circuit design computations for microstrip or stripline circuitry, the designer usually assumes isotropy of the electrical properties of laminated materials.

3.4.6.2 Copper foil adhesion to the laminate (courtesy of Rogers Corporation).

A concern sometimes associated with the use of PTFE materials is the condition of the copper foil adhesion to the PTFE laminate after processing at soldering temperatures. This concern exists due to the infrequent but damaging effects of soldering irons used in some assembly and repair procedures. The potential damage and loss of material integrity were addressed by a test that demonstrated the conditions under which damage could occur. Copper-clad laminates based on PTFE are normally manufactured by directly fusing the PTFE component of the laminate to the surface of the metallic foil. The nodular bonding surface of the more popular ED foil made during the electrodeposition process develops a mechanically interlocked bond with the PTFE. There have been occasions when problems with bond integrity during soldering have been encountered. This has been particularly true of cases where poorly controlled hand soldering tools have been used to apply shear or lifting forces during assembly or rework operations. To determine the effect of temperature on the bond strength of copper foil to a PTFE laminate, a laboratory study was performed with RT/duroid* 5870 laminate (1.59 mm) thick, clad with 1-oz/ft^2 (34-mm thick) ED copper foil by Rogers Corporation. This study involved exposure of specimens to a series of solder temperatures ranging from 182 to 371°C (360 to 700°F). The specimens consisted of 3-mm wide copper lines with a 0.250-in pad on one end etched onto 2-in square coupons of the laminate. Three measurements were made:

1. Cold-peel strength was measured as the force at a 90° angle to peel a 0.125-in (3.2-mm)-wide conductor strip from the base at an ambient temperature of 21°C after the specimen was floated on solder for 30 s.

2. Hot-peel strength was measured as the force required at a 90° angle to peel the 0.125-in (3.2-mm)-wide line from the base while the specimen was immersed in solder at temperature, after a stabilization time of about 1 min.

3. Shear force to destroy the bond was measured on the 0.250-in square pad at the end of the 0.125-in (3.2-mm)-wide strip while the specimen was immersed in solder at temperature after a stabilization time of about 1 min. To determine this force the 0.125-in (3.2-mm) strip was peeled away from the substrate up to the pad area. The force required to separate the pad from the base was measured by pulling on the 0.125-in (3.2-mm) strip parallel to the specimen surface.

---

*RT/duroid is a licensed trademark of Rogers Corporation.

Averages of test results are shown in Table 3.17. The cold-peel test after solder exposure shows that a bond is not destroyed until a temperature above the 327°C crystalline melt point of the PTFE is approached. At this temperature the copper begins to oxidize rapidly upon contact with the air. The protection from air afforded by the PTFE bond to the foil surface is significantly diminished as the PTFE begins its phase change. The hot peel and shear failure values show that the bond is susceptible to mechanical damage at elevated temperatures. Cold-peel failure up to 288°C was into the substrate (cohesive). All other peel failures were between foil and substrate (adhesive). The cold-peel value for material not floated on solder was 16.0 lb/in in width with failure into the substrate (cohesive). The 16.0-lb/in and larger hot shear failures were breakage of the 0.125-in wide copper strip used to apply force to the pad. Care must be exercised in hand soldering work to avoid extremes of force or exposure to temperatures in excess of 300°C.

3.4.6.3 Prevention and removal of absorbed chemicals in microwave laminates (courtesy of Rogers Corporation). A typical high performance microwave laminate is a polytetrafluoroethylene (PTFE) composite containing a large volume fraction (>50%) of ceramic filler particles along with glass microfibers. The high filler loading provides the laminate with a low z axis coefficient of thermal expansion (CTE) for excellent plated-through hole reliability and an in-plane CTE closely matched to copper for good dimensional stability. The material's high filler loading also contributes to approximately 5% vol porosity in the form of microvoids. These microvoids in the composite appear to exist at the filler interface but are not detectable in cross sections, even with scanning electron microscopy. Because of the low surface energy of PTFE and the treated ceramic filler, the microvoids do not result in high water absorption. However, low surface-tension liquids, such as organic solvents and surfactant-laden aqueous solutions, can penetrate the pores.

TABLE 3.17 Summary of Average Bond Strength, 1-oz Electrodeposited Copper on RT/Duroid Laminate

Solder temperature [°C (°F)]	Cold-peel 30-s float (lb/in width)	Hot-peel in solder (lb/in width)	Hot shear in solder (lb/0.25-in pad)
182 (360)	14.8	5.4	16.0+
204 (400)	14.8	3.2	16.0+
232 (450)	14.8	1.6	16.0+
260 (500)	14.4	0.8	12.8
288 (550)	12.7	0.1	8.0
316 (600)	11.9	0	6.4
343 (650)	6.3	0	6.4
371 (700)	6.4	0	1.6

SOURCE: The Rogers Corporation.

Since PTFE and the ceramic filler are inert to most process chemicals, liquid uptake simply fills microvoids but does not change the physical properties of the laminate. However, it is critical that volatile materials that have penetrated the composite be removed prior to exposing the boards to high temperatures (for example, bonding operations, solder reflow, and assembly processes). Parts should also be thoroughly rinsed immediately following exposure to process chemicals to ensure that nonvolatile dissolved materials are not left behind when the parts are baked. Failure to remove volatiles prior to high temperature operations such as bonding or reflow can result in dielectric blistering or delamination. The following baking procedures have been found to eliminate volatile materials-related problems during high temperature exposure.

Basic guidelines for baking

1. *Prior to bonding.* Bake fabricated inner layers for at least ½ h at 300°F under vacuum or nitrogen prior to bonding. If the boards are bonded in an autoclave, the bake cycle can be included at the front end of the bond cycle. No pressure should be applied.

2. *Prior to electroless copper application.* Bake boards for at least 1 h at 300°F under vacuum or nitrogen immediately prior to electroless copper plating. This bake is critical since absorbed glycol ethers from commercially available sodium etchants and alcohol from rinses are very difficult to remove once the multilayer board edges and machined features have been covered by electroless copper.

3. *Prior to reflow.* Bake boards for at least 2 h at 300°F under vacuum or nitrogen prior to reflow or hot-air solder leveling (HASL). Following the bake, exposure to flux must be kept to a minimum (<30 s) and the bake cycle must be repeated if the boards need to be reworked. When baking parts in a nitrogen purged bag, nitrogen flow out of the bag is necessary to ensure that volatiles are removed from the bag. Similarly, care must be taken when utilizing a vacuum bag to ensure that vacuum lines do not become plugged by the bagging materials. If volatiles remain in the bag, they will condense on the parts as the parts cool. This can significantly reduce the effectiveness of the bake. The time required for the oven to reach temperature should be added to the recommended bake time if the oven isn't preheated.

Dielectric contamination. Failure to adequately rinse the laminate following exposure to process chemicals can sometimes result in dielectric staining or increased dielectric loss. These problems can be prevented by minimizing exposure to very low surface energy solvents which contain nonvolatile components and by utilizing good rinsing procedures. For example, parts should not be allowed to soak in resist stripper solutions laden with dissolved photoresist any longer than is necessary to strip the resist. Also, parts must be thoroughly rinsed immediately following resist stripping.

Basic guidelines to prevent dielectric contamination

1. Minimize exposure to low surface energy solvents containing non-volatile components.
2. Dump rinse frequently to prevent accumulation of nonvolatile materials.
3. If the dielectric surface is exposed to low surface energy aqueous solutions or water-soluble organic solutions containing nonvolatile materials, soak the part in hot (>70°F) deionized water for 15 min immediately following exposure.
4. If the dielectric surface is exposed to water insoluble solvents containing nonvolatile materials, promptly soak the part in a water-soluble organic solvent, such as methanol, ethanol, or isopropyl alcohol, for 15 min and then soak it for 15 min in hot deionized water.

3.4.6.4 Thick metal cladding on microwave circuit dielectric laminates (courtesy of Rogers Corporation). In addition to conventional one- or two-sided copper foil–clad substrates, RT/duroid laminates can be found with a thick metal laminated on one side and copper foil on the other. The thick metal is typically in the 0.020- to 0.500-in thickness range. It serves as ground plane for stripline and microstrip circuit boards. Some applications show design, process, and performance advantages over laminates clad on both sides with thin copper foil, since the thicker metal ground plane can act as a structural member as well as a heat path. Figure 3.16 shows this construction.

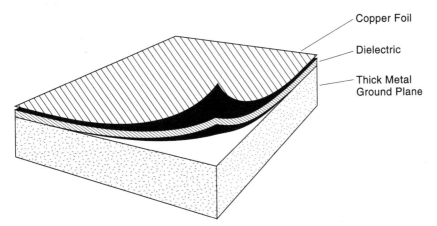

Copper Foil

Dielectric

Thick Metal
Ground Plane

Figure 3.16 Thick metal ground plane attached to RT/duroid foil-clad material. (*Courtesy of Rogers Corporation.*)

Heat sinking. Glass fiber–reinforced PTFE composites, such as RT/duroid 5880, or ceramic PTFE, such as RT/duroid 6010, have relatively low thermal conductivity values (typically 0.26 and 0.41 Wm^{-1}-K^{-1}, respectively, compared to about 220 Wm^{-1}-K^{-1} for aluminum metal or 9.7 to 15.6 Wm^{-1}-K^{-1} for alumina). In higher power applications, the designer must accommodate the heat dissipated by active devices, conductor resistance losses, and dielectric loss. Thick metal ground planes provide high heat flow, keeping component temperature under control.

Active and passive components, such as diodes and resistors, are mounted directly on the ground plane in slots or blind holes machined through the dielectric. This design provides maximum heat dissipation at the high heat-producing sites. For high current-carrying circuit traces, close proximity and direct bond of the heat sink ground plane and dielectric minimize the heat flow path. Also, thick metal-clad laminates eliminate the thermal barrier of solder or epoxy bonding to external heat sinks used with conventional copper foil–clad laminates.

Connector mounting. When the circuit board and housing are separate components of a package, reliability problems can arise with the use of coax transitions to microstrip or stripline circuits by edge-mounted launchers. CTE mismatches between the board and the housing induce thermal cycling fatigue failures where the connector pin attaches to the board and can even cause alignment problems during assembly. There are two ways that thick metal clad laminates can solve this:

1. The composite x,y plane CTE of the laminate is dominated by the stiffer thick metal layer. CTE mismatch is avoided by selecting the same metal for housing and thick metal cladding.

2. The connector body can edge mount directly to the rigid, integral chassis formed by the ground plane of the circuit board. Vertical slots, rather than holes, in the flange for the mounting screws ease alignment of the connector body. The pin soldered to the trace on the board forms a reliable strainfree assembly, as illustrated in Fig. 3.17.

Component mounting. Both active and passive circuit components may be mounted directly onto the ground plane in slots or blind holes milled through the dielectric. For example, conventional die attach methods using eutectic silicon gold or conductive epoxies may be used to bond transistors to the plated ground plane. Electrical connections are made by wire bonding, ribbon bonding, or soldering. These connections can be between components bonded into, or traces ending at the edge of, slots or blind holes. In addition to the heat sinking advantage previously mentioned, the thick metal ground plane provides a rigid base for

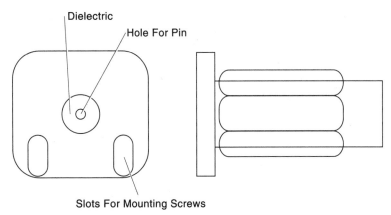

Figure 3.17 Edge launch connector with slots for mounting screws. (*Courtesy of Rogers Corporation.*)

attachment. Selecting a metal, such as Kovar,* to match the CTE of the component improves the reliability in extreme thermal cycling environments. Selecting a thick metal cladding with matching CTE could benefit the reliability of surface-mounted components by lowering strain on bonded or soldered leads during temperature excursions.

Thermal stress cracking of conductors. Extreme thermal shock or cycling environments can cause stress crack failure of copper conductors on boards with only thin copper foil cladding. The x,y or in-plane direction CTE mismatch between copper at about 17 ppm/°C versus PTFE glass composite averaging about 40 ppm/°C from -100 to $+150$°C is resolved with much of the strain being borne by the thin copper foil. Copper tends to work-harden and ultimately crack when the strain is great enough. A thick copper ground plane will put virtually all the strain on the soft dielectric to minimize the problem.

Common thick metal options. Aluminum, brass, and copper are often chosen as thick metal claddings, with aluminum being the most common. Metal and laminate properties are given in Tables 3.18 and 3.19. Other metals could be used depending on specific properties required for particular applications. Examples include Kovar and Invar,[†] a copper/Invar/copper composite for matching CTE to components; or stainless steel for strength, rigidity, and corrosion resistance. Metal thickness options are given in Table 3.20. These thicknesses and tolerances are considered industry standards, allowing for reduced lead times. These thicknesses can be specified to a tolerance of 0.002 in.

---

*Kovar is a registered trademark of Carpenter Technology Corporation.

[†]Invar is a registered trademark of Soc. Mtl. rgq D'Imphy.

TABLE 3.18   Typical Properties of Thick Metal Options

Property/units	Aluminum	Copper	Brass
Alloy number	6061	110	Cartridge
Tensile strength (kips/in^2)	20	35	45
Specific gravity	2.7	8.9	8.5
Specific heat (J/kg/K)	960	385	375
Thermal conductivity (W/m-K)	180	390	120
Thermal expansion (ppm/K)	24	17	20
Resistivity ($\mu\Omega$-cm)	5	2	6
Relative cost/area	1	1.5	1.4

SOURCE: The Rogers Corporation.

TABLE 3.19   Typical Properties of Thick Metal-Clad Laminates

Property, units	Value
Ground plane/dielectric interface	
Peel strength (lb/in width)*	
RT/duroid 6010, 0.025 in thick	24
RT/duroid 5880, 0.031 in thick	30
Shear strength (lb/in)*	
RT/duroid 6010	1350
RT/duroid 5880	1500
Typical surface roughness (μin) RMS[†]	70
Ground plane thickness tolerance (in)	
Standard	Various
Tight tolerance	±0.002
Parallelism (in)[†]	0.002 maximum

*Values when interface roughness is 70 μ in.
†Both sides of the ground plane normally have the same roughness.
SOURCE: The Rogers Corporation.

TABLE 3.20   Thickness and Tolerances of Metal Ground Planes

Copper	Brass	Aluminum
0.020±0.0035	0.020±0.003	0.015±0.003
0.027±0.004	0.025±0.003	0.020±0.003
0.032±0.004	0.032±0.004	0.025±0.003
0.040±0.005	0.040±0.004	0.032±0.0035
0.050±0.006	0.050±0.005	0.040±0.004
0.064±0.006	0.064±0.005	0.050±0.005
0.093±0.007	0.093±0.006	0.063±0.005
0.125±0.007	0.125±0.006	0.071±0.005
0.187±0.008	0.187±0.007	0.080±0.006
0.250±0.010	0.250±0.009	0.090±0.006
0.500±0.013	0.375±0.012	0.100±0.007

SOURCE: The Rogers Corporation.

RF losses. At high frequencies, conductor losses, including both the circuit trace and return (ground) path, can become significant. Conductor losses are a function of electrical resistivity and surface roughness. RF losses can be minimized by specifying a smoother dielectric/ground plane interface, but the trade-off here is in reduced bond strength. Copper or aluminum is preferred over brass for high frequency, low loss applications, because of the higher electrical resistivity of brass.

3.4.6.5 Low cost approaches to high performance PWBs for high frequency applications. The most easily manufactured laminate material is FR-4. Its dielectric constant and dissipation factor, however, are less than those desired for many microwave, RF, and high speed applications. In those cases, a clear need exists for low cost materials that would have the properties of Teflon laminates without the cost. Several materials being offered are designed to meet those requirements.

Rogers Corporation offers its RO4003, which is a thermosetting resin laminate with a ceramic filler. The dielectric constant of RO4003 is 3.38 and its dissipation factor is 0.0022. A flame-retardant version, RO4350, has a dielectric constant of 3.38 and a dissipation factor of 0.004. To fulfill the need for multilayer laminates, Rogers Corporation has made available a prepreg material, RO4403, that is compatible with FR-4 lamination processes. Allied Signal offers its FR-408 with a dielectric constant of 3.8 and a dissipation factor of 0.01. GE Electromaterials offers its GETEK laminates that have a dielectric constant of 3.8 and a dissipation factor of 0.008. PCL-LD-621, manufactured by Polyclad, is also compatible with FR-4 processing and has a dielectric constant of 3.5 with a dissipation factor of 0.005.

Bibliography

Brown, Richard: "Materials and Processes for Microwave Hybrids," a technical monograph of the ISHM (IMAPS), IMAPS, Reston, VA, 1993.

Crum, Susan: "Wireless Devices Set New Criteria for High Frequency Materials," *Electronic Packaging and Production*, 39(3):56, March 1999.

Harper, C. A., ed.: *Electronic Packaging and Interconnection Handbook*, 3rd ed., McGraw-Hill, New York, 2000, Chap. 12.

Konsowski, S. G., and A. R. Helland: *Electronic Packaging of High Speed Circuitry*, McGraw-Hill, New York.

Product Data Sheets RT 532, RT 495, RT 241, RT 2121, Rogers Corporation, Microwave Materials Division, Chandler, AZ, reproduced in part with permission.

Woermbke, James D.: "Soft Substrates Conquer Hard Designs," *Microwaves*, 21(1):89, January 1982.

Advanced Ceramic Substrates for Microelectronic Systems

Jerry E. Sergent

Director of Technology, TCA, Inc.
Williamsburg, Kentucky

4.1 Introduction

A *substrate* in this context is considered to be the component upon which
the interconnection scheme for an electronic circuit is fabricated. The
substrate is the foundation of an electronic circuit and must perform
many functions. It acts as the platform to mount components and must
be compatible with the processes intended to metallize the substrate and
to attach the components to the metallized traces. The substrate may
also be an integral part of the overall circuit package.

The desirable properties of a substrate for electronic applications
include

- *High electrical resistivity.* A substrate must have high electrical
  resistivity in order to isolate adjacent circuitry.

- *High thermal conductivity.* High thermal conductivity assists in
  transporting the heat generated by electronic components during
  normal operation away from the components.

- *Resistance to temperature.* Many of the processes used to metallize substrates and to assemble the components take place at elevated temperatures.

- *Inert to chemical corrosion.* Solvents, fluxes, and the like are harsh and must not attack the chemical structure of the substrate.

- *Cost.* The cost of the substrate material must be compatible with the cost of the end product.

The properties of ceramic substrates are advantageous for many microelectronic systems, ranging from simple, inexpensive circuits used in throwaway toys to elaborate multilayer structures used in space or medical applications. Only by understanding the properties of ceramic substrates can they be optimally utilized in this wide variety of applications.

Ceramics are crystalline in nature, with a dearth of free electrons. They have a high electrical resistivity, are very stable, chemically and thermally, and have a high melting point. They are formed by the bonding of a metal and a nonmetal and may exist as oxides, nitrides, carbides, or silicides. An exception is diamond, which consists of pure carbon subjected to high temperature and pressure. Diamond substrates meet the criteria for ceramics and may be considered as such in this context.

The primary bonding mechanism in ceramics is ionic bonding. An ionic bond is formed by the electrostatic attraction between positive and negative ions. Atoms are most stable when they have eight electrons in the outer shell. Metals have a surplus of electrons in the outer shell, which are loosely bound to the nucleus and readily become free, creating positive ions. Similarly, nonmetals have a deficit of electrons in the outer shell, and readily accept free electrons, creating negative ions. Figure 4.1 illustrates an ionic bond between a magnesium ion with a charge of $+2$ and an oxygen ion with a charge of -2, forming magnesium oxide (MgO). Ionically bonded materials are crystalline in nature, and have both a high electrical resistance and a high relative dielectric constant. Due to the strong nature of the bond, they have a high melting point and do not readily break down at elevated temperatures. By the same token, they are very stable chemically and are not attacked by ordinary solvents and most acids.

A degree of covalent bonding may also be present, particularly in some of the silicon- and carbon-based ceramics. The sharing of electrons in the outer shell forms a covalent bond. A covalent bond is depicted in Fig. 4.2, illustrating the bond between oxygen and hydrogen to form water. A covalent bond is also a very strong bond, and may be present in liquids, solids, or gases.

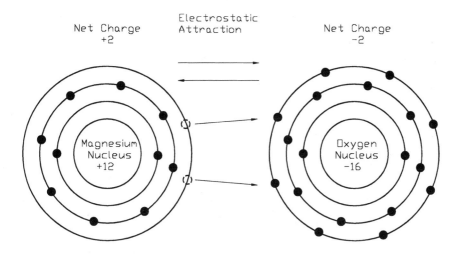

MgO Ionic Bond

Figure 4.1 Magnesium oxide ionic bond.

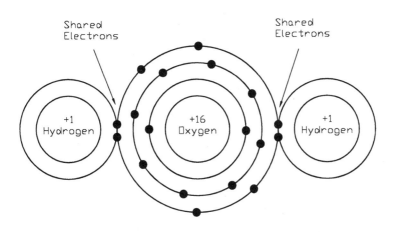

Covalent Bond - Water

Figure 4.2 Covalent bond between oxygen and hydrogen to form water.

This chapter considers the properties of substrates used in micro-electronic applications, including aluminum oxide (alumina, Al_2O_3), beryllium oxide (beryllia, BeO), aluminum nitride (AlN), boron nitride (BN), diamond (C), and silicon carbide (SiC). Two electrically conductive composite materials, aluminum silicon carbide (AlSiC) and Dymalloy, a diamond/copper structure, are also described. Although the conductive nature of this material prevents it from being used as a conventional substrate, it has a high thermal conductivity and may be used in applications where the relatively low electrical resistance is not a consideration.

4.2   Substrate Fabrication

It is difficult to manufacture ceramic substrates in the pure form. The melting point of most ceramics is very high, as shown in Table 4.1, and most are also very hard, limiting the ability to machine the ceramics. For these reasons, ceramic substrates are typically mixed with fluxing and binding glasses which melt at a lower temperature and make the finished product easier to machine.

The manufacturing process for Al_2O_3, BeO, and AlN substrates is very similar. The base material is ground into a fine powder, several micrometers in diameter, and mixed with various fluxing and bonding glasses, including magnesia and calcia, also in the form of powders. An organic binder, along with various plasticizers, is added to the mixture and the resultant slurry is ball-milled to remove agglomerates and to make the composition uniform.

The slurry is formed into a sheet, the so-called "green state," by one of several processes, as shown in Fig. 4.3,[1] and sintered at an elevated temperature to remove the organics and to form a solid structure:

Roll compaction. The slurry is sprayed onto a flat surface and partially dried to form a sheet with the consistency of putty. The sheet is fed through a pair of large parallel rollers to form a sheet of uniform thickness.

Tape casting. The slurry is dispensed onto a moving belt that flows under a knife edge to form the sheet. This is a relatively low pressure process compared to the others.

Powder pressing. The powder is forced into a hard die cavity and subjected to very high pressure (up to 20,000 lb/in^2) throughout the sintering process. This produces a very dense part with tighter as-fired tolerances than other methods, although pressure variations may produce excessive warpage.

TABLE 4.1   Melting Points of Selected Ceramics

Material	Melting point (°C)
SiC	2700
BN	2732
AlN	2232
BeO	2570
Al_2O_3	2000

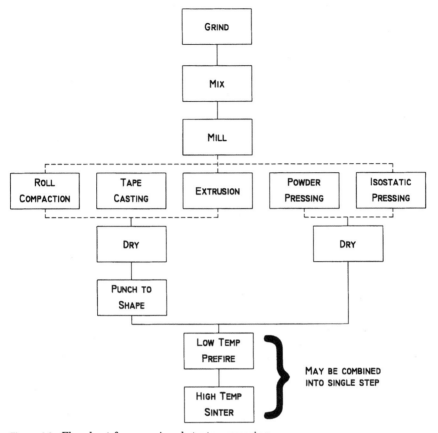

Figure 4.3 Flowchart for ceramic substrate processing.

Isostatic powder pressing. This process utilizes a flexible die surrounded with water or glycerin and compressed with up to 10,000 lb/in^2. The pressure is more uniform and produces a part with less warpage.

Extrusion. The slurry, which is less viscous than for other processes, is forced through a die. Tight tolerances are hard to obtain, but the process is very economical and produces a thinner part than is attainable by other methods.

In the green state, the substrate is approximately the consistency of putty and may be punched to the desired size. Holes and other geometries may also be punched at this time.

Once the part is formed and punched, it is sintered at a temperature above the glass melting point to produce a continuous structure. The temperature profile is very critical and the process may actually be performed in two stages—one stage to remove the volatile organic materials and a second stage to remove the remaining organics and to sinter the glass/ceramic structure. The peak temperature may be as high as several thousand degrees Celsius and may be held for several hours, depending on the material and the type and amount of binding glasses. For example, pure alumina substrates formed by powder processing with no glasses are sintered at 1930°C.

It is essential that all the organic material be removed prior to sintering. Otherwise, the gases formed by the organic decomposition may leave serious voids in the ceramic structure and cause serious weakening. The oxide ceramics may be sintered in air. In fact, it is desirable to have an oxidizing atmosphere to aid in removing the organic materials by allowing them to react with the oxygen to form CO_2. The nitride ceramics must be sintered in the presence of nitrogen to prevent oxides of the metal from being formed. In this case, no reaction of the organics takes place; they are evaporated and carried away by the nitrogen flow.

During sintering, a degree of shrinkage takes place as the organic is removed and the fluxing glasses activate. Shrinkage may be as low as 10% for powder processing to as high as 22% for sheet casting. The degree of shrinkage is highly predictable and may be considered during design.

Powder pressing generally forms boron nitride substrates. Various silica and/or calcium compounds may be added to lower the processing temperature and improve machinability. Diamond substrates are typically formed by chemical vapor deposition (CVD). Composite substrates, such as AlSiC, are fabricated by creating a spongy structure of SiC, and forcing molten aluminum into the crevices.

4.3   Surface Properties of Ceramic Substrates

The surface properties of interest, surface roughness and camber, are highly dependent on the particle size and method of processing. *Surface roughness* is a measure of the surface microstructure and *camber* is a measure of the deviation from flatness. In general, the smaller the particle size, the smoother the surface.

Surface roughness may be measured by electrical or optical means. Electrically, surface roughness is measured by moving a fine-tipped stylus across the surface. The stylus may be attached to a piezoelectric crystal or to a small magnet that moves inside a coil, inducing a voltage proportional to the magnitude of the substrate variations. The stylus must have a resolution of 25.4 nm (1 μin) in order to read accurately in the most common ranges. Optically, a coherent light beam from a laser diode or other source is directed onto the surface. The deviations in the substrate surface create interference patterns that are used to calculate the roughness. Optical profilometers have a higher resolution than the electrical versions and are used primarily for very smooth surfaces. For ordinary use, the electric profilometer is adequate and is widely used to characterize substrates in both manufacturing and laboratory environments.

The output of an electric profilometer is plotted in schematic form in Fig. 4.4 and in actual form in Fig. 4.5. A quantitative interpretation of surface roughness can be obtained from this plot in one of two ways—by the root-mean-square (rms) value and by the arithmetic average.

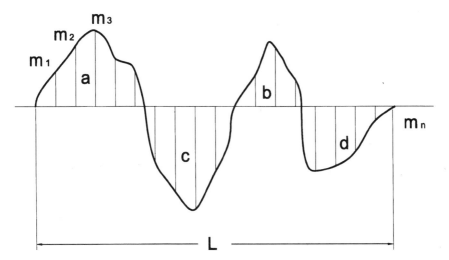

Figure 4.4   Schematic of surface trace.

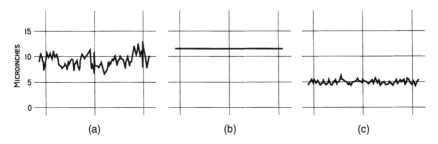

Figure 4.5   Surface trace of three substrate surfaces. (a) 23-μin surface. (b) 0.3-μin surface. (c) 7-μin surface.

The rms value is obtained by dividing the plot into n small, even increments of distance and measuring the height, m, at each point, as shown in Fig. 4.4. The rms value is calculated by

$$\text{rms} = \sqrt{\frac{m_1^2 + m_2^2 + \cdots + m_n^2}{n}} \qquad (4.1)$$

and the average value [usually referred to as the centerline average (CLA)] is calculated by

$$\text{CLA} = \frac{a_1 + a_2 + a_3 + >>> + a_n}{L} \qquad (4.2)$$

where $a_1, a_2, a_3,...$ = areas under the trace segments (Fig. 4.4)

L = length of travel

For systems where the trace is magnified by a factor, M, Eq. (4.2) must be divided by the same factor.

For a sine wave, the average value is 0.636 × peak and the rms value is 0.707 × peak, which is 11.2% larger than the average. The profilometer trace is not quite sinusoidal in nature. The rms value may be greater than the CLA value by 10 to 30%.

Of the two methods, the CLA is the preferred method of use because the calculation is more directly related to the surface roughness, but it also has several shortcomings:

- The method does not consider surface waviness or camber, as shown in Fig. 4.6.[2]

- Surface profiles with different periodicities and the same amplitudes yield the same results, although the effect in use may be somewhat different.

- The value obtained is a function of the tip radius.

Surface roughness has a significant effect on the adhesion and performance of thick- and thin-film depositions. For adhesion purposes, it is desirable to have a high surface roughness to increase the effective interface area between the film and the substrate. For stability and repeatability, the thickness of the deposited film should be much greater than the variations in the surface. For thick films, which have a typical thickness of 10 to 12 μin, surface roughness is not a consideration, and a value of 25 μm (625 nm) is desirable. For thin films, however, which may have a thickness measured in angstroms, a much smoother surface is required. Figure 4.7 illustrates the difference in a thin film of tantalum nitride (TaN) deposited on both a 1-μin surface

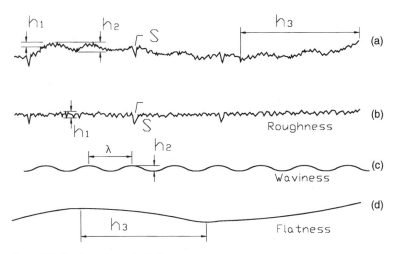

Figure 4.6 Surface characteristics.

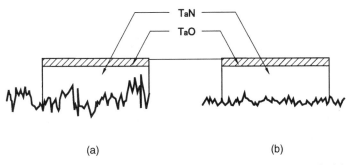

Figure 4.7 TaN resistor with TaO passivation on substrates with different surface roughness. (*a*) 5-μin surface. (*b*) 1-μin surface.

and a 5-μin surface. Tantalum nitride is commonly used to fabricate resistors in thin-film circuits and is stabilized by growing a layer of tantalum oxide, which is nonconductive, over the surface by baking the resistors in air. Note that the oxide layer in the rougher surface represents a more significant percentage of the overall thickness of the film in areas where the surface deviation is the greatest. The result is a wider variation in both the initial and poststabilization resistor values and a larger drift in value with time.

Camber and waviness are similar in form in that they are variations in flatness over the substrate surface. Referring to Fig. 4.6, camber can be considered as an overall warpage of the substrate, while waviness is more periodic in nature. Both of these factors may occur as a result of uneven shrinkage during the organic removal/sintering process or as a result of nonuniform composition. Waviness may also occur as a result of a "flat spot" in the rollers used to form the green sheets.

Camber is measured in units of length/length, interpreted as the deviation from flatness per unit length, and is measured with reference to the longest dimension by placing the substrate through parallel plates that are set a specific distance apart. Thus, a rectangular substrate would be measured along the diagonal. A typical value of camber is 0.003 in/in (also 0.003 mm/mm), which, for a 2 × 2-in substrate, represents a total deviation of 0.003 × 2 × 1.414 in = 0.0085 in. For a substrate that is 0.025 in thick, a common value, the total deviation represents one-third of the overall thickness!

The nonplanar surface created by camber adversely affects subsequent metallization and assembly processes. In particular, screen printing is made more difficult due to the variable snapoff distance. Torsion bar printing heads on modern screen printers can compensate to a certain extent but not entirely. A vacuum holddown on the screen printer platen also helps, but it only flattens the substrate temporarily during the actual printing process. Camber can also create excessive stresses and a nonuniform temperature coefficient of expansion. At temperature extremes, these factors can cause cracking, breaking, or even shattering of the substrate.

Camber is measured by first measuring the thickness of the substrate and then placing the substrate between a series of pairs of parallel plates set specific distances apart. Camber is calculated by subtracting the substrate thickness from the smallest distance that the substrate will pass through and dividing by the longest substrate dimension. A few generalizations can be made about camber:

- Thicker substrates will have less camber than thinner.
- Square shapes will have less camber than rectangular.
- The pressed methods of forming will produce substrates with less camber than the sheet methods.

4.4 Thermal Properties of Substrate Materials

4.4.1 Thermal conductivity

The *thermal conductivity* of a material is a measure of the ability to carry heat and is defined as

$$q = -k \frac{dT}{dx} \tag{4.3}$$

where k = thermal conductivity (W/m-°C)

q = heat flux (W/cm^2)

dT/dx = temperature gradient (°C/m) in steady state

The negative sign denotes that heat flows from areas of higher temperature to areas of lower temperature.

There are two mechanisms that contribute to thermal conductivity—the movement of free electrons and lattice vibrations, or phonons. When a material is locally heated, the kinetic energy of the free electrons in the vicinity of the heat source increases, causing the electrons to migrate to cooler areas. These electrons undergo collisions with other atoms, losing their kinetic energy in the process. The net result is that heat is drawn away from the source toward cooler areas. In a similar fashion, an increase in temperature increases the magnitude of the lattice vibrations, which, in turn, generate and transmit phonons, carrying energy away from the source. The thermal conductivity of a material is the sum of the contributions of these two parameters:

$$k = k_p + k_e \qquad (4.4)$$

where k_p = contribution due to phonons

k_e = contribution due to electrons

In ceramics, the heat flow is primarily due to phonon generation, and the thermal conductivity is generally lower than that of metals. Crystalline structures, such as alumina and beryllia, are more efficient heat conductors than amorphous structures such as glass. Organic materials used to fabricate printed circuit boards or epoxy attachment materials are electrical insulators and highly amorphous, and tend to be very poor thermal conductors.

Impurities or other structural defects in ceramics tend to lower the thermal conductivity by causing the phonons to undergo more collisions, lowering the mobility and lessening their ability to transport heat away from the source. This is illustrated by Table 4.2, which lists the thermal conductivity of alumina as a function of the percentage of glass. Although the thermal conductivity of the glass binder is lower than that of the alumina, the drop in thermal conductivity is greater than expected from the addition of the glass alone. If the thermal conductivity is a function of the ratio of the materials alone, it should have the form

$$k_T = P_1 k_1 + P_2 k_2 \qquad (4.5)$$

where k_T = net thermal conductivity

P_1 = percentage of material 1 in decimal form

k_1 = thermal conductivity of material 1

P_2 = percentage of material 2 in decimal form

k_2 = thermal conductivity of material 2

TABLE 4.2   Thermal Conductivity of Alumina Substrates with Different Concentrations of Alumina

Percentage of alumina	Thermal conductivity (W/m-°C)
85	16.0
90	16.7
94	22.4
96	24.7
99.5	28.1
100	31.0

In pure form, alumina has a thermal conductivity of about 31 W/m-°C, and the binding glass has a thermal conductivity of about 1 W/m-°C. Equation (4.5) and the parameters from Table 4.2 are plotted in Fig. 4.8.

By the same token, as the ambient temperature increases, the number of collisions increases, and the thermal conductivity of most materials decreases. A plot of the thermal conductivity versus temperature for several materials is shown in Fig. 4.9.[3] One material not plotted in this graph is diamond. The thermal conductivity of diamond varies widely with composition and the method of preparation, and is much higher than those materials listed. Diamond will be discussed in detail in a later section. Selected data from Fig. 4.9 was analyzed and extrapolated into binomial equations that quantitatively describe the thermal conductivity versus temperature relationship. This data is summarized in Table 4.3.

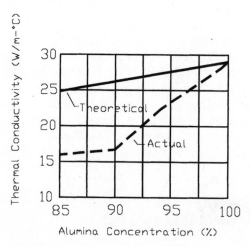

Figure 4.8 Thermal conductivity of alumina versus concentration—theoretical and actual.

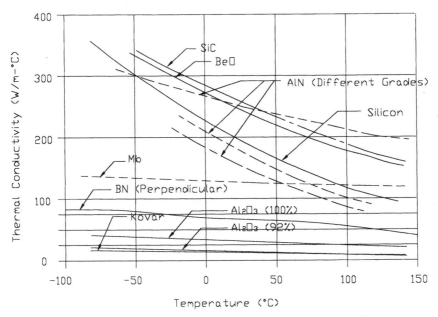

Figure 4.9 Thermal conductivity versus temperature for selected materials.

TABLE 4.3 **Approximate Thermal Conductivity Versus Temperature for Selected Ceramic Materials (Binomial Relationship)**

Material	Constant	T coefficient	T^2 coefficient
SiC	285	−1.11	1.55×10^{-3}
BeO	275	−1.10	1.06×10^{-3}
AlN (pure)	271	−0.60	3.81×10^{-4}
AlN (grade 1)	210	−1.39	3.45×10^{-3}
AlN (grade 2)	185	−1.37	4.40×10^{-3}
BN (perpendicular)	73	−0.06	2.17×10^{-4}
Al₂O₃ (99%)	34	−0.12	2.00×10^{-4}
Al₂O₃ (96%)	17	−0.07	5.70×10^{-5}

4.4.2 Specific heat

The specific heat of a material is defined as

$$c = \frac{dQ}{dT} \tag{4.6}$$

where c = specific heat (W-s/g-°C)
 Q = energy (W-s)
 T = temperature (K)

The specific heat, c, is defined in a similar manner and is the amount of heat required to raise the temperature of 1 g of material by

1°C, with units of watt-second/gram-degree Celsius. The quantity "specific heat" in this context refers to the quantity, c_v, which is the specific heat measured with the volume constant, as opposed to c_p, which is measured with the pressure constant. At the temperatures of interest, these numbers are nearly the same for most solid materials. The specific heat is primarily the result of an increase in the vibrational energy of the atoms when heated, and the specific heat of most materials increases with temperature up to a temperature called the Debye temperature, at which point it becomes essentially independent of temperature. The specific heat of several common ceramic materials as a function of temperature is shown in Fig. 4.10.

The heat capacity, C, is similar in form, except that it is defined in terms of the amount of heat required to raise the temperature of 1 mol of material by 1°C and has the units of watt-second/mol-degree Celsius.

4.4.3 Temperature coefficient of expansion

The temperature coefficient of expansion (TCE) arises from the asymmetrical increase in the interatomic spacing of atoms as a result of increased heat. Most metals and ceramics exhibit a linear, isotropic relationship in the temperature range of interest, while certain plastics may be anisotropic in nature. The TCE is defined as

$$\alpha = \frac{l\,(T_2) - l\,(T_1)}{l\,(T_1)\,(T_2 - T_1)} \tag{4.7}$$

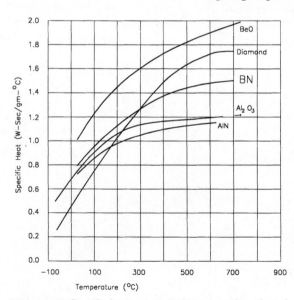

Figure 4.10 Specific heat versus temperature for selected materials.

where α = temperature coefficient of expansion (ppm/$^\circ$C^{-1})
T_1 = initial temperature in-$^\circ$C
T_2 = final temperature in-$^\circ$C
$l\,(T_1)$ = length at initial temperature
$l\,(T_2)$ = length at final temperature

The TCE of most ceramics is isotropic. For certain crystalline or single-crystal ceramics, the TCE may be anisotropic, and some may even contract in one direction and expand in the other. Ceramics used for substrates do not generally fall into this category, as most are mixed with glasses in the preparation stage and do not exhibit anisotropic properties as a result. The temperature coefficient of expansion of several ceramic materials is shown in Table 4.4.

4.5 Mechanical Properties of Ceramic Substrates

The mechanical properties of ceramic materials are strongly influenced by the strong interatomic bonds that prevail. Dislocation mechanisms, which create slip mechanisms in softer metals, are relatively scarce in ceramics, and failure may occur with very little plastic deformation. Ceramics also tend to fracture with little resistance.

4.5.1 Modulus of elasticity

The TCE phenomenon has serious implications in the applications of ceramic substrates. When a sample of material has one end fixed, which may be considered to be a result of bonding to another material that has a much smaller TCE, the net elongation of the hotter end per unit length, or "strain" (E), of the material is calculated by

$$E = \text{TCE} \times \Delta T \qquad (4.8)$$

TABLE 4.4 Temperature Coefficient of Expansion of Selected Ceramic Substrate Materials

Material	TCE (ppm/$^\circ$C)
Alumina (96%)	6.5
Alumina (99%)	6.8
BeO (99.5%)	7.5
BN:	
Parallel	0.57
Perpendicular	−0.46
Silicon carbide	3.7
Aluminum nitride	4.4
Diamond, type IIA	1.02
AlSiC (70% SiC loading)	6.3

where E = strain in length/length
ΔT = temperature differential across the sample

Elongation develops a stress, S, per unit length in the sample as given by Hooke's law:

$$S = E\,Y \tag{4.9}$$

where S = stress [lb/in² (N/m²/m)]
Y = modulus of elasticity [lb/in² (N/m²)]

When the total stress, as calculated by multiplying the stress/unit length by the maximum dimension of the sample, exceeds the strength of the material, mechanical cracks will form in the sample that may even propagate to the point of separation. The small elongation that occurs before failure is referred to as *plastic deformation*. This analysis is somewhat simplistic in nature, but serves as a basic understanding of the mechanical considerations. The modulus of elasticity of selected ceramics is summarized in Table 4.5, along with other mechanical properties.

4.5.2 Modulus of rupture

Ordinary stress-strain testing is not generally used to test ceramic substrates since they do not exhibit elastic behavior to a great degree. An alternate test, the modulus of rupture (bend strength) test, as described in Fig. 4.11, is preferred. A sample of ceramic, either circular or rectangular, is suspended between two points, a force is applied in the center, and the elongation of the sample is measured. The stress is calculated by

$$\sigma = \frac{Mx}{I} \tag{4.10}$$

where σ = stress (MPa)
M = maximum bending moment (N-m)
x = distance from center to outer surface (m)
I = moment of inertia (N-m²)

The expressions for σ, M, x, and I are summarized in Table 4.6. When these are inserted into Eq. (4.10), the result is

$$\sigma = \frac{3\,F\,L}{2 \times y^{2}} \qquad \text{rectangular cross section} \tag{4.11}$$

$$\sigma = \frac{FL}{\pi R^{3}} \qquad \text{circular cross section} \tag{4.12}$$

TABLE 4.5 Mechanical Properties of Selected Ceramics

Material	Modulus of elasticity (GPa)	Tensile strength (MPa)	Compressive strength (MPa)	Modulus of rupture (MPa)	Flexural strength (MPa)	Density (g/cm^3)
Alumina (99%)	370	500	2,600	386	352	3.98
Alumina (96%)	344	172	2,260	341	331	3.92
Beryllia (99.5%)	345	138	1,550	233	235	2.87
Boron nitride (normal)	43	2,410	6,525	800	53.1	1.92
Aluminum nitride	300	310	2,000	300	269	3.27
Silicon carbide	407	197	4,400	470	518	3.10
Diamond (type IIA)	1,000	1,200	11,000	940	1,000	3.52

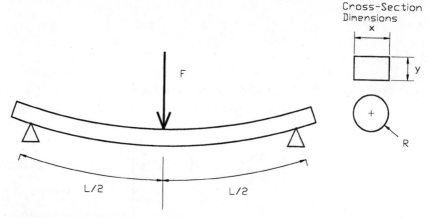

Figure 4.11 Modulus of rupture test setup.

TABLE 4.6   Parameters of Stress in Modulus of Rupture Test[5]

Cross section	M	c	I
Rectangular	$F\,L/4$	$y/2$	$x\,y^3/12$
Circular	$F\,L/4$	R	$\pi R^2/4$

where F = applied force (N)
    x = long dimension of rectangular cross section (m)
    y = short dimension of rectangular cross section (m)
    L = length of sample (m)
    R = radius of circular cross section (m)

The modulus of rupture is the stress required to produce fracture and is given by

$$\sigma_r = \frac{3\,F_r\,L}{2 \times y^2} \qquad \text{rectangular} \qquad (4.13)$$

$$\sigma_r = \frac{F_r\,L}{\pi R^3} \qquad \text{circular} \qquad (4.14)$$

where σ_r = modulus of rupture (N/m²)
    F_r = force at rupture

The modulus of rupture for selected ceramics is shown in Table 4.5.

4.5.3 Tensile and compressive strength

A force applied to a ceramic substrate in a tangential direction may produce tensile or compressive forces. If the force is tensile, in a direction such that the material is pulled apart, the stress produces plastic deformation as defined in Eq. (4.9). As the force increases past a value referred to as the *tensile strength,* breakage occurs. Conversely, a force applied in the opposite direction creates compressive forces until a value referred to as the *compressive strength* is reached, at which point breakage also occurs. The compressive strength of ceramics is, in general, much larger than the tensile strength. The tensile and compressive strength of selected ceramic materials are shown in Table 4.5.

In practice, the force required to fracture a ceramic substrate is much lower than predicted by theory. The discrepancy is due to small flaws or cracks residing within these materials as a result of processing. For example, when a substrate is sawed, small edge cracks may be created. Similarly, when a substrate is fired, trapped organic material may outgas during firing, leaving a microscopic void in the bulk. The result is an amplification of the applied stress in the vicinity of the void that may exceed the tensile strength of the material and create a fracture. If the microcrack is assumed to be elliptical with the major axis perpendicular to the applied stress, the maximum stress at the tip of the crack may be approximated by[4]

$$S_M = 2\,S_O \leq \left(\frac{a}{\rho_t} \right)^{1/2} \tag{4.15}$$

where S_M = maximum stress at the tip of the crack
S_O = nominal applied stress
a = length of the crack as defined in Fig. 4.11
ρ_t = radius of the crack tip

The ratio of the maximum stress to the applied stress may be defined as

$$K_t = \frac{S_M}{S_O} = 2 \left(\frac{a}{\rho_t} \right)^{1/2} \tag{4.16}$$

where K_t = stress concentration factor

For certain geometries, such as a long crack with a small tip radius, K_t may be much larger than 1, and the force at the tip may be substantially larger than the applied force.

Based on this analysis, a material parameter called the *plain strain fracture toughness,* a measure of the ability of the material to resist fracture, can be defined as

$$K_{IC} = Z \, S_C \, \sqrt{\pi \, a} \qquad (4.17)$$

where K_{IC} = plain strain fracture toughness [lb/in²-in$^{1/2}$ (MPa-m$^{1/2}$)]
  Z = dimensionless constant, typically 1.2 [Ref. 4]
  S_C = critical force required to cause breakage

From Eq. (4.17), the expression for the critical force can be defined as

$$S_C = Z \, \frac{K_{IC}}{\sqrt{\pi a}} \qquad (4.18)$$

When the applied force on the die due to TCE or thermal differences exceeds this figure, fracture is likely. The plain strain fracture toughness for selected materials is presented in Table 4.7. It should be noted that Eq. (4.18) is a function of thickness up to a point, but is approximately constant for the area-to-thickness ratio normally found in substrates.

4.5.4 Hardness

Ceramics are among the hardest substances known and the hardness is correspondingly difficult to measure. Most methods rely on the ability of one material to scratch another and the measurement is presented on a relative scale. Of the available methods, the Knoop method is the most frequently used. In this approach, the surface is highly polished and a pointed diamond stylus under a light load is allowed to impact on the material. The depth of the indentation formed by the stylus is measured and converted to a qualitative scale called the "Knoop" or "HK" scale. The Knoop hardness of selected ceramics is given in Table 4.8.

TABLE 4.7 Fracture Toughness for Selected Materials

Material	Fracture toughness (MPa-m$^{1/2}$)
Silicon	0.8
Alumina (96%)	3.7
Alumina (99%)	4.6
Silicon carbide	7.0
Molding compound	2.0

TABLE 4.8   Knoop Hardness for Selected Ceramics

Material	Knoop hardness (100 g)
Diamond	7000
Aluminum oxide	2100
Aluminum nitride	1200
Beryllium oxide	1200
Boron nitride	5000
Silicon carbide	2500

4.5.5 Thermal shock

Thermal shock occurs when a substrate is exposed to temperature extremes in a short period of time. Under these conditions, the substrate is not in thermal equilibrium and internal stresses may be sufficient to cause fracture. Thermal shock can be liquid-to-liquid or air-to-air, with the most extreme exposure occurring when the substrate is transferred directly from one liquid bath to another. The heat is more rapidly absorbed or transmitted, depending on the relative temperature of the bath, due to the higher specific of the liquid as opposed to air.

The ability of a substrate to withstand thermal shock is a function of several variables, including the thermal conductivity, the coefficient of thermal expansion, and the specific heat. Winkleman and Schott[5] developed a parameter called the *coefficient of thermal endurance* that qualitatively measures the ability of a substrate to withstand thermal stress:

$$F = \frac{P}{\alpha\,Y}\sqrt{\frac{k}{\rho c}} \qquad (4.19)$$

where F = coefficient of thermal endurance
P = tensile strength (MPa)
α = thermal coefficient of expansion (1/K)
Y = modulus of elasticity (MPa)
k = thermal conductivity (W/m-K)
ρ = density (kg/m^3)
c = specific heat (W-s/kg-K)

The coefficient of thermal endurance for selected materials is shown in Table 4.9. The phenomenally high coefficient of thermal endurance for BN is primarily a result of the high tensile strength to modulus of elasticity ratio as compared to other materials. Diamond is also high, primarily due to the high tensile strength, the high thermal conductivity, and the low TCE.

TABLE 4.9   Thermal Endurance Factor for Selected Materials at 25°C

Material	Thermal endurance factor
Alumina (99%)	0.640
Alumina (96%)	0.234
Beryllia (99.5%)	0.225
Boron nitride (a axis)	648
Aluminum nitride	2.325
Silicon carbide	1.40
Diamond (type IIA)	30.29

The thermal endurance factor is a function of temperature in that several of the variables, particularly the thermal conductivity and the specific heat, are functions of temperature. From Table 4.9, it is also noted that the thermal endurance factor may drop rapidly as the alumina-to-glass ratio drops. This is due to the difference in the thermal conductivity and TCE of the alumina and glass constituents that increase the internal stresses. This is true of other materials as well.

4.6   Electrical Properties of Ceramic Substrates

The electrical properties of ceramic substrates perform an important task in the operation of electronic circuits. Depending on the applications, the electrical parameters may be advantageous or detrimental to circuit function. Of most interest are the resistivity, the breakdown voltage, or dielectric strength, and the dielectric properties, including the dielectric constant and the loss tangent.

4.6.1   Resistivity

The electrical resistivity of a material is a measure of the ability of that material to transport a charge under the influence of an applied electric field. More often, this ability is presented in the form of the electrical conductivity, the reciprocal of the resistivity as defined in Eq. (4.20):

$$\sigma = \frac{1}{\rho} \qquad (4.20)$$

where σ = conductivity (S/unit length)
ρ = resistivity (Ω-unit length)

The conductivity is a function primarily of two variables—the concentration of charge and the mobility, the ability of that charge to be transported through the material. The current density and the applied field are related by the expression defined in Eq. (4.21):

$$\mathbf{J} = \sigma \, \mathbf{E} \tag{4.21}$$

where $\mathbf{J}$ = current density (A/unit area)
$\mathbf{E}$ = electric field (V/unit length)

It should be noted that both the current density and the electric field are vectors since the current is in the direction of the electric field. The current density may also be defined as

$$\mathbf{J} = n v_d \tag{4.22}$$

where n = free carrier concentration (C/unit volume)
v_d = drift velocity of electrons (unit length/s)

The drift velocity is related to the electric field by

$$v_d = \mu \, \mathbf{E} \tag{4.23}$$

where μ = mobility (length^2/V-s)

In terms of the free carrier concentration and the mobility, the current density is

$$\mathbf{J} = n \mu \mathbf{E} \tag{4.24}$$

Comparing Eq. (4.19) with Eq. (4.23) the conductivity can be defined as

$$\sigma = n \, \mu \tag{4.25}$$

The free carrier concentration may be expressed as

$$n = n_t + n_i \tag{4.26}$$

where n_t = free carrier concentration due to thermal activity
n_i = free carrier concentration due to field injection

The thermal charge density, n_t, in insulators is a result of free electrons obtaining sufficient thermal energy to break the interatomic bonds, allowing them to move freely within the atomic lattice. Ceramic materials characteristically have few thermal electrons as a result of the strong ionic bonds between atoms. The injected charge density, n_i, occurs when a potential is applied and is a result of the inherent capacity of the material. The injected charge density is given by

$$n_i = \varepsilon \, \mathbf{E} \tag{4.27}$$

where ε = dielectric constant of the material (F/unit length)

Inserting Eq. (4.27) and Eq. (4.26) into Eq. (4.24), the result is

$$\mathbf{J} = \mu\, n_t\, \mathbf{E} + \mu\, \varepsilon\, \mathbf{E}^2 \qquad (4.28)$$

For conductors, $n_t \gg n_i$ and Ohm's law applies. For insulators, $n_i \gg n_t$, and the result is a square law relationship between the voltage and the current:[6]

$$\mathbf{J} = \mu\, \varepsilon\, \mathbf{E}^2 \qquad (4.29)$$

The conductivity of ceramic substrates is extremely low. In practice, it is primarily due to impurities and lattice defects, and may vary widely from batch to batch. The conductivity is also a strong function of temperature. As the temperature increases, the ratio of thermal to injected carriers increases. As a result, the conductivity increases and the *V-l* relationship follows Ohm's law more closely. Typical values of the resistivity of selected ceramic materials are presented in Table 4.10.

4.6.2   Breakdown voltage

The term *breakdown voltage* is very descriptive. While ceramics are normally very good insulators, the application of excessively high potentials can dislodge electrons from orbit with sufficient energy to allow them to dislodge other electrons from orbit, creating an "avalanche effect." The result is a breakdown of the insulation proper-

TABLE 4.10   Electrical Properties of Selected Ceramic Substrates

	Property			
Material	Electrical resistivity (Ω-cm)	Breakdown voltage (AC kV/mm)	Dielectric constant	Loss tangent (@ 1 MHz)
Alumina (96%):				
25°C	$>10^{14}$		9.0	
500°C	4×10^9	8.3	10.8	0.0002
1000°C	1×10^6			
Alumina (99.5%):				
25°C	$>10^{14}$			
500°C	2×10^{10}	8.7	9.4	0.0001
1000°C	2×10^6		10.1	
Beryllia:				
25°C	$>10^{14}$	6.6	6.4	0.0001
500°C	2×10^{10}		6.9	0.0004
Aluminum nitride	$>10^{13}$	14	8.9	0.0004
Boron nitride	$>10^{14}$	61	4.1	0.0003
Silicon carbide*	$>10^{13}$	0.7	40	0.05
Diamond (type II)	$>10^{14}$	1000	5.7	0.0006

*Depends on method of preparation. May be substantially lower.

ties of the material, allowing current to flow. This phenomenon is accelerated by elevated temperature, particularly when mobile ionic impurities are present.

The breakdown voltage is a function of numerous variables, including the concentration of mobile ionic impurities, grain boundaries, and the degree of stoichiometry. In most applications, the breakdown voltage is sufficiently high as to not be an issue. However, there are two cases where it must be a consideration:

1. At elevated temperatures created by localized power dissipation or high ambient temperature, the breakdown voltage may drop by orders of magnitude. Combined with a high potential gradient, this condition may be susceptible to breakdown.

2. The surface of most ceramics is highly "wettable," in that moisture tends to spread rapidly. Under conditions of high humidity, coupled with surface contamination, the effective breakdown voltage is much lower than the intrinsic value.

4.6.3 Dielectric properties

Two conductors in proximity with a difference in potential have the ability to attract and store electric charge. Placing a material with dielectric properties between them enhances this effect. A dielectric material has the capability of forming electric dipoles, displacements of electric charge, internally. At the surface of the dielectric, the dipoles attract more electric charge, thus enhancing the charge storage capability, or capacitance, of the system. The relative ability of a material to attract electric charge in this manner is called the *relative dielectric constant*, or *relative permittivity*, and is usually given the symbol, K. The relative permittivity of free space is 1.0 by definition and the absolute permittivity is

$$\varepsilon_0 = \text{permittivity of free space}$$

$$\varepsilon_0 = \frac{1}{36\,\pi} \times 10^{-9} \quad \frac{F}{\text{meter}} \tag{4.30}$$

The relationship between the polarization and the electric field is

$$\overline{P} = \varepsilon_0\,(K-1)\,\overline{E}\ \frac{Q}{m^2} \tag{4.31}$$

where P = polarization (C/m^2)
$\mathbf{E}$ = electric field (V/m)

There are four basic mechanisms that contribute to polarization:

1. *Electronic polarization.* In the presence of an applied field, the cloud of electrons is displaced relative to the positive nucleus of the atom or molecule, creating an induced dipole moment. Electronic polarization is essentially independent of temperature and may occur very rapidly. The dielectric constant may, therefore, exist at very high frequencies, up to 10^{17} Hz.

2. *Molecular polarization.* Certain molecular structures create permanent dipoles that exist even in the absence of an electric field. These may be rotated by an applied electric field, generating a degree of polarization by orientation. Molecular polarization is inversely proportional to temperature and occurs only at low to moderate frequencies. Molecular polarization does not occur to a great extent in ceramics, and is more prevalent in organic materials and liquids such as water.

3. *Ionic polarization.* Ionic polarization occurs in ionically bonded materials when the positive and negative ions undergo a relative displacement to each other in the presence of an applied electric field. Ionic polarization is somewhat insensitive to temperature and occurs at high frequencies, up to 10^{13} Hz.

4. *Space charge polarization.* Space charge polarization exists as a result of charges derived from contaminants or irregularities that exist within the dielectric. These charges exist to a greater or lesser degree in all crystal lattices and are partly mobile. Consequently, they will migrate in the presence of an applied electric field. Space charge polarization occurs only at very low frequencies.

In a given material, more than one type of polarization can exist and the net polarization is given by

$$\bar{P}_t = \bar{P}_e + \bar{P}_m + \bar{P}_i + \bar{P}_s \qquad (4.32)$$

where P_t = total polarization
P_e = electronic polarization
P_m = molecular polarization
P_i = ionic polarization
P_s = space charge polarization

Normally, the dipoles are randomly oriented in the material and the resulting internal electric field is zero. In the presence of an external applied electric field, the dipoles become oriented as shown in Fig. 4.12.

There are two common ways to categorize dielectric materials—polar or nonpolar and paraelectric or ferroelectric. Polar materials include those which are primarily molecular in nature, such as water,

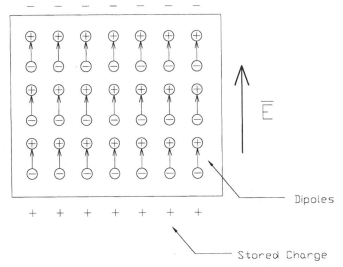

Figure 4.12   Orientation of dipoles in an electric field.[1]

and nonpolar materials include both electronically and ionically polarized materials. Paraelectric materials are polarized only in the presence of an applied electric field and lose their polarization when the field is removed. Ferroelectric materials retain a degree of polarization after the field is removed. Materials used as ceramic substrates are usually nonpolar and paraelectric in nature. An exception is silicon carbide, which has a degree of molecular polarization.

In the presence of an electric field that is changing at a high frequency, the polarity of the dipoles must change at the same rate as the polarity of the signal in order to maintain the dielectric constant at the same level. Some materials are excellent dielectrics at low frequencies, but the dielectric qualities drop off rapidly as the frequency increases. Electronic polarization, which involves only displacement of free charge and not ions, responds more rapidly to the changes in the direction of the electric field, and remains viable up to about 10^{17} Hz. The polarization effect of ionic displacement begins to fall off at about 10^{13} Hz, and molecular and space charge polarizations fall off at still lower frequencies. The frequency response of the different types is shown in Fig. 4.13, which also illustrates that the dielectric constant decreases with frequency.

Changing the polarity of the dipoles requires a finite amount of energy and time. The energy is dissipated as internal heat, quantified by a parameter called the *loss tangent* or *dissipation factor*. Further, dielectric materials are not perfect insulators. These phenomena may

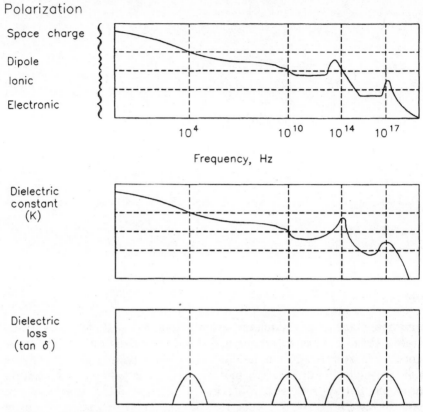

Figure 4.13 Frequency effects on dielectric materials. (From Ref. 1, Fig. 8.4.)

be modeled as a resistor in parallel with a capacitor. The loss tangent, as expected, is a strong function of the applied frequency, increasing as the frequency increases.

In alternating current applications, the current and voltage across an ideal capacitor are exactly 90° out of phase, with the current leading the voltage. In actuality, the resistive component causes the current to lead the voltage by an angle less than 90°. The loss tangent is a measure of the real or resistive component of the capacitor and is the tangent of the difference between 90° and the actual phase angle. Thus

$$\text{Loss tangent} = \tan(90° - \delta) \qquad (4.33)$$

where δ = phase angle between voltage and current. The loss tangent is also referred to as the *dissipation factor* (DF).

The loss tangent may also be considered as a measure of the time required for polarization. It requires a finite amount of time to change the polarity of the dipole after an alternating field is applied. The resulting phase retardation is equivalent to the time indicated by the difference in phase angles.

4.7 Metallization of Ceramic Substrates

There are three fundamental methods of metallizing ceramic substrates—thick film, thin film, and copper, which includes direct bond copper (DBC), plated copper, and active metal braze (AMB). Not all of these processes are compatible with all substrates. The selection of a metallization system depends on both the application and the compatibility with the substrate material.

4.7.1 Thick film

The *thick-film process* is an additive process by which conductive, resistive, and dielectric (insulating) patterns in the form of a viscous paste are screen printed, dried, and fired onto a ceramic substrate at an elevated temperature to promote the adhesion of the film. In this manner, by depositing successive layers, as shown in Fig. 4.14, multilayer interconnection structures can be formed which may contain integrated resistors, capacitors, or inductors.

The initial step is to generate 1:1 artworks corresponding to each layer of the circuit. The screen is a stainless-steel mesh with a mesh count of 80 to 400 wires/in. The mesh is stretched to the proper tension and mounted to a cast aluminum frame with epoxy. It is coated with a photosensitive material and exposed to light through one of the artworks. The unexposed portion is rinsed away, leaving openings in the screen mesh corresponding to the pattern to be printed.

Thick-film materials in the fired state are a combination of glass ceramic and metal, referred to as "cermet" thick films, and are designed to be fired in the range 850 to 1000°C. A standard cermet thick-film paste has four major ingredients—an active element, which establishes the function of the film; an adhesion element, which provides the adhesion to the substrate and a matrix which holds the active particles in suspension; an organic binder, which provides the proper fluid properties for screen printing; and a solvent or thinner, which establishes the viscosity of the vehicle phase.

4.7.1.1 The active element. The active element within the paste dictates the electrical properties of the fired film. If the active element is a metal, the fired film will be a conductor; if it is a conductive metal oxide, a resistor; and, if it is an insulator, a dielectric. The active element is most commonly found in powder form ranging from 1 to 10 μm in size, with a mean diameter of about 5 μm.

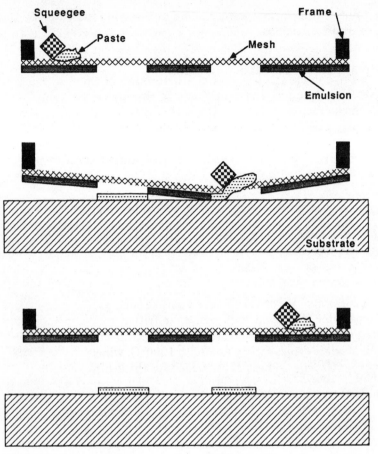

Figure 4.14 Screen printing process for material deposition onto a substrate.
(From Ref. 1, Fig. 1.2.)

4.7.1.2 The adhesion element. There are two primary constituents
used to bond the film to the substrate—glass and metal oxides, which
may be used singly or in combination. Films that use a glass, or "frit,"
are referred to as *fritted materials,* and have a relatively low melting
point (500 to 600°C). There are two adhesion mechanisms associated
with the fritted materials, a chemical reaction and a physical reaction.
In the chemical reaction, the molten glass chemically reacts with the
glass in the substrate to a degree. In the physical reaction, the glass
flows into and around the irregularities in the substrate surface. The
total adhesion is the sum of the two factors. The physical bonds are
more susceptible to degradation by thermal cycling or thermal storage
than the chemical bonds, and are generally the first to fracture under
stress. The glass also creates a matrix for the active particles, holding

them in contact with each other to promote sintering and to provide a series of three-dimensional continuous paths from one end of the film to the other. Principal thick film glasses are based on B_2O_3–SiO_2 network formers with modifiers such as PbO, Al_2O_3, Bi_2O_3, ZnO, BaO, and CdO added to change the physical characteristics of the film, such as melting point, viscosity, and coefficient of thermal expansion. Bi_2O_3 also has excellent wetting properties, both to the active element and to the substrate, and is frequently used as a flux. The glass phase may be introduced as a prereacted particle or formed in situ by using glass precursors such as boric oxide, lead oxide, and silicon. Fritted conductor materials tend to have glass on the surface, making subsequent component assembly processes more difficult.

A second class of materials utilizes metal oxides to provide the adhesion to the substrate. In this case, a pure metal, such as copper or cadmium, is mixed with the paste and reacts with oxygen atoms on the surface of the substrate to form an oxide. The conductor adheres to the oxide and to itself by sintering, which takes place during firing. During firing, the oxides react with broken oxygen bonds on the surface of the substrate, forming a Cu or Cd spinel structure such as $CuAl_2O_4$. Pastes of this type offer improved adhesion over fritted materials, and are referred to as *fritless, oxide-bonded,* or *molecular-bonded materials.* Fritless materials typically fire at 900 to 1000°C, which is undesirable from a manufacturing aspect. Ovens used for thick-film firing degrade more rapidly and need more maintenance when operated at these temperatures for long periods of time.

A third class of materials utilizes both reactive oxides and glasses. The oxides in these materials react at lower temperatures, but are not as strong as copper. A lesser concentration of glass than found in fritted materials is added to supplement the adhesion. These materials, referred to as *mixed bonded systems,* incorporate the advantages of both technologies and fire at a lower temperature.

The selection of a binding material is strongly dependent on the substrate material. For example, the most common glass composition used with alumina is a lead/bismuth borosilicate composition. When this glass is used in conjunction with aluminum nitride, however, it is rapidly reduced at firing temperatures.[1] Alkaline earth borosilicates must be used with AlN to promote adhesion.

4.7.1.3 Organic binder. The organic binder is generally a thixotropic fluid and serves two purposes—it holds the active and adhesion elements in suspension until the film is fired and it gives the paste the proper fluid characteristics for screen printing. The organic binder is usually referred to as the *nonvolatile organic* since it does not evaporate, but begins to burn off at about 350°C. The binder must oxidize cleanly during firing, with no residual carbon which could contaminate the film. Typical materials used in this application are ethyl cellulose and various acrylics.

For nitrogen-fireable films, where the firing atmosphere can contain only a few parts per million of oxygen, the organic vehicle must decompose and thermally depolymerize, departing as a highly volatile organic vapor in the nitrogen blanket provided as the firing atmosphere, since oxidation into CO_2 or H_2O is not feasible due to the oxidation of the copper film.

4.7.1.4 Solvent or thinner. The organic binder in the natural form is too thick to permit screen printing, which necessitates the use of a solvent or thinner. The thinner is somewhat more volatile than the binder, evaporating rapidly above about 100°C. Typical materials used for this application are terpineol, butyl carbitol, or certain of the complex alcohols into which the nonvolatile phase can dissolve. The low vapor pressure at room temperature is desirable to minimize drying of the pastes and to maintain a constant viscosity during printing. Additionally, plasticizers, surfactants, and agents that modify the thixotropic nature of the paste are added to the solvent to improve paste characteristics and printing performance.

To complete the formulation process, the ingredients of the thick-film paste are mixed together in proper proportions and milled on a three-roller mill for a sufficient period of time to ensure that they are thoroughly mixed and that no agglomeration exists.

Thick-film conductor materials may be divided into two broad classes—air fireable and nitrogen fireable. Air-fireable materials are made up of noble metals that do not readily form oxides, gold and silver in the pure form, or alloyed with palladium and/or platinum. Nitrogen-fireable materials include copper, nickel, and aluminum, with copper being the most common.

Thick-film resistors are formed by adding metal oxide particles to glass particles and firing the mixture at a temperature/time combination sufficient to melt the glass and to sinter the oxide particles together. The resulting structure consists of a series of three-dimensional chains of metal oxide particles embedded in a glass matrix. The higher the metal oxide–to–glass ratio, the lower the resistivity and vice versa. The most common materials used are ruthenium-based, such as ruthenium dioxide RuO_2, and bismuth ruthenate, $BiRu_2O_7$.

Thick-film dielectric materials are used primarily as insulators between conductors, either as simple cross-overs or in complex multilayer structures. Small openings, or "vias," may be left in the dielectric layers so that adjacent conductor layers may interconnect. In complex structures, as many as several hundred vias per layer may be required.

In this manner, complex interconnection structures may be created. Although the majority of thick-film circuits can be fabricated with only three layers of metallization, others may require several more. If more than three layers are required, the yield begins dropping dramatically with a corresponding increase in cost.

Dielectric materials used in this application must be of the "devitrifying" or "recrystallizable" type. Dielectrics in the paste form are a mixture of glasses that melt at a relatively low temperature. During firing, when they are in the liquid state, they blend together to form a uniform composition with a higher melting point than the firing temperature. Consequently, on subsequent firings they remain in the solid state, which maintains a stable foundation for firing sequential layers. By contrast, vitreous glasses always melt at the same temperature and would be unacceptable for layers to either "sink" and short to conductor layers underneath, or "swim" and form an open circuit. Additionally, secondary loading of ceramic particles is used to enhance devitrification and to modify the TCE.

Dielectric materials have two conflicting requirements in that they must form a continuous film to eliminate short circuits between layers while, at the same time, they must maintain openings as small as 0.010 in. In general, dielectric materials must be printed and fired twice per layer to eliminate pinholes and prevent short circuits between layers.

The TCE of thick-film dielectric materials must be as close as possible to that of the substrate to avoid excessive "bowing," or warpage, of the substrate after several layers. Excessive bowing can cause severe problems with subsequent processing, especially where the substrate must be held down with a vacuum or where it must be mounted on a heated stage. In addition, the stresses created by the bowing can cause the dielectric material to crack, especially when it is sealed within a package. Thick-film material manufacturers have addressed this problem by developing dielectric materials that have an almost exact TCE match with alumina substrates. Where a serious mismatch exists, matching layers of dielectric must be printed on the bottom of the substrate to minimize bowing, which obviously increases the cost.

4.7.2 Thin film

The *thin-film technology* is a subtractive technology in that the entire substrate is coated with several layers of metallization and the unwanted material is etched away in a succession of selective photoetching processes. The use of photolithographic processes to form the patterns enables much finer and more well-defined lines than can be formed by the thick-film process. This feature promotes the use of thin-film technology for high density and high frequency applications.

Thin-film circuits typically consist of three layers of material deposited on a substrate. The bottom layer serves two purposes—it is the resistor material and also provides the adhesion to the substrate. The adhesion mechanism of the film to the substrate is an oxide layer that forms at the interface between the two. The bottom layer must, therefore, be a material which oxidizes readily. The most common types of resistor material are nichrome (NiCr) and tantalum nitride (TaN). Gold and silver, for example, are noble metals and do not adhere well to ceramic surfaces.

The middle layer acts as an interface between the resistor layer and the conductor layer, either by improving the adhesion of the conductor or by preventing diffusion of the resistor material into the conductor. The interface layer for TaN is usually tungsten (W), and for NiCr, a thin layer of pure Ni is used.

Gold is the most common conductor material used in thin-film circuits because of the ease of wire and die bonding and the high resistance of the gold to tarnish and corrosion. Aluminum and copper are also frequently used in certain applications. It should be noted that copper and aluminum will adhere directly to ceramic substrates, but gold requires one or more intermediate layers since it does not form the necessary oxides for adhesion.

The term *thin film* refers more to the manner in which the film is deposited onto the substrate as opposed to the actual thickness of the film. Thin films are typically deposited by one of the vacuum deposition techniques, sputtering or evaporation, or by electroplating.

Sputtering is the prime method by which thin films are applied to substrates. In ordinary DC triode sputtering, as shown in Fig. 4.15, a current is established in a conducting plasma formed by striking an arc in an inert gas, such as argon, with a partial vacuum of approximately 10 μm pressure. A substrate at ground potential and a target material at high potential are placed in the plasma. The potential may be AC or DC. The high potential attracts the gas ions in the plasma to the point where they collide with the target with sufficient kinetic energy to dislodge microscopically sized particles with enough residual kinetic energy to travel the distance to the substrate and adhere.

The adhesion of the film is enhanced by presputtering the substrate surface by random bombardment of argon ions prior to applying the potential to the target. This process removes several atomic layers of the substrate surface, creating a large number of broken oxygen bonds and promoting the formation of the oxide interface layer. The oxide formation is further enhanced by the residual heating of the substrate as a result of the transfer of the kinetic energy of the sputtered particles to the substrate when they collide.

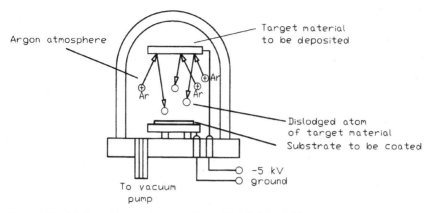

Figure 4.15   A DC sputtering chamber. (From Ref. 1, Fig. 4.4.)

DC triode sputtering is a very slow process, requiring hours to produce films with a usable thickness. By utilizing magnets at strategic points, the plasma can be concentrated in the vicinity of the target, greatly speeding up the deposition process. In most applications, an RF potential at a frequency of 13.56 MHz is applied to the target. The RF energy may be generated by a conventional electronic oscillator or by a magnetron. The magnetron is capable of generating considerably more power with a correspondingly higher deposition rate.

By adding small amounts of other gases, such as oxygen and nitrogen to the argon, it is possible to form oxides and nitrides of certain target materials on the substrate. It is this technique, called *reactive sputtering,* that is used to form tantalum nitride, a common resistor material.

Evaporation of a material into the surrounding area occurs when the vapor pressure of the material exceeds the ambient pressure and can take place from either the solid state or the liquid state. In the thin-film process, the material to be evaporated is placed in the vicinity of the substrate and heated until the vapor pressure of the material is considerably above the ambient pressure. The evaporation rate is directly proportional to the difference between the vapor pressure of the material and the ambient pressure and is highly dependent on the temperature of the material.

There are several techniques by which evaporation can be accomplished. The two most common are resistance heating and electron-beam (E-beam) heating.

Evaporation by resistance heating, as depicted in Fig. 4.16, usually takes place from a boat made with a refractory metal, a ceramic crucible wrapped with a wire heater, or a wire filament coated with the evaporant. A current is passed through the element and the heat generated heats the evaporant. It is somewhat difficult to monitor the temperature of the melt by optical means due to the propensity of the evaporant to coat the inside of the chamber, and control must be done by empirical means. Closed-loop systems exist that can control the deposition rate and the thickness, but these are quite expensive. In general, adequate results can be obtained from the empirical process if proper controls are used.

The E-beam evaporation method takes advantage of the fact that a stream of electrons accelerated by an electric field tend to travel in a circle when entering a magnetic field. This phenomenon is utilized to direct a high energy stream of electrons onto an evaporant source. The kinetic energy of the electrons is converted into heat when they strike the evaporant. E-beam evaporation is somewhat more controllable since the resistance of the boat is not a factor and the variables controlling the energy of the electrons are easier to measure and control. In addition, the heat is more localized and intense, making it possible to evaporate metals with higher 10^{-2} torr temperatures and lessening the reaction between the evaporant and the boat.

While evaporation provides a more rapid deposition rate, there are certain disadvantages when compared with sputtering:

1. It is difficult to evaporate alloys, such as nichrome (NiCr), due to the difference between the 10^{-2} torr temperatures. The element with the lower temperature tends to evaporate somewhat faster, causing the composition of the evaporated film to be different from the composition of the alloy. To achieve a particular film composition, the composition of the melt must contain a higher portion of the material with

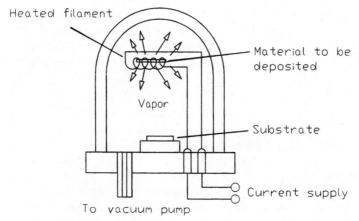

Figure 4.16 A thermal vacuum evaporation system. (From Ref. 1, Fig. 4.1.)

the higher 10^{-2} torr temperature and the temperature of the melt must be tightly controlled. By contrast, the composition of a sputtered film is identical to that of the target.

2. Evaporation is limited to the metals with lower melting points. Refractory metals, ceramics, and other insulators are virtually impossible to deposit by evaporation.

3. Reactive deposition of nitrides and oxides is very difficult to control.

Electroplating is accomplished by applying a potential between the substrate and the anode, which are suspended in a conductive solution of the material to be plated. The plating rate is a function of the potential and the concentration of the solution. In this manner, most metals can be plated to a metal surface.

In the thin-film technology, it is a common practice to sputter a film of gold a few angstroms thick and to build up the thickness of the gold film by electroplating. This is considerably more economical and results in much less target usage. For added savings, photoresist can be applied to the substrate and gold electroplated only where actually required by the pattern.

The interconnection and resistor patterns are formed by selective photoetching. The substrate is coated with a photosensitive material exposed to ultraviolet light through a pattern formed on a glass plate. The photoresist may be of the positive or negative type, with the positive type being prevalent due to its inherently higher resistance to the etchant materials. The unwanted material that is not protected by the photoresist may be removed by "wet," or chemical, etching or by "dry," or sputter etching.

In general, two masks are required—one corresponding to the conductor pattern and one corresponding to a combination of both the conductor and resistor patterns, generally referred to as the "composite" pattern. As an alternate to the composite mask, a mask that contains only the resistor pattern plus a slight overlap onto the conductor to allow for misalignment may be used. The composite mask is preferred since it allows a second gold etch process to be performed to remove any bridges or extraneous gold which might have been left from the first etch.

Sputtering may also be used to etch thin films. In this technique, the substrate is coated with photoresist and the pattern exposed in exactly the same manner as with chemical etching. The substrate is then placed in a plasma and connected to a potential. In effect, the substrate acts as the target during the sputter etching process, with the unwanted material being removed by the impingement of the gas ions on the exposed film. Since the photoresistive film is considerably thicker than the sputtered film, it is not affected.

4.7.3 Copper metallization technologies

The thick-film and thin-film technologies are limited in their ability to deposit films with a thickness greater than 1 mil (25 μm). This factor directly affects the ohmic resistance of the circuit traces and affects their ability to handle large currents or high frequencies. The copper metallization technologies provide conductors with greatly increased conductor thickness which offer improved circuit performance in many applications. There are three basic technologies available to the hybrid designer—direct bond copper (DBC), active metal braze (AMB), and the various methods of plating copper directly to ceramic.

4.7.3.1 Direct bond copper. Copper may be directly bonded to alumina ceramic by placing a film of copper in contact with the alumina and heating to about 1065°C, just below the melting point of copper, which is 1083°C. At this temperature, a combination of 0.39% O_2 and 99.61% Cu form a liquid that can melt, wet, and bond tightly to the surfaces in contact with it when cooled to room temperature. In this process, the copper remains in the solid state during the bonding process and a strong bond is formed between the copper and the alumina with no intermediate material required. The metallized substrate is slowly cooled to room temperature at a controlled rate to avoid quenching. To prevent excessive bowing of the substrate, copper must be bonded to both sides of the substrate to minimize stresses due to the difference in TCE between copper and alumina.

In this manner, a film of copper from 5 to 25 mils thick can be bonded to a substrate and a metallization pattern formed by photolithographic etching. For subsequent processing, the copper is usually plated with several hundred microinches of nickel to prevent oxidation. The nickel-plated surface is readily solderable, and aluminum wire bonds to nickel is one of the most reliable combinations.[7] Aluminum wire bonded directly to copper is not as reliable and may result in failure on exposure to heat and/or moisture.[8]

Multilayer structures of up to four layers have been formed by etching patterns on both sides of two substrates and bonding them to a common alumina substrate. Interconnections between layers are made by inserting oxidized copper pellets into holes drilled or formed in the substrates prior to firing. Vias may also be created by using one of the copper plating processes.

The line and space resolution of DBC is limited due to the difficulty of etching thick layers of metal without substantial undercutting. Special design guidelines must be followed to allow for this factor.[1] While the DBC technology does not have a resistor system, the thick-film technology can be used in conjunction with DBC to produce integrated resistors and areas of high density interconnections.

Aluminum nitride can also be used with copper, although the consistency of such factors as grain size and shape are not as good as aluminum oxide at this time. Additional preparation of the AlN surface is required to produce the requisite layer of oxide necessary to produce the bond. This can be accomplished by heating the substrate to about 1250°C in the presence of oxygen.

4.7.3.2 Plated copper technology. The various methods of plating copper to a ceramic all begin with the formation of a conductive film on the surface. This film may be vacuum deposited by thin-film methods, screen printed by thick-film processes, or deposited with the aid of a catalyst. A layer of electroless copper may be plated over the conductive surface, followed by a layer of electrolytic copper to increase the thickness.

A pattern may be generated in the plated surface by one of two methods. Conventional photolithographic methods may be used to etch the pattern, but this may result in undercutting and loss of resolution when used with thicker films. To produce more precise lines, a dry-film photoresist may be utilized to generate a pattern on the electroless copper film that is the negative of the one required for etching. The traces may then be electroplated to the desired thickness using the photoresist pattern as a mold. Once the photoresist pattern is removed, the entire substrate may be immersed in an appropriate etchant to remove the unwanted material between the traces. Plated copper films created in this manner may be fired at an elevated temperature in a nitrogen atmosphere to improve the adhesion.

4.7.3.3 Active metal brazing copper technology. The active metal brazing (AMB) process utilizes one or more of the metals in the IV-B column of the periodic table, such as titanium, hafnium, or zirconium, to act as an activation agent with ceramic. These metals are typically alloyed with other metals to form a braze that can be used to bond copper to ceramic. One such example is an alloy of 70% Ti/15% Cu/15% Ni that melts at 960 to 1000°C. Numerous other alloys can also be used.[2]

The braze may be applied in the form of a paste, a powder, or a film. The combination is heated to the melting point of the selected braze in a vacuum to minimize oxidation of the copper. The active metal forms a liquidus with the oxygen in the system that acts to bond the metal to the ceramic. After brazing, the copper film may be processed in much the same manner as DBC.

4.8 Substrate Materials

The characteristics of various substrate materials have been summarized in previous sections. However, there may be substantial variations in the parameters due to processing, composition, stoichiometry, or other factors. This section covers the materials in more detail and also describes some common applications.

4.8.1 Aluminum oxide

Aluminum oxide, Al_2O_3, commonly referred to as alumina, is by far the most common substrate material used in the microelectronics industry because it is superior to most other oxide ceramics in mechanical, thermal, and electrical properties. The raw materials are plentiful, low in cost, and amenable to fabrication by a wide variety of techniques into a wide variety of shapes.

Alumina is hexagonal close-packed with a corundum structure. Several metastable structures exist, but they all ultimately irreversibly transform to the hexagonal alpha phase.

Alumina is stable in both oxidizing and reducing atmospheres up to 1925°C. Weight loss in a vacuum over the temperature range 1700 to 2000°C can vary from 10^{-7} to 10^{-6} g/cm^2-s. It is resistant to attack by all gases, except wet fluorine, to at least 1700°C. Alumina is attacked at elevated temperatures by alkali metal vapors and halogen acids, especially the lower purity alumina compositions that may contain a percentage of glasses.

Alumina is used extensively in the microelectronics industry as a substrate material for thick- and thin-film circuits, for circuit packages, and as multilayer structures for multichip modules. Compositions exist for both high and low temperature processing. High temperature cofired ceramics (HTCC) use a refractory metal, such as tungsten or molybdenum/manganese, as a conductor and fire at about 1800°C. The circuits are formed as separate layers, laminated together, and fired as a unit. Low temperature cofired ceramics (LTCC) use conventional gold or palladium silver as conductors and fire as low as 850°C. Certain power metal-oxide-semiconductor field-effect transistor (MOSFETs) and bipolar transistors are mounted on alumina substrates to act as electrical insulators and thermal conductors.

The parameters of alumina are summarized in Table 4.11.

4.8.2 Beryllium oxide

Beryllium oxide (BeO, beryllia) is cubic close-packed and has a zinc-blende structure. The alpha form of BeO is stable to above 2050°C. BeO is stable in dry atmospheres and is inert to most materials. It hydrolyzes at temperatures greater than 1100°C with the formation and volatilization of beryllium hydroxide. BeO reacts with graphite at high temperature forming beryllium carbide.

TABLE 4.11   Typical Parameters of Aluminum Oxide

Parameter	Test	Percentage (%)			
		85	90	96	99.5
Density (g/cm^3)	ASTM C20	3.40	3.60	3.92	3.98
Elastic modulus (GPa)	ASTM C848	220	275	344	370
Poisson's ratio	ASTM C848	0.22	0.22	0.22	0.22
Compressive strength (MPa)	ASTM C773	1930	2150	2260	2600
Fracture toughness (MPa-m$^{0.5}$)	Notched beam	3.1	3.3	3.7	4.6
Thermal conductivity (W/m-°C)	ASTM C408	16	16.7	24.7	31.0
TCE (10^{-6}/°C)	ASTM C372	5.9	6.2	6.5	6.8
Specific heat (W-s/g-°C)	ASTM E1269	920	920	880	880
Dielectric strength (AC kv/mm)	ASTM D116	8.3	8.3	8.3	8.7
Loss tangent (1 MHz)	ASTM D2520	0.0009	0.0004	0.0002	0.0001
Volume resistivity (Ω-cm)					
25°C	ASTM D1829	>10^{14}	>10^{14}	>10^{14}	>10^{14}
500°C		4×10^8	4×10^8	4×10^9	2×10^{10}
1000°C			5×10^5	1×10^6	2×10^6

Beryllia has an extremely high thermal conductivity, higher than aluminum metal, and is widely used in applications where this parameter is critical. The thermal conductivity drops rapidly above 300°C, but is suitable for most practical applications.

Beryllia is available in a wide variety of geometries formed using a variety of fabrication techniques. While beryllia in the pure form is perfectly safe, care must be taken when machining BeO, however, as the dust is toxic if inhaled.

Beryllia may be metallized with thick-film, thin-film, or by one of the copper processes. However, thick-film pastes must be specially formulated to be compatible. Laser or abrasive trimming of BeO must be performed in the presence of a vacuum to remove the dust.

The properties of 99.5% beryllia are summarized in Table 4.12.

4.8.3 Aluminum nitride

Aluminum nitride is covalently bonded with a wurtzite structure and decomposes at 2300°C under 1 atm of argon. In a nitrogen atmosphere of 1500 lb/in^2, melting may occur in excess of 2700°C. Oxidation of AlN

TABLE 4.12   Typical Parameters for 99.5% Beryllium Oxide

Parameter	Value
Density (g/cm^3)	2.87
Hardness (Knoop 100 g)	1200
Melting point (°C)	2570
Modulus of elasticity (GPa)	345
Compressive strength (MPa)	1550
Poisson's ratio	0.26
Thermal conductivity (W/m-K)	
25°C	250
500°C	55
Specific heat (W-s/g-K)	
25°C	1.05
500°C	1.85
TCE (10^{-6}/K)	7.5
Dielectric constant	
1 MHz	6.5
10 GHz	6.6
Loss tangent	
1 MHz	0.0004
10 GHz	0.0004
Volume resistivity (Ω-cm)	
25°C	$>10^{14}$
500°C	2×10^{10}

in even a low concentration of oxygen ($< 0.1\%$) occurs at temperatures above 700°C. A layer of aluminum oxide protects the nitride to a temperature of 1370°C, above which the protective layer cracks, allowing oxidation to continue. Aluminum nitride is not appreciably affected by hydrogen, steam, or oxides of carbon to 980°C. It dissolves slowly in mineral acids and decomposes slowly in water. It is compatible with aluminum to 1980°C, gallium to 1300°C, iron or nickel to 1400°C, and molybdenum to 1200°C.

Aluminum nitride substrates are fabricated by mixing AlN powder with compatible glass powders containing additives, such as CaO and Y_2O_3, along with organic binders, and casting the mixture into the desired shape. Densification of the AlN requires very tight control of both atmosphere and temperature. The solvents used in the preparation of substrates must be anhydrous to minimize oxidation of the AlN powder and prevent the generation of ammonia during firing.[9] For maximum densification and maximum thermal conductivity, the substrates must be sintered in a dry reducing atmosphere to minimize oxidation.

Aluminum nitride is primarily noted for two very important properties—a high thermal conductivity and a TCE closely matching that of silicon. There are several grades of aluminum nitride with different thermal conductivities available. The prime reason is the oxygen content of the material. It is important to note that even a thin surface layer of oxidation on a fraction of the particles can adversely affect the thermal conductivity. Only with a high degree of material and process control can AlN substrates be made consistent.

The thermal conductivity of AlN does not vary as widely with temperature as that of BeO. Considering the highest grade of AlN, the cross-over temperature is about 20°C. Above this temperature, the thermal conductivity of AlN is higher; below 20°C, BeO is higher.

The TCE of AlN closely matches that of silicon, an important consideration when mounting large power devices. The second level of packaging is also critical. If an aluminum nitride substrate is mounted directly to a package with a much higher TCE, such as copper, the result can be worse than if a substrate with an intermediate, although higher, TCE were used. The large difference in TCE builds up stresses during the mounting operation that can be sufficient to fracture the die and/or the substrate.

Thick-film, thin-film, and copper metallization processes are available for aluminum nitride. Certain of these processes, such as direct bond copper (DBC), require oxidation of the surface to promote adhesion. For maximum thermal conductivity, a metallization process should be selected that bonds directly to AlN to eliminate the relatively high thermal resistance of the oxide layer.

Thick-film materials must be formulated to adhere to AlN. The lead oxides prevalent in thick-film pastes that are designed for alumina and beryllia oxidize AlN rapidly, causing blistering and a loss of adhesion. Thick-film resistor materials are primarily based on RuO_2 and MnO_2.

Thin-film processes available for AlN include NiCr/Ni/Au, Ti/Pt/Au, and Ti/Ni/Au.[10] Titanium, in particular, provides excellent adhesion by diffusing into the surface of the AlN. Platinum and nickel are transition layers to promote gold adhesion. Solders, such as Sn60%/Pb40% and Au80%/Sn20% can also be evaporated onto the substrate to facilitate soldering.

Multilayer circuits can be fabricated with W or MbMn conductors. The top layer is plated with nickel and gold to promote solderability and bondability. Ultrasonic milling may be used for cavities, blind vias, and through vias. Laser machining is suitable for through vias as well.

Direct bond copper may be attached to AlN by forming a layer of oxide over the substrate surface, which may require several hours at temperatures above 900°C. The DBC forms a eutectic with aluminum oxide at about 963°C. The layer of oxide, however, increases the thermal resistance by a significant amount, partially negating the high thermal conductivity of the aluminum nitride. Copper foil may also be brazed to AlN with one of the compatible braze compounds. Active metal brazing (AMB) does not generate an oxide layer. The copper may also be plated with nickel and gold.

The properties of aluminum nitride are summarized in Table 4.13.

4.8.4 Diamond

Diamond substrates are primarily grown by chemical vapor deposition (CVD). In this process, a carbon-based gas is passed over a solid surface and activated by a plasma, a heated filament, or a combustion flame. The surface must be maintained at a high temperature, above 700°C, to sustain the reaction. The gas is typically a mixture of methane (CH_4) and hydrogen (H_2) in a ratio of 1 to 2% CH_4 by volume.[11] The consistency of the film in terms of the ratio of diamond to graphite is inversely proportional to the growth rate of the film. Films produced by plasma have a growth rate of 0.1 to 10 μm/hr and are very high quality, while films produced by combustion methods have a growth rate of 100 to 1000 μm/hr and are of lesser quality.

The growth begins at nucleation sites and is columnar in nature, growing faster in the normal direction than in the lateral direction. Eventually, the columns grow together to form a polycrystalline struc-

TABLE 4.13   Typical Parameters for Aluminum Nitride (Highest Grade)

Parameter	Value
Density (g/cm^3)	3.27
Hardness (Knoop 100g)	1200
Melting point (°C)	2232
Modulus of elasticity (GPa)	300
Compressive strength (MPa)	2000
Poisson's ratio	0.23
Thermal conductivity (W/m-K)	
25°C	270
150°C	195
Specific heat (W-s/g-K)	
25°C	0.76
150°C	0.94
TCE (10^{-6}/K)	4.4
Dielectric constant	
1 MHz	8.9
10 GHz	9.0
Loss tangent	
1 MHz	0.0004
10 GHz	0.0004
Volume resistivity (Ω-cm)	
25°C	$>10^{12}$
500°C	2×10^8

ture with microcavities spread throughout the film. The resulting substrate is somewhat rough, with a 2- to 5-μm surface. This feature is detrimental to the effective thermal conductivity and the surface must be polished for optimum results. An alternative method is to use an organic filler[12] on the surface for planarization. This process has been shown to have a negligible effect on the overall thermal conductivity from the bulk, and dramatically improves heat transfer. Substrates as large as 10 cm^2 and as thick as 1000 μm have been fabricated.

Diamond can be deposited as a coating on refractory metals, oxides, nitrides, and carbides. For maximum adhesion, the surface should be a carbide-forming material with a low TCE.[13]

Diamond has an extremely high thermal conductivity, several times that of the next highest material. The primary application is, obviously, in packaging power devices. Diamond has a low specific heat, however, and works best as a heat spreader in conjunction with a heat sink. For maximum effectiveness[13]

$$t_D = 0.5 \rightarrow 1 \times r_h$$
$$r_D = 3 \times r_h$$

where t_D = thickness of diamond substrate
r_h = radius of heat source
r_D = radius of diamond substrate

Applications of diamond substrates include heat sinks for laser diodes and laser diode arrays. The low dielectric constant of diamond coupled with the high thermal conductivity makes it attractive for microwave circuits as well. As improved methods of fabrication lower the cost, the use of diamond substrates is expected to expand rapidly.

The properties of diamond are summarized in Table 4.14.

4.8.5 Boron nitride

There are two basic types of boron nitride (BN). Hexagonal (alpha) BN is soft and is structurally similar to graphite. It is white in color and is sometimes called *white graphite*. Cubic (beta) BN is formed by subjecting hexagonal BN to extreme heat and pressure, similar to the process used to fabricate synthetic industrial diamonds. Melting of either phase is possible only under nitrogen at high pressure.

Hot-pressed BN is very pure (>99%), with the major impurity being boric oxide (BO). Boric oxide tends to hydrolyze in water, degrading the dielectric and thermal shock properties. Calcium oxide (CaO) is frequently added to tie up the BO to minimize the water absorption. When exposed to temperatures above 1100°C, BO forms a thin coating on the surface, slowing further oxide growth.

TABLE 4.14   Typical Parameters for CVD Diamond

Parameter	Value
Density (g/cm^3)	3.52
Hardness (Knoop 100g)	7,000
Modulus of elasticity (GPa)	1,000
Compressive strength (MPa)	11,000
Poisson's ratio	0.148
Thermal conductivity (W/m-K)	
Normal	2,200
Tangential	1,610
Specific heat (W-s/g-K)	
25°C	0.55
150°C	0.90
TCE (10^{-6}/K)	1.02
Dielectric constant	
1 MHz	5.6
10 GHz	5.6
Loss tangent	
1 MHz	0.001
10 GHz	0.001
Volume resistivity (Ω-cm)	
25°C	$>10^{13}$
500°C	2×10^{11}

Boron nitride in the hot-pressed state is easily machinable and may be formed into various shapes. The properties are highly anisotropic and vary considerably in the normal and tangential directions of the pressing force. The thermal conductivity in the normal direction is very high and the TCE is very low, making BN an attractive possibility for a substrate material. However, it has not yet been proven possible to metallize BN,[14] thereby limiting the range of applications. It can be used in contact with various metals, including, copper, tin, and aluminum, and may be used as a thermally conductive electrical insulator. Applications of BN include microwave tubes and crucibles.

The properties of boron nitride are summarized in Table 4.15.

4.8.6   Silicon carbide

Silicon carbide (SiC) has a tetrahedral structure and is the only known alloy of silicon and carbon. Both elements have four electrons in the outer shell, with an atom of one bonded to four atoms of the other. The result is a very stable structure not affected by hydrogen or nitrogen up to 1600°C. In air, SiC begins decomposing above 1000°C. As with other compounds, a protective oxide layer forms over the silicon, reducing the rate of decomposition. Silicon carbide is highly resistant to both acids and bases. Even the so-called "white etch" (hydrofluoric acid mixed with nitric and sulfuric acids) has no effect.

TABLE 4.15   Typical Parameters for Boron Nitride

Parameter	Value
Density (g/cm^3)	1.92
Hardness (Knoop 100g)	5000
Modulus of elasticity (GPa)	
Normal	43
Tangential	768
Compressive strength (MPa)	
Normal	110
Tangential	793
Poisson's ratio	0.05
Thermal conductivity (W/m-K)	
Normal	73
Tangential	161
Specific heat (W-s/g-K)	
25°C	0.84
150°C	1.08
TCE (10^{-6}/K)	
Normal	0.57
Tangential	−0.46
Dielectric constant @ 1 MHz	4.1
Loss tangent @ 1 MHz	0.0003
Volume resistivity (Ω-cm)	
25°C	1.6×10^{12}
500°C	2×10^{10}

Silicon carbide structures are formed by hot pressing, dry and isostatic pressing (preferred), by CVD, or by slip casting. Isostatic pressing using gas as the fluid provides optimum mechanical properties.

Silicon carbide in pure form is a semiconductor and the resistivity depends on the impurity concentration. In the intrinsic form, the resistivity is less than 1000 Ω-cm, which is unsuitable for ordinary use. The addition of a small percentage (<1%) of BeO during the fabrication process[15] increases the resistivity to as high as 10^{13} Ω-cm by creating carrier-depleted layers around the grain boundaries.

Both thick and thin films can be used to metallize SiC, although some machining of the surface to attain a higher degree of smoothness is necessary for optimum results. The two parameters that make SiC attractive as a substrate are the exceptionally high thermal conductivity, second only to diamond, and the low TCE, which matches that of silicon to a higher degree than any other ceramic. SiC is also less expensive than either BeO or AlN. A possible disadvantage is the high dielectric constant, 4 to 5 times higher than other substrate materials. This parameter can result in cross-coupling of electronic signals or in excessive transmission delay.

The parameters of SiC are summarized in Table 4.16.

TABLE 4.16   **Typical Parameters for Silicon Carbide**

Parameter	Value
Density (g/cm^3)	3.10
Hardness (Knoop 100 g)	500
Modulus of elasticity (GPa)	407
Compressive strength (MPa)	4400
Fracture toughness (MPa-m$^{1/2}$)	7.0
Poisson's ratio	0.14
Thermal conductivity (W/m-K)	
25°C	290
150°C	160
Specific heat (W-s/g-K)	
25°C	0.64
150°C	0.92
TCE (10^{-6}/K)	3.70
Dielectric constant	40
Loss tangent	0.05
Volume resistivity (Ω-cm)	
25°C	$>10^{13}$
500°C	2×10^9

4.9   Composite Materials

Ceramics typically have a low thermal conductivity and a low TCE, while metals have a high thermal conductivity and a high TCE. It is a logical step to combine these properties to obtain a material with a high thermal conductivity and a low TCE. The ceramic in the form of particles or continuous fibers is mixed with the metal to combine the desirable properties of both. The resultant structure is referred to as a metal matrix composite (MMC).

The most common metals used in this application are aluminum and copper, with aluminum being more common due to lower cost. Fillers include SiC, AlN, BeO, graphite, and diamond. Compatibility of the materials is a prime consideration. Graphite, for example, has an electrolytic reaction with aluminum but not with copper.[14] Two materials will be described here—AlSiC, a composite made up of aluminum and silicon carbide, and Dymalloy, a combination of copper and diamond.

4.9.1   Aluminum silicon carbide (AlSiC)

AlSiC is produced by forcing liquid aluminum into a porous SiC preform. The preform is made by any of the common ceramic processing technologies, including dry pressing, slip molding, and tape casting. The size and shape of the preform is selected to provide the desired volume fraction of SiC. The resulting combination has a thermal conductivity almost as high as pure aluminum, with a TCE as low as 6.1

ppm/°C. AlSiC is also electrically conductive, prohibiting its use as a conventional substrate.

The mechanical properties of the composite are determined by the ratio of SiC to aluminum, as shown in Fig. 4.17.[14] A ratio of 70 to 73% of SiC by volume provides the most optimum properties for electronic packaging.[16] This ratio gives a TCE of about 6.5 ppm/°C, which closely matches that of alumina and beryllia. This allows AlSiC to be used as a baseplate for ceramic substrate, using its high thermal conductivity to maximum advantage.

Since AlSiC is electrically conductive, it may be readily plated with aluminum to provide a surface for further processing. The aluminum coating may be plated with nickel and gold to permit soldering, or may be anodized where an insulating surface is required.[17] An alternate approach is to flame-spray the AlSiC with various silver alloys for solderability.

Two other advantages of AlSiC are strength and weight. The aluminum is somewhat softer than SiC and reduces the propagation of cracks. The density is only about one-third that of Kovar, and the thermal conductivity is over 12 times greater.

AlSiC has been used to advantage in the fabrication of hermetic single-chip and multichip packages and as heat sinks for power devices and circuits. While difficult to drill and machine, AlSiC can be formed into a variety of shapes in the powder state. It has been successfully integrated with patterned AlN to form a power module package.[17]

The TCE linearly increases with temperature up to about 350°C, and then begins to decrease. At this temperature, the aluminum matrix softens and the SiC matrix dominates. This factor is an important feature for power packaging.

The parameters of AlSiC are summarized in Table 4.17.

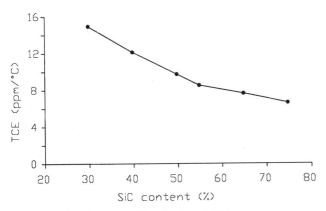

Figure 4.17   TCE versus SiC content for AlSiC.

TABLE 4.17   Typical Parameters for AlSiC (70% SiC by volume)

Parameter	Value
Density (g/cm^3)	3.02
Modulus of elasticity (GPa)	224
Flexural strength (MPa)	270
Tensile strength (MPa)	192
Thermal conductivity (25°C) (W/m-K)	218
TCE (10^{-6}/K)	7.0
Volume resistivity (25°C) ($\mu\Omega$-cm)	34

4.9.2   Dymalloy

Dymalloy is a matrix of type I diamond and Cu20%/Ag80% alloy.[18] The diamond is ground into a powder in the 6- to 50-μm range. The powder is coated with W74/Rh26 to form a carbide layer approximately 100 Å thick followed by a 1000-Å coating of copper. The copper is plated to a thickness of several micrometers to permit brazing.

The powder is packed into a form and filled in a vacuum with Cu20%/Ag80% alloy that melts at approximately 800°C. This material is selected over pure copper, which melts at a much higher temperature, to minimize graphization of the diamond. The diamond loading is approximately 55% by volume.

The parameters of Dymalloy are summarized in Table 4.18.

4.10   Multilayer Substrate Structures

In all but the simplest electronic circuits, it is necessary to have a method for fabricating multilayer interconnection structures to enable all the necessary points to be connected. The thick-film technology is limited to three layers for all practical purposes due to yield and planarity considerations, and thin-film multilayer circuits are quite expensive to fabricate. The copper technologies are limited to a single layer due to processing limitations.

TABLE 4.18   Typical Parameters for Dymalloy (55% Diamond by Volume)

Parameter	Value
Density (g/cm^3)	6.4
Tensile strength (MPa)	400
Specific heat* (W-s/g-°C)	$0.316 + 8.372 \times 10^{-4}\,T$
Thermal conductivity (W/m-K)	360
TCE† (10^{-6}/K)	$5.48 + 6.5 \times 10^{-3}\,T$

*Temperature in degrees Celsius from 25 to 75°C.
†Temperature in degrees Celsius 25 to 200°C.

The ceramic technology may be used to economically fabricate very complex cofired multilayer structures consisting of many layers. There are three basic classes of materials—high temperature cofired ceramic (HTCC), low temperature cofired ceramic (LTCC), and aluminum nitride.

4.10.1   High temperature cofired ceramic

HTCC multilayer circuits are based on alumina mixed with an appropriate glass. The alumina and glass in powder form are ball milled and formed into a slurry by mixing with an organic binder, a plasticizer, and a solvent. The slurry is formed into a sheet by a doctor blade and dried. The resultant film, approximately 25 μm thick, is referred to as a *green tape* or *greensheet*. At this stage, the green tape has the consistency of putty.

The next step is blanking and punching. The green tape is blanked into sheets of uniform size and holes are punched where vias and alignment holes are required. The metal patterns are printed and dried next. Refractory metals, such as tungsten and molybdenum, are used due to the high firing temperature. Via fills may be accomplished during conductor printing or during a separate printing operation. The process is repeated for each layer.

The individual layers are aligned and laminated under heat and pressure to form a monolithic structure in preparation for firing. The structure is heated to approximately 600°C to remove the organic materials. Carbon residue is removed by heating to approximately 1200°C in a wet hydrogen atmosphere. Sintering and densification take place at approximately 1600°C.

During firing, HTCC circuits shrink anywhere from 14 to 17%, depending on the organic content. With careful control of the material properties and processing parameters, the shrinkage can be controlled to within 0.1%. Shrinkage must be taken into consideration during the design, punching, and printing processes. The artwork enlargement must exactly match the shrinkage factor associated with a particular lot of green tape.

Processing of the substrate is completed by plating the outer layers with nickel and gold for component mounting and wire bonding. The gold is plated to a thickness of 25 μin for gold wire and 5 μin for aluminum wire. Gold wire bonds to the gold plating, while aluminum wire bonds to the nickel underneath. The gold plating in this instance is simply to protect the nickel surface from oxidation or corrosion.

The properties of HTCC materials are summarized in Table 4.19.

TABLE 4.19 Properties of Multilayer Ceramic Materials

	Low temperature cofired ceramic (LTCC)	High temperature cofired ceramic (HTCC)	Aluminum nitride
Material	Cordierite: MgO, SiO_2, Al_2O_3 Glass-filled composites: SiO_2, B_2O_3, Al_2O_3, PbO, SiO_2, CaO, Al_2O_3 Crystalline phase ceramics: Al_2O_3, CaO, SiO_2, MgO, B_2O_3	88–92% alumina	AlN, yttria, CaO
Firing temperature (°C)	850–1050	1500–1600	1600–1800
Conductors	Au, Ag, Cu, PdAg	W, MoMn	W, MoMn
Conductor resistance ($M\Omega$/[square])	3–20	8–12	8–12
Dissipation factor	$15–30 \times 10^{-4}$	$5–15 \times 10^{-4}$	$20–30 \times 10^{-3}$
Relative dielectric constant	5–8	9–10	8–9
Resistor values	$0.1\,\Omega - 1\,M\Omega$	N/A	N/A
Firing shrinkage (%):			
x,y	12.0 ± 0.1	12–18	15–20
z	17.0 ± 0.5	12–18	15–20
Repeatability (%)	0.3–1	0.3–1	0.3–1
Line width (μm)	100	100	100
Via diameter (μm)	125	125	125
Number of metal layers	33	63	8
CTE (ppm/°C)	3–8	6.5	4.4
Thermal conductivity (W/M·°C)	2–6	15–20	180–200

4.10.2   Low temperature cofired ceramic

LTCC circuits consist of alumina mixed with glasses with the capability to simultaneously sinter and crystallize.[19] These structures are often referred to as *glass ceramics*. Typical glasses are listed in Table 4.19. The processing steps are similar to those used to fabricate HTCC circuits with two exceptions[20]—the firing temperature is much lower, 850 to 1050°C, and the metallization is gold-based or silver-based thick film formulated to be compatible with the LTCC material. Frequently, silver-based materials are used in the inner layers with gold on the outside for economic reasons. Special via fill materials are used between the gold and the silver layers to prevent electrolytic reaction.

The shrinkage of LTCC circuits during firing is in the range of 12 to 18%. If the edges are restrained, the lateral shrinkage can be held to 0.1% with a corresponding increase in the vertical shrinkage.

One advantage that LTCC has over the other multilayer technologies is the ability to print and fire resistors. Where trimming is not required, the resistors can be buried in intermediate layers with a corresponding saving of space. It is also possible to bury printed capacitors of small value.

The properties of LTCC materials are summarized in Table 4.19.

4.10.3   Aluminum nitride

Aluminum nitride multilayer circuits are formed by combining AlN powder with yttria or calcium oxide.[19] Glass may also be added. Sintering may be accomplished in three ways:

1. Hot pressing during sintering
2. High temperature (>1800°C) sintering without pressure
3. Low temperature (<1650°C) sintering without pressure

Tungsten or molybdenum pastes are used to withstand the high firing temperatures. Control of the processing parameters during sintering is critical if optimum properties are to be attained. A carbon atmosphere helps in attaining a high thermal conductivity by preventing oxidation of the AlN particles. Shrinkage is in the range of 15 to 20%.

The properties of aluminum nitride multilayer materials are summarized in Table 4.19.

References

1. Jerry Sergent and Charles Harper, *Hybrid Microelectronics Handbook,* 2d ed., McGraw-Hill, New York, 1995.
2. Richard Brown, "Thin Film Substrates," in Leon Maissel and Reinhard Glang, eds., *Handbook of Thin Film Technology,* McGraw-Hill, New York, 1971.

3. Philip Garrou and Arne Knudsen, "Aluminum Nitride for Microelectronic Packaging," *Advancing Microelectronics,* vol. 21, no. 1, January–February 1994.
4. C. G. M. Van Kessel, S. A. Gee, and J. J. Murphy, "The Quality of Die Attachment and Its Relationship to Stresses and Vertical Diecracking," *Proceedings, IEEE Components Conference,* 1983.
5. A. Winkleman and O. Schott, *Annals of Physics in Chemistry,* vol. 51, 1984.
6. Jerry Sergent and H. Thurman Henderson, "Double Injection in Semi-Insulators," *Proceedings,* Solid State Materials Conference, 1973.
7. George Harman, "Wire Bond Reliability and Yield," *ISHM Monograph,* 1989.
8. Craig Johnston et al., "Temperature Dependent Wear-Out Mechanism for Aluminum/Copper Wire Bonds," *Proceedings,* ISHM Symposium, 1991.
9. Ellice Y. Yuh, et al., "Current Processing Capabilities for Multilayer Aluminum Nitride," *International Journal of Microelectronics and Electronic Packaging,* vol. 16, no. 2, Second Quarter, 1993.
10. Nobuyiki Karamoto, "Thin Film and Co-Fired Metallization on Shapal Aluminum Nitride," *Advancing Microelectronics,* vol. 21, no. 1, January/February 1994.
11. Paul W. May, "CVD Diamond—A New Technology for the Future?" *Endeavor Magazine,* vol. 19, no. 3, 1995.
12. Ajay P. Malshe et al., "Diamond for MCMs," *Advanced Packaging,* September/October 1995.
13. Thomas Moravec and Arjun Partha, "Diamond Takes the Heat," *Advanced Packaging,* Special Issue, October 1993.
14. S. Fuchs and P. Barnwell, "A Review of Substrate Materials for Power Hybrid Circuits," *The IMAPS Journal of Microcircuits and Electronic Packaging,* vol. 20, no. 1, First Quarter, 1997.
15. Mitsuru Ura and Osami Asai, Internal Report, Hitachi Research Laboratory, Hitachi Industries, Ltd.
16. M. K. Premkumar and R. R. Sawtell, "Alcoa's AlSiC Cermet Technology for Microelectronics Packaging," *Advancing Microelectronics,* July/August 1995.
17. M. K. Premkumar and R. R. Sawtell, "Aluminum-Silicon Carbide," *Advanced Packaging,* September/October 1996.
18. J. A. Kerns et al., "Dymalloy: A Composite Substrate for High Power Density Electronic Components," *International Journal of Microcircuits and Electronic Packaging,* vol. 19, no. 3, Third Quarter, 1996.
19. Philip E. Garrou and Iwona Turlik, *Multichip Module Technology Handbook,* McGraw-Hill, New York, 1998.
20. Jerry E. Sergent, "Materials for Multichip Modules," *Electronic Packaging and Production,* December 1996.

Flexible Printed Circuits

Marty Jawitz

Jaw—Mac Enterprises
Las Vegas, Nevada

5.1 Introduction

Flexible printed circuits are printed circuits that are fabricated on a thin flexible-based material and are usually constructed with a nonreinforced polymeric material. The ability of the core material to bend allows them to be used as interconnects in place of wire harnesses or in places where a connection is needed to a moving part. They were also designed to replace bundles of wire and cables of conventional round insulated wire for space and weight savings. There are two distinct advantages of using flexible circuits:

1. Once the circuit layout has been proven, it is almost impossible to make a wrong solder connection.
2. Thinness of the flexible circuits allows them to be used in locations where the thickness of even a ribbon cable cannot be accommodated.

Etched flexible circuit can also provide for a two-dimensional conductor geometry that can mate with a connector pin array and provide a more elaborate wire harness.

5.2   Films

5.2.1   General

In order to reduce stresses that are or could be built up during production processing, the flexible printed wiring films should have

- High tensile strength
- High tensile modulus
- High melting point
- High T_g
- Good thermal stability
- Low coefficient of thermal expansion
- Low residual stresses

In a finished flexible circuit, the desirable film properties should include

- Low dielectric constant
- Low dissipation factor
- High volume resistivity
- High surface resistivity

Every flexible-clad laminate consists of a conductive foil and a flexible dielectric base. Copper is usually the choice for the conductor material while most flexible circuits are fabricated on either a polyimide- or polyester-based polymer film. For special applications, aramids or fluorocarbons can be used.

The reason for choosing a particular film type is

1. *Military or high performance flexible circuits.* These circuits are fabricated using polyimide films because they offer the best overall performance for the cost.
2. *Commercial.* Cost-sensitive circuits are fabricated from polyester films (similar properties to polyimide films but at a lower cost). The polyester film also has a lower thermal resistance.
3. *Aramid fibers, nonwoven.* This material is inexpensive and has excellent electrical and mechanical properties, but features excessive moisture and residual chemistry absorption.
4. *Fluorocarbons.* These films are expensive and hard to handle. They have superior dielectric properties, which makes them best suited for controlled impedance boards.

Flexible circuits are used in a multitude of applications ranging from the lowest end consumer products to the highest end military and commercial systems. It is no coincidence, therefore, that the choice of materials that could be used to fabricate these circuits is as diverse in performance as the range of products in which they are used.

In order to determine the best material for the intended application, one must first understand the major performance characteristics of each material. A good reference knowledge of the mechanical, electrical, thermal, and chemical properties will allow for an effective material choice to be made for each flexible circuit application.

Mechanical	Electrical	Thermal	Chemical
Flexibility	Dielectric strength	Coefficient of thermal expansion	Moisture absorption
Tear strength	Dielectric constant	Service temperature range	Resistance to acid and alkali
Tear propagation strength	Dissipation factor		Resistance to automotive fluids
Dimensional stability	Volume resistivity		

These properties must be matched to meet the required design characteristics. Besides the important properties that are listed here, the dimensional stability and tear resistance of the film are critical when selecting a particular material.

5.2.1.1 Dimensional stability. Dimensional stability is crucial in flexible circuit manufacturing. The primary difference between flexible and rigid circuits is that the flexible film laminates are used in place of reinforced rigid-glass cloth laminates. The major consequence of this difference is that flexible laminates inherently expand and shrink more during exposure to processing conditions.

The stability of a flexible laminate is a complex composite of film properties, degraded by the properties of the cured adhesive and processing conditions that were used to form the laminate. Careful laminate manufacturing, using low web tension, vacuum-assisted lamination, and thermally stabilized films, helps to minimize shrinkage. Polyimide films with a high tensile strength and modulus help to improve stability. While after-etch shrinkage of 0.1% is achievable with a high performance film, shrinkage for laminates made with conventional polyimide films is more commonly 0.15%. These shrinkage factors may seem trivial, but economical manufacturing of flexible circuits requires the use of large panels (18 × 24 in). The distance from tooling pin to etch feature can be as great as 9 in (assuming that the tooling pins are at the border of an 18-in panel) across this span; 0.1% shrinkage translates to a 9-mil shift, tolerable and predictable if not accompanied by other errors, but undesirable and costly to neutralize.

5.2.1.2 Tear resistance. Unfortunately, film characteristics that promote better stability also lower initiating and propagating tear values, since distributing and dissipating tear energy requires a softer film with greater elongation and yield before failure.

Flexible circuits generally have complex outlines with multiple stress concentration points that makes tear resistance an important property. Tear resistance is a characteristic in which an adhesive can enhance laminate performance, since most flexible adhesives have better tear resistance than the polyimide film.

5.2.2 Fluorocarbons

The first flexible circuits were supported and insulated by fusion bonding high performance fluorocarbons approximately 15 years before polyimide films came on the scene. Unmatched chemical inertness, high thermal resistance plus outstanding dielectric properties, combined with tough mechanical properties, suggest that fluorocarbons would be ideal for flexible circuits. Early experience with the fusion bonding process quickly identified dimensional stability as a fatal flaw in the fusion-bonded fluorocarbon dielectric. Lamination at the required temperature of approximately 260 to 288°C created stresses on the molten dielectric that could destroy the conductive patterns.

The tear resistance of the fluorocarbons is extremely high and, as a result, it is sometimes used to reinforce weak corners on polyimide films. Inert to all common chemistries and inherently incombustible, fluorocarbon laminates do not require pampering during fabrication, assembly, or usage.

On the downside, fluorocarbons do not lend themselves to the plated-through hole process because of their excellent resistance to most chemicals. Electroless copper deposits do not generally adhere to the drilled wall of the fluorocarbon laminate, thus requiring the use of additional steps (that is, tetra-etch). Today, for ease of circuit and laminate manufacturing, fluorocarbons can be assembled with adhesives instead of using the fusion process. Dimensional stability is much improved although not at the level of the polyimide film. Provided the adhesive is kept as thin as possible, a circuit so constructed will display some of the excellent electrical characteristics of a fusion-made

TABLE 5.1 Flexible-Based Dielectric Fluorocarbon (FEP)*

Property	Requirements
Tensile strength, minimum (lb/in^2)	2500
Elongation, minimum (%)	175
Initiation tear strength, minimum (g-m-s)	NA
Propagation tear strength, minimum (g-m-s)	
<0.038 mm	75
>0.038 to 0.1 mm	90
>0.1mm	90
Dimensional stability, maximum (%)	5.0
Chemical resistance (lb/in^2)	
Tensile strength, minimum (lb/in^2)	2500
Elongation, minimum (%)	175
Dielectric constant, maximum @ 1 MHz	2.2
Dissipation factor, maximum @ 1 MHz	0.0007
Volume resistivity, minimum (MΩ-cm)	10^7
Surface resistance, minimum (MΩ)	10^7
Dielectric strength, minimum (V/mil)	2500
Moisture absorption, maximum (%)	0.1
Fungus resistance	Nonnutrient
Flammability, minimum (min)	UL94

*Data for this material were reproduced (with permission) from IPC specification IPC-FC-231.

circuit but at a lower cost and higher yield. The properties of the fluorocarbon film is shown in Table 5.1.

The fluorocarbon films today incorporate an adhesive rather than use the fusion process in order to bond the copper foil to the flurocarbon dielectric or the covercoat to the conductive traces and/or used as a bonding layer. The properties for these materials are given in Tables 5.2 and 5.3.

5.2.3 Polyethylene terephthalate (polyester or PET)

Polyethylene terephthalate (PET), commonly referred to as *polyester*, has been used extensively in low cost, high production, flexible printed circuitry for many years. PET films are mostly used in the automotive field for behind-the-dash cluster circuits and also for membrane touch switches.

The polyethylene terephthalate polymer is the reaction product formed through the polymerization of ethylene glycol with either terephthalic acid or dimethyl terephthalate.

Roughly 20% of the flexible printed wiring boards produced in the United States today are fabricated using a combination of polyester laminates and cover layers. This film can be easily processed in roll form.

TABLE 5.2   Flexible Adhesive Bonding Film (Adhesive on One or Both Sides of the Dielectric Film)*

Property	FEP/ acrylic	FEP/ epoxy	FEP/ polyester
Peel strength, minimum			
As received (lb/in)	2.5	5.0	4.0
After solder float (lb/in)	2.0	5.0	4.0
After temperature cycling (lb/in)	2.5	5.0	4.0
Tensile strength, minimum (lb/in^2)	2500		
Elongation, minimum (%)	200		
Low temperature flexibility	Pass (5 cycles)		
Flow, maximum (mil/mil)	5	0.127	0.076
Volatile content, maximum (%)	2	2	
Flammability, minimum (%O_2)	15		
Solder float	Pass	Pass	
Chemical resistance (%)	80	80	90
Dielectric constant, maximum at 1 MHz	3.0	3.0	N/A
Dissipation factor, maximum at 1 MHz	0.025	0.025	N/A
Volume resistivity, minimum (MΩ-cm)	10^6	10^6	10^6
Surface resistance, minimum (MΩ)	10^5	10^4	10^6
Volatile content, maximum (%)	2		2.0
Dimensional stability, maximum (%)	1.0	1.0	1.0
Dielectric strength, minimum (V/mil)	2000	2000	2000
Insulation resistance, minimum (MΩ)	10^4		
Moisture and insulation resistance, minimum (MΩ)	10^3	10^3	10^4
Moisture absorption, maximum (%)	4.0	4.0	2.0
Fungus resistance	Non-nutrient	Non-nutrient	Non-nutrient

*Data for these materials were reproduced (with permission) from IPC specification IPC-FC-233.

Compared to polyimide films, the polyester films have

- Lower dielectric constant
- Higher insulation resistance
- Greater tear strength
- Lower cost
- Lower moisture absorption (less than 1%)
- Better dimensional stability
- Lower lamination temperatures

Polyester films come up short only in the area of thermal resistance.

Taking advantage of the plastic memory properties of the polyester film allows us to easily fabricate *retractiles,* which are long strip circuits folded in half and then coiled so that the ends can be pulled apart; when released, the circuits coil up again. The centerfold gives the

TABLE 5.3   **Flexible Metal-Clad Dielectric Fluorocarbon***

Property	FEP/acrylic	FEP/epoxy
Peel strength		
(1 oz or greater), minimum		
As received (lb/in)	2.5	8.0
After solder float (lb/in)	2.0	7.0
After temperature (lb/in)	2.5	8.0
Tensile strength, minimum (lb/in)	2500	2500
Elongation, minimum (%)	200	200
Initiation tear strength, minimum (g)	200	200
Propagation tear strength,		
(>0.0015 in), minimum (g-m-s)	125	125
Flexural endurance,		
(<0.002 in), minimum (cycles)	1000	100
Low temperature flexibility,		
minimum (5 cycles)	Pass	Pass
Dimensional stability, maximum (%)	0.8	0.8
Flammability, minimum (%O_2)	15	20
Solder float	Pass	Pass
Solderability	Pass	Pass
Chemical resistance (%)	80	80
Dielectric constant, maximum @ 1 MHz	3.0	3.0
Dissipation factor, maximum @ 1 MHz	0.025	0.025
Volume resistivity, minimum (MΩ	10^6	106
Surface resistance, minimum (MΩ)	10^5	104
Dielectric strength, minimum (V/mil)	2000	2000
Insulation resistance, minimum (MΩ)	10^4	104
Moisture and insulation resistance,		
minimum (MΩ)	10^3	103
Moisture absorption, maximum (%)	4.0	4.0
Fungus resistance	Non-nutrient	Non-nutrient

*Data for these materials were reproduced (with permission) from IPC specification IPC-FC-241.

tracks a noninductive bifilar coil that makes excellent connections for slide-out draw-type units. The two major disadvantages of the polyester films are

1. The base material melts below soldering temperatures

2. They have a low T_g of 80°C, which make them difficult to install and service without risk and damage or difficult to bond to other laminates.

Various special forms of connectors were devised which pierced the polyester insulation and made pressure contact with the copper conductors. However, the polyester films were largely superseded by the polyimide films which can safely withstand soldering and bonding temperatures. The properties of the polyethylene terephthalate films are shown in Table 5.4 while the properties of the polyester copper-clad laminates are shown in Table 5.7.

5.2.4 Polyethylene Naphthalate (PEN)

Although polyethylene naphthalate (PEN) films have been around for quite some time, they are only now making in roads into the field of flexible circuitry. PEN is currently finding its niche in the more demanding applications where either heat, such as that used for sur-face-mount applications, or environmental conditions are too harsh for the polyester films.

The polyethylene naphthalate polymer is the reaction product of 2,6 naphthalene dicarboxylate with ethylene glycol to form polyethylene 2,6 naphthalate, which is more commonly referred to as PEN. The resulting film is more crystalline in nature and exhibits a T_g of 120°C versus a T_g of 80°C for polyester. PEN films have a UL rating of 160°C

TABLE 5.4   Dielectric-Base Materials Polyethylene Terephthalate (PET)* and Polyethylene Naphthalate (PEN)

Properties	Polyester (PEN)	PET
Tensile strength (lb/in^2)		
MD	28,000	32,000
TD	33,000	32,000
Yield strength (lb/in^2)		
MD	15,000	21,000
TD	15,000	22,500
Elongation at break (%)		
MD	150	60
TD	110	65
Tensile modulus (lb/in^2)		
MD	600,000	870,000
TD	700,000	870,000
Initiation tear strength (g/mil)		
MD	1,000	1,000
TD	900	
Propagation tear strength (g/mil)		
MD	16	11.5
TD	12	12.5
Folding endurance (cycles)		
MD	200,000	750
TD	50,000	850
Dielectric strength, minimum (kV/mil)	7	12
Dielectric constant @ 1 MHz	3.3	
Volume resistivity (Ω-cm)	10^{17}	10^{16}
Surface resistivity (Ω/®)	10^{15}	10^{15}

*Data for these materials were reproduced (with permission) from IPC spec-ification IPC-FC-231.

mechanical and 180°C electrical. These films can withstand short-term exposure of 5 to 10 s at soldering temperatures of 260°C. The properties of polyethylene naphthalate films are shown in Table 5.4.

5.2.5 Polyimide

Polyimide films can be easily adopted for use in flexible circuitry. However, because laminate behavior is determined by the combined properties of the adhesive and supporting film, it is important to look beyond the film's properties.

The main reason for the high volume use of polyimide films (approximately 80% of all flexible circuits fabricated in the United States) is probably due to its ability to withstand the heat of manual and automatic soldering. Polyimide films have excellent thermal resistance. As a class, they have a continuous use rating of 300°C that would quickly destroy copper foils and solder joints through oxidation and intermetallic growth.

Polyimide films are inherently nonburning and, when combined with specially compounded fire-retardant adhesive products, produce laminates that are UL-94 V0 rated. However, many flexible circuit adhesives have much less resistance. Although they are able to stand up briefly to soldering temperatures, these adhesives are the weak link in the polyimide laminate. Some polyimide films absorb a great deal of moisture. Polyimide laminates must be baked at least 1 h at 100°C or better (for a single-sided flex and longer for a double-sided) prior to exposing the laminates to elevated temperatures such as that required for soldering. Because moisture reabsorption is very rapid, the laminate should be stored under dry conditions if not ready for immediate use.

Polyimide films are hard to surpass as insulators and high voltage barriers. However, in flex laminates, the primary insulation (the material controlling the conductor runs) is an adhesive with its own thermal and dielectric properties, thus the designer has to look carefully at all the properties of the laminate and not just that of the film itself when laying out a conductor run.

Polyimide films and their associated adhesives have poor electrical properties when used for controlled impedance flex applications due to a dielectric constant of 3.7 or greater combined with a dissipation factor greater than 0.03. It is recommended that another material (like fluorocarbons) should be used. The properties of the polyimide polymer films are given in Table 5.5 and the properties of the copper-clad polyimide laminates are shown in Table 5.7.

5.2.6 Aramids

Although polyimides and polyester films are the two most widely used insulating materials for fabricating flexible circuits, there are other

TABLE 5.5   Flexible-Based Dielectric Polyimide*

Property	Requirements
Tensile strength, minimum (lb/in^2)	
>0.025 mm	24,000
<0.025 mm	20,000
Elongation, minimum (%)	
>0.025 mm	45
<0.025 mm	35
Initiation tear strength, minimum (g)	
>0.025 mm	500
<0.025 mm	100
Propagation tear strength, minimum (g)	
<0.025 mm	1
0.025 mm	4
>0.025 to <0.05 mm	12
Dimensional stability, maximum (%)	
>0.025 mm	0.10
<0.025 mm	N/A
Chemical resistance	
IPC-TM650 method 2.5.2, method B	
Tensile strength, minimum (lb/in^2)	
>0.025 mm	24,000
<0.025 mm	20,000
Elongation, minimum (%)	
>0.025 mm	40
<0.025 mm	28
Dielectric constant, maximum @ 1 MHz	
>0.025 mm	3.8
<0.025 mm	4.0
Dissipation factor, maximum @ 1 MHz	0.012
Volume resistivity, minimum (MΩ-cm)	10^6
Surface resistance, minimum (MΩ)	10^5
Dielectric strength, minimum (V/mil)	2,500
Moisture absorption, maximum (%)	3.9
Fungus resistance	Non-nutrient
Temperature index, minimum (UL 746B)	200
Flammability, minimum (UL 94)	V-0

*Data for these materials were reproduced (with permission) from IPC specification IPC-FC-231.

films, such as the random fiber aramids, that could be used. The aramid paper isn't a film in the true sense but functions as a mechanical backbone and adhesive carrier just as the polymer films do. The aramid paper has a very low initiating and propagation tear strength, but it also has low dielectric constant (1.6 to 2.0) and dissipation factor (0.0015) with good dimensional stability. They are rated for continuous usage at 220°C and are nonburning (UL rated SE-0) and, when coated with a suitable laminating adhesive, form the basis of a very good flexible laminate.

Aramid films have very good tensile and tear strength values as well as dimensional stability and fall short only in the area of moisture absorption. The aramid fibers are very hydroscopic (approximately 3.7%) and like the polyimide laminates, if the laminate is not thoroughly dried before soldering the assembly and then kept dry throughout the process, it can blister seriously. Both sides of the paper should be thoroughly coated with a resin before these laminates are subjected to any liquid processing or the processing chemicals may wick into the aramid fibers, leaving a permanent stain and potential insulation-resistance problem. Although these fibers have many desirable properties and are inexpensive, their shortcomings make them difficult to use.

Flexible laminates based on aramid papers with an epoxy resin have been used to fabricate military flex circuits with coextensive "flexible" layers. The x-y coefficient of thermal expansion is very low (although the z axis is very high), thus providing very good dimensional stability for layer-to-layer alignment. These fibers, when coated with a flame-retardant resin, produce a laminate that is nonflammable. The properties of the aramid films are shown in Table 5.6 while the properties of the copper-clad aramid laminates are shown in Table 5.7.

TABLE 5.6   Flexible-Based Dielectrics Aramid Paper*

Property	Requirements
Tensile strength, minimum (lb/in^2)	4000
Elongation, minimum (%)	4
Initiation tear strength, minimum (g)	N/A
Propagation tear strength, minimum (g)	
0.050 mm	50
0.076 mm	70
Dimensional stability, maximum (%)	0.65
Chemical resistance	
IPC-TM650 method 2.5.2, method B	
Tensile strength, minimum (lb/in^2)	4000
Elongation, minimum (%)	4
Dielectric constant, maximum, @ 1 MHz	3.0
Dissipation factor, maximum @ 1 MHz	0.013
Volume resistivity, minimum (MΩ-cm)	10^6
Surface resistance, minimum (MΩ)	10^5
Dielectric strength, minimum (V/mil)	390
Moisture absorption, maximum (%)	13.0
Fungus resistance	Non nutrient
Temperature index, minimum (UL 746B)	220
Flammability, minimum (UL 94)	V O

*Data for this material were reproduced (with permission) from IPC specification IPC-FC-231.

TABLE 5.7 Flexible Metal-Clad Dielectrics*

Property	Polyimide/ acrylic	Polyimide/ epoxy	Polyester/ polyester	Polyester/ epoxy	Polyimide/ butryal: phenolic
Peel strength, minimum (lb/in^2)					
As received:					
<0.001	4.0	4.0	4.0	4.0	3.0
>0.001	8.0	8.0	5.0	8.0	5.0
After solder float:					
<0.001	3.0	3.0	N/A	3.0	2.5
>0.001	7.0	7.0	N/A	7.0	4.5
After temperature cycling:					
<0.001	4.0	4.0	4.0	8.0	3.0
>0.001	8.0	8.0	4.0		5.0
Tensile strength, minimum (lb/in^2)					
>0.001	24,000	24,000	20,000	5,000	24,000
<0.001	15,000	15,000			15,000
Elongation, minimum (%)	40	40	90	2.5	15
Initiation tear strength, minimum (g-m-s)	500	500	800	3,500	500
Propagation tear strength, minimum (g-m-s)					
<0.001	5.0	5.0	10.0	250	5.0
>0.001 to <0.002	10.0	10.0	20.0		10.0
>0.002 to <0.04	15.0	15	50.0		15.0

Property					
Flexural endurance, minimum (cycles)	1,000	1,000	1,000	15	1,000
Low temperature flexibility	Pass	Pass	Pass	Pass	Pass
Dimensional stability, maximum (%)	(5 cycles)	(5 cycles)	(5 cycles)	(5 cycles)	(5 cycles)
Flammability, minimum (%O_2)	15	20	20	28	20
Solder float	0.15	0.15	0.15	0.08	0.15
Solderability	Pass	Pass	N/A	Pass	Pass
Chemical resistance (%)	Pass	Pass	Pass	Pass	Pass
Dielectric constant, minimum @ 1 MHz	80	80	90	70	85
Dissipation factor, maximum @ 1 MHz	4.0	4.0	4.0		
Volume resistivity, minimum (mΩ-cm)	0.04	0.04	0.02		
Surface resistance, minimum (MΩ)	10^6	10^6	10^6	10^6	10^6
Dielectric strength, minimum (V/mil)	10^4	10^4	10^4	10^5	10^5
Insulation resistance, minimum (MΩ)	2,000	2,000	2,500	1,500	2,000
Moisture and insulation resistance, minimum (MΩ)	10^4	10^4	10^5	10^6	10^5
	10^3	10^3	10^4	10^5	10^5
Moisture absorption, maximum (%)	4.0	4.0	2.0	1.5	4.0
Fungus resistance	Non nutrient	Non nutrient	Non nutrient	Non nutrient	Non nutrient

*Data for these materials were obtained (with permission) from IPC specification IPC-FC-241.

5.3   Conductive Materials

5.3.1   General

Copper foil is by far the leading choice for flexible printed wiring board conductor material. Although aluminum is superior in conductivity, it is chemically more active and, therefore, presents corrosion problems when joined to other metals. Overall, copper is preferred as it is the standard for conductivity and current carrying capacity. It is easy to etch and, if necessary, can be used to create a high density circuit pattern.

5.3.2   Electrodeposited (ED) copper foil

The electrodeposited (ED) copper process starts with a copper solution being plated at very high deposition rates; a copper foil is built up on a highly polished steel drum. It produces a foil that has a smooth and shiny finish on one surface (the surface facing the drum is called the "drum side") and a rough and dull surface on the other side (the side facing the anodes). The copper grains have a vertical orientation, resulting in a foil with a high yield and tensile strength. The vertical grain structure provides smooth, even etching properties that aid in the production of thin conductor runs on close centers. Because of their inherently rough surfaces, conventional ED copper foils bond well to the flexible-based films where their reduced flexural properties are irrelevant. Table 5.8 shows the properties of electrodeposited foils.

5.3.3   Rolled annealed (RA) copper

Rolled copper foils are different in every way from foils produced by the electrodeposition process. The rolled foil process starts with copper ingots that are hot rolled to an intermediate gauge. At this stage, all the surfaces of the foil are milled and annealed prior to being rolled to the final gauge thickness in a specially designed rolling machine. The grains produced by this process have a horizontal grain orientation with smooth surfaces on both sides. Roll reduction quickly causes work hardening, with the result that rolled foils must be periodically annealed. These foils are sold in several degrees of hardness ranging from "as-rolled" to "dead soft anneal." Rolled anneal is a standard anneal condition which provides good flexural endurance and resistance to fracturing in dynamic applications. Table 5.9 shows the properties of rolled annealed foils.

5.3.4   Low temperature anneal (LTA) foils

Modern electrodeposition technology has altered the logic for choosing rolled annealed foils for flexible circuitry. With careful adjustment of

TABLE 5.8 Copper Foil, Electrodeposited*

Property	IPC-MF-150/1 CU-E1 (standard electrodeposited) (oz/ft²)			IPC-MF-150/2 CU-E2 (high ductility electrodeposited) (oz/ft²)			IPC-MF-150/3 CU-E3 (high temperature elongation) (oz/ft²)			IPC-MF-150/4 CU-E4 (annealed electrodeposited) (oz/ft²)		
	$\frac{1}{2}$	1	2	$\frac{1}{2}$	1	2	$\frac{1}{2}$	1	2	$\frac{1}{2}$	1	2
Tensile strength at 23°C [kips/in² (MPa)]	30 (207)	40 (276)	40 (276)	15 (103)	30 (207)	30 (207)	30 (207)	40 (276)	40 (276)		20 (138)	20 (138)
Ductility at 23°C (%) Fatigue ductility	N/A	N/A	N/A	N/A	N/A	N/A	N/A	N/A	N/A	N/A	N/A	N/A
Elongation	2	3	3	5	10	15	2	3	3		10	15
Tensile strength at 180°C [kips/in² (MPa)]											15 (103)	15 (103)
Elongation at 180°C (%)											4	8

*Data for these materials were reproduced (with permission) from IPC specification IPC-MF-150.

TABLE 5.9 Copper Foil Wrought*

Property	IPC-MF-150/5 CU-W5 (as-rolled wrought) (oz/ft²)			IPC-MF-150/6 CU-W6 (light cold-rolled wrought) (oz/ft²)			IPC-MF-150/7 CU-W7 (annealed wrought) (oz/ft²)			IPC-MF-150/8 CU-W8 (as-rolled wrought low temperature anneal) (oz/ft²)		
	1/2	1	2	1/2	1	2	1/2	1	2	1/2	1	2
Tensile strength at 23°C [kips/in² (MPa)]	50 (345)	50 (345)	50 (345)		25–50 (177–345)	25–50 (177–345)	15 (103)	20 (138)	25 (172)	15 (103)	20 (138)	20 (138)
Ductility (%)												
Fatigue ductility	30	30	30		65–30	65–30	65	65	65	25	25	25
Elongation	0.5	0.5	1.0		10–0.05	20–1	5	10	20	5	10	10
Tensile strength at 180°C [kips/in² (MPa)]		20 (138)	40 (276)					14 (97)	22 (152)			
Elongation at 180°C (%)	2	2	3					6	11			

*Data for these materials were reproduced (with permission) from IPC specification IPC-MF-150.

the plating current density and bath chemistries, high quality ED foils can be produced which have competitive, if not superior, flexural properties than the rolled anneal foils.

Early foils produced by the electrodeposition process were very brittle and usually caused the plated-through holes to crack when the flexible circuit was subjected to thermal stress testing. RA metallurgy provides good flexural strength but, due to its softness, these foils are easily distorted as a result of handling during the manufacturing processing. An alternative form to rolled foil that offers better handling and flexural properties is a type called *low temperature anneal (LTA) foils*. These foils are sold with high levels of residual rolling stresses (the grain structure has an unusual elongation). High residual stresses are the result of a fast anneal when the foil is heated. LTA foils can be a good choice as these foils can resist foil damage because initially they have a very high yield strength and, therefore, resist denting and crazing, but anneal during cover-coat lamination to a softer more flexarally forgiving grain size.

5.4 Adhesives

The earliest experience in fabricating flexible circuits incorporated a thermoplastic polymer as its base dielectric insulation. These thermoplastic films were laminated to conductive traces using a combination of high heat and pressure. The conductive traces were then encapsulated by laminating a cover coat over the conductors using the same film as that used for the base. In this process, both films were fused together to provide a uniform homogeneous insulation. FEP, Teflon, Aclar, and polyvinyl chloride–insulated flexible circuits were made using this process. The fluorocarbon polymer provided excellent dielectric properties and environmental protection. As experience with this polymer was gained, a number of problems became evident:

1. The fluorocarbon film was unstable at elevated temperatures.
2. During the cover-coat lamination process, the copper conductors were either distorted or "swam" due to the high heat and pressure that was required for this process.
3. The fluorocarbon film had memory.
4. Copper conductors required a special treatment in order to bond to the film.
5. Low modulus of elasticity of the plastic film transferred the stresses to the copper conductors, which resulted in the conductors being easier to stretch and break.

Research to improve substrate strength and dimensional stability led to the development of flexible circuits based on biaxially oriented polyester and polyimide films. These films were stronger and more dimensionally stable than the fluorocarbons but these films could not be fusion bonded. In order to make a copper-clad laminate, an adhesive had to be used to bond either the copper conductors to the base dielectric or the cover coat to the copper conductors.

The function of an adhesive in a flexible circuit application is to bond the cover coats and/or the copper foils to the dielectric base material and, in multilayer and rigid flex printed wiring boards, the adhesive is used to bond the inner layers together. A flexible laminate's performance depends upon the combined properties of the adhesive and supporting dielectric film. Adhesives, such as acrylics, epoxies, modified epoxies, polyester, and butryal phenolics, have long been used with varying degrees of success as bonding agents in flexible circuitry applications.

A conventional flexible printed wiring circuit consists of five distinct layers of materials: a base dielectric film, a conductor layer, a cover-coat film, and two layers of adhesives. One of the adhesive layers is used to bond the conductor and base together to form the copper-clad laminate and the second adhesive layer joins the cover coat to the etched circuitry. Adhesives are the known weak link in the flexible circuit. However, something must hold the structure together. The adhesives largely determine the flexible circuits' electrical, thermal, and chemical performance and are also a critical factor in a controlled impedance board design. The adhesive is the dielectric which surrounds the etched copper conductors and, in a typical construction, accounts for approximately 50% or more of the total flexible circuit thickness.

The base dielectric films classify flexible printed wiring boards but the adhesive dominates the performance for the following reasons:

1. Because the adhesive surrounds the conductors, it is the primary insulation in flexible printed wiring and determines the electrical properties such as volume and surface resistivity, dielectric constant, and dissipation factor. Film properties are secondary.
2. Flammability is adhesive driven. Virtually all films are self-extinguishing. Most adhesives support combustion unless they are formulated to include suppressors.
3. Adhesives are the controlling factors in critical areas such as thermal ratings.
4. Bond strength is an adhesive property. Selection of a proper adhesive can prevent assembly damage, especially during soldering and connector terminations.
5. Chemical resistance is also an adhesive function.

Two important mechanical properties to consider in selecting an adhesive system are tensile strength and elongation. A high tensile strength suggests that the material has good mechanical stability and resistance to handling stresses resulting from repeated connecting and disconnecting electrical connectors. Materials with a high elongation have the ability to flex repeatedly and move easily. However, these two properties tend to negate each other. A laminate with a high tensile strength will have a low elongation, resulting in a stiffer substrate without the capability of dynamic flexing. A laminate with a high elongation would be excellent for dynamic flexing, but may not be mechanically durable.

5.4.1 Adhesive forms

Flexible printed wiring adhesives are used in several forms: applied to one side of the dielectric film, applied to both sides of the film, as a bond ply, and as an unsupported cast film:

- *Base and cover coat.* These materials, when used together, are generally identical in polymer and adhesive formulation.

- *Bond ply and cast films.* These materials are used to join layers of flexible printed wiring together to form multilayer and rigid flex structures. Bond plies provide an assured dielectric barrier between facing conductive layers and act as a cover coat for both while cast adhesives allow for a thinner construction.

5.4.2 Applying the adhesive coating

Adhesives that are used for flexible printed wiring products are produced from mixtures of solvents, polymers, and curing agents. These mixtures are applied to the film in a horizontal coating machine and coat either one or both sides of the film to form a base, a cover coat, bond ply, or onto a temporary carrier film to create a cast film adhesive. The solvent is removed in a horizontal oven leaving a tack-free coating (semicured) that is protected by a release film. This release film is introduced at the last rewind stage of the coating machine and remains on the adhesive coating until lay-up. The release film is sometimes removed prior to drilling in order to remove any stresses that could upset the dimensional stability. Stripping off the release film can create a large static charge that can contribute to foreign matter buildup on the material.

5.4.3 Requirements for adhesive coatings

The basic requirements for an adhesive coating are

1. Adhesive coating must be as tack-free as possible so as to minimize foreign material pickup and to allow for ease of registration and alignment of cover coat and bond plies for lamination.

2. The cast films that are used are thin (0.001 to 0.002 in) and very fragile. The films must be sufficiently cohesive but not brittle, so that they may be lifted from the carrier sheet and positioned in a lay-up without being distorted. Most cast films are weak and limp, making accurate alignment of drilled or punched holes with an etched pattern extremely difficult and labor intensive.

3. Adhesive coating must have a low residual volatile content (less than 1%) so as to minimize lamination voids and possible poor adhesion.

4. The cured adhesive must be resistant to attack by processing chemicals and, as a conflicting requirement, drilling residue must also be removed from the pad edges prior to plating the drilled holes.

5. The adhesives are all thermosetting to some degree; therefore, it is good practice to protect them from elevated temperatures. Storage of the material in a refrigerator or freezer until ready for use is preferred.

6. The adhesive should be cured in a reasonable time and at moderate temperatures.

7. All adhesives have a limited shelf life. Therefore, these materials must be monitored to ensure freshness and expected performance.

5.4.4 Adhesive options

In order to determine the best material combination for a given application, you must first understand the major performance characteristics of each material. A good reference knowledge of the mechanical, electrical, thermal, and chemical properties will allow for an effective material choice for each flexible printed circuit board application.

Shown in the following table are the adhesives that could be used to laminate the copper foil or cover coat to the base dielectric:

Dielectric material	Adhesive
Polyester	Polyester
Polyester	Flame-retardant polyester
Polyester	Modified epoxy
Polyimide	Acrylic
Polyimide	Flame-retardant acrylic
Polyimide	Modified epoxy
Polyimide	Polyester
Polyimide	Polyimide
Polyimide	Adhesiveless
Aramid	Acrylic
Aramid	Modified epoxy
Composite	Modified epoxy
Composite	Polyester
Glass	Epoxy

5.4.5 Acrylic adhesives

Acrylic adhesives, which offer high heat resistance and good electrical properties, have been used successfully on polyimide films for many years. The modulus of the polyimide/acrylic adhesive system has made this laminate the preferred choice for dynamic flex applications. However, they are now finding that the acrylic thickness and its high z-axis expansion have limited this material's usefulness. The polyimide/acrylic adhesive laminates are also vulnerable to attack by some solvents used in the photoresist process and to alkaline solutions used in the plating and etching processes. Absorbed solvents are especially difficult to remove prior to multilayer lamination and may result in a delamination and/or blistering problems. Dimensional stability and small hole–drilling problems (especially drill smear from the acrylic adhesive) can decrease yields, especially with high density designs. Plasma desmearing, which is used to clean out debris from the drilled holes, will show a higher etch-back rate for the acrylic adhesive than for the polyimide film. If not controlled, plated-through hole failures can occur. The properties of the acrylic adhesives are shown in Table 5.10.

5.4.6 Polyimide and epoxy adhesives

Polyimide substrates can also be coated with a polyimide adhesive. The polyimide adhesives' electrical properties are at least as good as, if not better than, the acrylic adhesives. Additionally, they offer better heat resistance than any other flexible adhesive system as the resins will cross-link during cure.

The reduced dynamic flexing ability of the polyimide, epoxy, or modified epoxy is not a serious limitation since the majority of flexible circuits produced are used in static flexing applications. The trade-off for this increased laminate stiffness is better dimensional stability, better processability, and lower overall adhesive thickness in multilayer and rigid flex boards. Properties of these adhesives are shown in Table 5.10.

5.4.7 Polyester and phenolics

Polyester-based substrates are primarily used in cost-sensitive consumer and automotive electronic applications. Adhesives used on polyester substrates are polyester, modified polyester, or phenolics. Polyester adhesives have excellent electrical properties, excellent flexibility, and fair resistance to heat. The heat resistance can be improved by partially cross-linking the adhesive.

TABLE 5.10 Flexible Adhesive Bonding Films*

Property	Adhesive/film						
	Acrylic/ polyimide	Epoxy/ polyimide	Polyester/ polyester	Acrylic/ aramid	Acrylic/ glass	Polyimide/ polyimide	Polyimide/ butryal: phenolic
Peel strength, minimum (lb/in)							
As received	8.0	8.0	4.0	5.0	8.0	875	875
After solder float	7.0	7.0	N/A	4.0	7.0	875	875
After temperature cycling	8.0	8.0	4.0	5.0		875	875
Tensile strength, minimum (lb/in²)							
>0.001	24,000	24,000	20,000	4,000			
<0.001	15,000	15,000					
Elongation, minimum (%)	40	40	90	7			
Low temperature flexibility	Passes (5 cycles)	Passes (5 cycles)	Passes (5 cycles)	Passes (5 cycles)	Passes (5 cycles)		
Flow, maximum (squeeze out mil/mil of adhesive)	5	10	10	5		0.127	0.127
Volatile content, maximum (%)	1.5	2.0	1.5	2.0	2.0	2.0	3
Flammability, minimum (%O_2)	15	20	20	15	15		
Solder float	Passes	Passes	N/A	Pass	N/A	Pass	Pass
Chemical resistance (%)	80	80	90	80	80	80	85
Dielectric constant, maximum @ 1 MHz	4.0	4.0	4.0		4.0	4.0	3.0
Dissipation factor, maximum @ 1 MHz	0.04	0.04	0.02		0.025	0.010	0.025
Volume resistivity, minimum (MΩ-cm)	10^6	10^6	10^6	10^6	10^6	10^6	10^6
Surface resistance, minimum (MΩ-cm)	10^5	10^4	10^4	10^5	10^5	10^5	10^4
Dielectric strength, minimum (V/mil)	2,000	2,000	2,500	500	1,500	3,200	2,000
Insulation resistance, minimum (MΩ)	10^4	10^4	10^5	10^4			
Moisture and insulation resistance, minimum (MΩ)	10^3	10^3	10^4	10^3		10^3	10^3
Moisture absorption, maximum (%)	6.0	4.0	2.0	15	6.0	3.0	2.0
Fungus resistance	Non-nutrient	Non-nutrient	Non-nutrient		Non-nutrient	Non-nutrient	Non-nutrient

*Data for this material were reproduced (with permission) from IPC specification IPC-FC-233.

5.4.8 Butryal phenolic

Butryal phenolics are more heat resistant than polyester adhesives, but the electrical properties are not as good and they are not as flexible as the acrylic adhesives. The addition of additives to the butryal phenolic adhesive can increase its flexibility.

5.5 Problems with Adhesive-Bonded Constructions

The disadvantage of adhesive-bonded construction arises from the fact that one is now dealing with a composite dielectric film. A layer of adhesive now exists between the insulating film and the copper foil. When the copper foil is etched to provide the base circuit, the exposed insulation is the adhesive rather than the insulating film. After cover-coat lamination, the primary insulation encapsulating the conductive traces is the adhesive rather than the base dielectric film. (The limitation with adhesive-bonded construction is shown later in this section). This results from the fact that adhesive formulations have not yet been developed that can provide properties comparable to the base dielectric films and covercoats.

Performance of the adhesive systems is lacking in

- Resistance to delamination during the soldering operation
- Poor moisture resistance
- Poor resistance to heat aging, discoloration, embrittlement, and outgasing
- Poor insulation at elevated temperatures

5.6 Adhesiveless Laminates

Traditionally, plastic films that were laminated with a metal foil have been utilized in the fabrication of flexible circuits for many, many years. Metal-clad laminates prepared on heat-resistant polymer films, using a variety of adhesives like epoxies, acrylics, polyesters, fluorocarbons, etc., have been the most commonly used materials in the higher performance flexible circuit board fabrication. Polyimide films were generally chosen as the dielectric substrate because of their high temperature stability, good dielectric behavior and properties. The nonpolyimide adhesives used in bonding the substrate and copper conductors together are often the weak link. Because the adhesives used have a poor thermal stability and high coefficient of thermal expansion, these adhesives often fail during the fabrication and/or assembly process.

Over the years, researchers have come up with several different versions of the adhesives, but have not provided a satisfactory solution to the functional needs. The latest approach to this problem was to develop an adhesiveless system made from polyimide resin and copper metal devoid of any nonpolyimide adhesives.

Significant performance advantages can be realized when adhesiveless-based laminates are used to make flexible circuits, high layer count multilayers, and rigid-flexes. Many of these advantages emanate from the inherent thinner circuits and the elimination of the mismatched characteristics of the adhesive in relation to the film and copper foil.

Adhesiveless laminates are thinner due to the absence of the 1 or 2 mils of adhesive that is used in the adhesive-clad laminates. Shown in the following table is a comparison of a double-sided circuit that was fabricated using adhesive versus an adhesiveless system. The savings in thickness is 4 mils.

Adhesive laminate	Thickness (mil)	Adhesiveless laminate	Thickness (mil)
1-mil copper (plated)	0.0010	1-oz. copper (plated)	0.0014
1-oz copper foil (RA)	0.0014	2-mil polyimide film	0.0020
1-mil adhesive	0.0010	1-oz copper (plated)	0.0014
2-mil polyimide film	0.0020		
1-mil adhesive	0.0010		
1-oz copper foil (RA)	0.0014		
1-mil copper (plate)	0.0010		
Total thickness	0.0088		0.0048

Some of the requirements that are redefining material characteristics today include

- Surface-mount technology

- Reduced thickness and weight

- Smaller plated-through holes

- Finer line width

- Higher operating temperature

- Smaller and denser assemblies

- Improved dimensional stability

- Material impedance characteristics

- Flame retardation (UL94 VO)

- Improved thermal management

- Matched coefficient of thermal expansion
- Increased number of flex cycles

Adhesiveless copper-clad laminates offer improved operating characteristics for single- and double-sided circuits as well as for multilayer and rigid-flex circuits. In the adhesiveless copper-clad laminates, the copper is bonded to the base material without the use of adhesives. Compared to adhesive-based laminates, adhesiveless laminates provide a thinner circuit, greater flexibility, and better thermal conductivity. Additionally, the thermal stress resistance of the higher layer count rigid flexes is significantly better.

5.6.1 Adhesiveless films

Adhesiveless laminates can be classified into four manufacturing technologies: (1) cast to foil, (2) vapor deposition on film, (3) sputter on film, and (4) plated on film. The properties of these adhesiveless laminates are shown in Table 5.11.

5.6.1.1 Cast to foil. The cast-to-foil process involves casting a liquid solution of polyamic acid onto the surface of a metal foil. The entire composition (foil and polyamic acid) is heated to a temperature that will evaporate the solvent and imidize the polyamic acid, resulting in either a polyimide or polyamide film. The cast-to-foil process is an efficient method for creating a single-sided laminate. However, to produce a double-sided laminate requires a thermal compression stage in which two single-sided cast-to-foil laminates are bonded together with a compatible adhesive (usually the same polymer as that of the film). Alternatively, a second sheet of copper foil may be bonded by thermal compression to the uncured B-stage cast laminate.

Although the adhesion of the copper foil to the polyimide film is good, this process is usually limited to a minimum copper foil thickness of 1 oz. Thinner copper foils are too difficult to handle. The thin foils (less than 1 oz) become highly annealed as a result of the high temperatures involved in making the case-to-foil laminate, which creates a problem in processing the laminate in such fabricating processes as routing and drilling.

5.6.1.2 Vapor deposition on film. In the vapor deposition on film process, the copper is vaporized in a vacuum chamber and the metal vapors are deposited on the polyimide film. A treatment on the surface of the polyimide film is normally required in order to enhance the copper adhesion. It is a relatively fast process and is usually limited to a copper thickness of 2 μm. Additional copper can be added by electrodeposition.

TABLE 5.11 Flexible Adhesiveless Films*

Property	Cast film on foil	Sputtered metal on film	Plated metal on film	Additive metal on film
Peel strength, minimum (lb/in)				
As received:				
<0.001	6.0	5.0	4.0	6.0
>0.001			4.0	6.0
After solder float:				
>0.001	6.0	6.0	6.0	6.0
<0.001			6.0	6.0
After temperature cycling:				
>0.001	6.0	6.0	6.0	6.0
<0.001			6.0	6.0
Initiation tear strength, minimum (g·m-s)	500	500	500	500
Propagation tear strength, minimum (g·m-s)	5	4	4	4
Flexural endurance, minimum (cycles)	N/A	1000		1000
Low temperature flexibility				Pass
Dimensional stability, maximum (%)	0.20	0.20	0.10	0.10
Solder float	Pass	Pass	Pass	Pass
Solderability	Pass	Pass	Pass	Pass
Chemical resistance, minimum (%)	80	80	90	80
Dielectric constant, maximum @ 1 MHz	4.0	4.0	4.0	3.5
Dissipation factor, maximum @ 1 MHz	0.01	0.012	0.012	0.008
Volume resistivity, minimum (MΩ-cm)	10^6	10^6	10^6	10^{10}
Surface resistance, minimum (MΩ)	10^5	10^5	10^5	10^8
Dielectric strength, minimum (V/mil)	3500	2500	2000	2500
Moisture and insulation resistance, minimum (MΩ)	10^3	10^2	10^4	10^2
Moisture absorption, maximum (%)	3.0	4.0	4.0	2.0
Fungus resistance	Non-nutrient	Non-nutrient	Non-nutrient	Non-nutrient

*Data for these materials were reproduced (with permission) from IPC specification IPC-FC-241.

5.6.1.3 Sputter to film. The sputtering-to-film method involves placing the polyimide film in a large vacuum chamber that has a copper cathode. The cathode is bombarded with positive ions, causing small particles of the charged copper to impinge on the film surface, resulting in an ultra-thin copper coating. The film is then electroplated to a specified thickness. The copper adhesion is not as good as that of the cast or plated method, and the dimensional stability does not compare favorably with the adhesive-based materials. Sputtered-to-film laminates require a base metal undercoating of chrome, nickel, or an oxide coating if subjected to elevated temperatures or humidity. Sputter coating is a slower process when compared to the vapor deposition method.

5.6.1.4 Plated on film. Another method of manufacturing adhesiveless laminates is to plate copper directly onto the polyimide film. The polyimide film is roll processed through the electroless metal chemistries to produce the "seed" layer. The polyimide film is then plated to a desired final thickness. The copper metal deposits can be controlled so as to provide a very thin copper foil. When compared to other adhesiveless laminates, the plated-to-film materials do not induce any thermal stresses due to their low processing temperatures. The electrodeposited copper foil laminates are much easier to handle and process.

5.6.2 Advantages of adhesiveless laminates

5.6.2.1 Electrical. Adhesiveless technology offers certain electrical advantages that are not instantly obvious. Thinner conductive coatings are readily available because adhesiveless materials are usually metallized using the roll processing technique. Time and current are the major controlling factors for metal thickness, so creating a thinner foil can be as simple as increasing roll speed or reducing the amperage from the plating rectifier.

From an electrical standpoint, thinner conductive coatings offer greater latitude to the electrical engineer in designing a controlled impedance board. In a microstrip or strip-line design with an adhesiveless material, the dielectric spacing is reduced which usually necessitates a thinner conductor in order to obtain the desired impedance factor. This may make the design less produceable and, subsequently, result in a prohibitive cost, but with options of using ¼-oz copper (½-oz finished), the cross-sectional area of the conductors can be reduced by 50% without impacting line width. This, coupled with a lower dielectric constant, allows for higher impedance values without seriously affecting image or etch production yields. A beneficial side effect of using adhesiveless materials for controlled impedance designs is a narrowing of the impedance range as noted on the TDR (time domain reflectometer) curve.

In an adhesive system, the adhesives have a tendency to allow embossing (conductors embedding themselves into the adhesive). However, in an adhesiveless system, the copper conductors cannot deform into the base material due to the high glass transition temperature of the polyimide film. Embossing can result from a variation in dielectric separation which is determined somewhat by the density of the conductive pattern. This is seen on the TDR curve as a greater delta (tolerance) between the minimum and maximum impedance values.

5.6.2.2   Thermal. In higher layer count rigid-flex, the advantages of the adhesiveless materials are well known. Acrylic adhesives with a z axis coefficient of thermal expansion (CTE) of 500 to 600 ppm/°C and a T_g of 30 to 40°C are the well-known weak links from a reliability standpoint. The more acrylic adhesive in a construction, the greater the chance of barrel cracking and/or pad lifting during either thermal cycling or soldering. Utilizing adhesiveless inner layers, combined with polyimide prepreg or bond films and polyimide cap layers, a high reliability (all polyimide) rigid flex of 20+ layers can be produced without plated-through hole reliability problems.

These constructions are excellent for military, aerospace, and other high reliability applications, but material costs could be prohibitive for most commercial applications.

5.6.2.3   Mechanical. A reduction in dielectric spacing typically one-half that of the traditional acrylic adhesive-based materials means that the metal foil layers are significantly closer to the neutral axis and, therefore, subjected to less stress and strain during flexing. This is especially important in dynamic (continual) flexing applications where increased life performance and reliability can be achieved.

5.7   Cover Coat

5.7.1   Cover-coat usage

The etched conductive traces on the flexible-based dielectric are usually protected from damage (either mechanical or environmental) by covering these traces with a protective film. Normally, the film material is of the same polymer as that of the base dielectric. The cover coat serves four distinct purposes:

1. Protects the circuit against corrosion and contamination.
2. Affords protection against mechanical damage which might occur if a moving flexible circuit touches another part of the equipment.

3. Puts the copper on a single-sided flex on the neutral bending axis.

4. Helps anchor the terminal pads to the base dielectric.

Thermally laminated cover-coat materials were the flex circuit standard for many years. The finished flex was suitable for both flex-to-install and for dynamic disk drive applications that required many millions of flex cycles. One of the basic reasons for the success of this type of cover-coat application was that the cover-coat thickness could be matched precisely to the base film thickness, thus placing the conductive traces in the neutral axis. This minimized the stresses and strains and also reduced work hardening of the conductive traces that could occur during continual flexing.

The copper traces have adequate adhesion to the base material, but this adhesion is reduced at soldering temperatures and, unless the lands are given some mechanical support or reinforcement, the risk exists that the lands could possibly lift, causing the flexible circuit to fail. Figure 5.1 shows three ways in which a cover coat can be used to hold down the lands.

The simplest way is to use a land larger than the hole in the cover coat (Fig. 5.1a). The land must be of sufficient size to ensure that the outer edges of the land will be securely encased under the cover coat even with the worst-case misregistration of either the conductive traces or lands to the bonded-on cover coat.

The best anchorage is obtained when the lands are provided with two, but preferably three, ears that are encased under the covercoat (Fig. 5.1b). The cover coat can be securely bonded to the base film between the ears, and the land sizes should be of normal size as appropriate to that hole diameter.

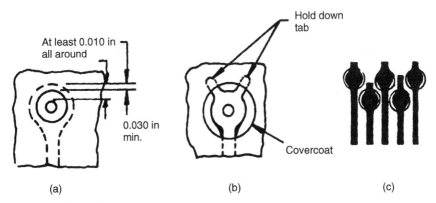

Figure 5.1 Covered exposure.

In cases such as in two row connectors, where the conductive traces must pass between fairly close spaced lands, a single ear has to suffice (Fig. 5.1c).

Flexible solder masks are available that could be used as a cover-coat material. These coatings are screen-printed onto the flexible circuit but offer no mechanical support, that is, anchoring down the lands or putting the conductive traces in the neutral axis during flexing or bending. Another problem arises when the screened-on solder mask is image defined. The liquid solder-mask coating is directly transferred through the photoimaged silk-screen stencil onto the flex circuit. Due to the flow characteristics of the liquid material, it is not possible to define openings around fine pitch devices or hold tight land tolerances. Screened-on solder masks cannot be used as an alternative for film cover coat.

For surface-mounted flex designs, a liquid solder mask can be used. It can be applied either by screen printing or spraying. Spray coatings offer some distinct advantages over screen printing in that greater copper thicknesses can be coated and the mask thickness variation from conductor knee to nonconductor is significantly less. This is critical when flex performance is considered because an increase in mask thickness will cause cracking and peeling during even limited flexing. Screen printing liquid photoimageable solder mask over dense circuit patterns can only really be successful when applied over less than 1 oz of copper. The copper thickness for a typical 1-oz copper (pattern plate) will yield conductors varying from 3.0 to 4.2 mils thick in isolated high current density areas. With a dense random conductor pattern, it is impossible to keep all of the conductors out of the plane parallel to the squeegee direction.

In a design that has 8-mil lines and spacing, a single pass of liquid solder mask cannot adequately cover these lines and spaces without leaving some areas void. This then requires a two-pass coverage. By using excessive pressure on the squeegee, a two-pass process can be used without adding solder-mask thickness to the top of the conductive traces, but it will add additional thickness in the spaces between the conductors.

As with screened solder mask, the major setback in polymer material cover coat is one of cover-coat–to–circuit definition. The problem is adhesive inflow around the drilled or punched holes and dimensional distortion during the cover-coat lamination process. The materials generally used for the cover-coat applications are either a polyimide or polyester polymer with an acrylic, epoxy, or polyester adhesive on one or both sides of the dielectric film. The properties of polymer-based covercoat materials are given in Tables 5.10 and 5.12.

TABLE 5.12 Flexible Bonding Films (Adhesive on One or Both Sides of the Dielectric Film)*

Property	Polyimide/acrylic	Polyimide/epoxy	Polyester/polyester	Aramid/acrylic
Peel strength, minimum (lb/in)				
As received	8.0	8.0	4.0	5.0
After solder float	7.0	7.0	N/A	4.0
After temperature cycling	8.0	8.0	4.0	5.0
Tensile strength, minimum (lb/in²)				
>0.001 in 24,000		24,000	20,000	4,000
<0.001 in 15,000		15,000		
Elongation, minimum (%)	40	40	90	7
Low temperature flexibility	Pass (5 cycles)	Pass (5 cycles)	Pass (5 cycles)	Pass (5 cycles)
Flow, maximum (mil/mil)	5	10	10	5
Volatile content, minimum (%)	1.5	2.0	1.5	2.0
Flammability, minimum (%O$_2$)	15	20	20	15
Solder float Pass	Pass	Pass	N/A	Pass
Chemical resistance (%)	80	80	90	80
Dielectric constant, maximum @ 1 MHz	4.0	4.0	4.0	
Dissipation factor, maximum @ 1 MHz	0.04	0.04	0.02	
Volume resistivity, minimum (MΩ-cm)	10^6	10^6	10^6	10^6
Surface resistance, minimum (MΩ)	10^5	10^4	10^4	10^5
Dielectric strength, minimum (V/mil)	2,000	2,000	2,500	500
Insulation resistance, minimum (MΩ)	10^4	10^4	10^5	10^4
Moisture and insulation resistance, minimum (MΩ)	10^3	10^3	10^4	10^3
Moisture absorption, maximum (%)	6.0	4.0	2.0	15
Fungus resistance	Non-nutrient	Non-nutrient	Non-nutrient	

*Data for these materials were reproduced (with permission) from IPC specification IPC-FC-232.

5.7.2 Cover-coat bonding

Cover-coat bonding is similar to normal multilayer lamination except that the cover-coat material is normally supplied with a low flow adhesive on one or both sides of the polymer film. Good land capture is shown in Fig. 5.1. The properties of the materials generally used for the cover-coat application are shown in Tables 5.10 and 5.12.

5.7.3 Cast Films

Cast films are used as an adhesive in both multilayer and rigid flex fabrication and also for adhering heat sinks to flex circuits. The properties of the cast films are shown in Table 5.13.

5.8 Neutral Axis

The neutral axis should be located at the center of the copper conductor. Several design techniques are popular for approximating this condition, such as using cover-coat and interdigitizing conductors front to back (Fig. 5.2).

5.9 Fabrication

TABLE 5.13   Unsupported Flexible Adhesive Bonding Films (Cast Films)*

Property	Polyester	Acrylic	Epoxy
Peel strength, minimum (lb/in)			
As received	875	1400	1400
After solder float	875	1400	1400
After temperature cycling	875	1400	1400
Flow, maximum (mil/mil)	0.254	0.127	0.127
Volatile content, maximum (%)	1.5	2	2
Chemical resistance, minimum (%)	90	80	80
Dielectric constant, maximum @ 1 MHz	4.6	4.0	4.0
Dissipation factor, maximum @ 1 MHz	0.13	0.05	0.06
Volume resistivity, minimum (MΩ-cm)	10^6	10^6	10^6
Surface resistance, minimum (MΩ)	10^5	10^5	10^4
Dielectric strength, minimum (V/mil)	1000	1000	500
Moisture and insulation resistance, minimum (MΩ)	10	N/A	10^3
Moisture absorption, maximum (%)	2.0	6.0	4.0
Fungus resistance	Non-nutrient	Non-nutrient	Non-nutrient

*Data for these materials were reproduced (with permission) from IPC specification IPC-FC-232.

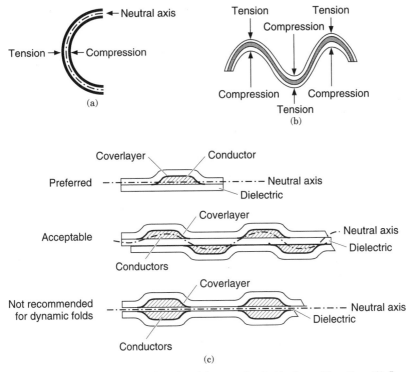

Figure 5.2 Neutral axis consideration; (*a*) neutral axis ideal consideration; (*b*) flexural cycles; (*c*) neutral axis conductor consideration.

5.9.1 Fabrication methods

Flexible printed wiring boards are made up of many materials and parts that are fabricated together to form the end product. These may consist of the etched and/or plated circuit patterns, cover-coat sheets, copper foil, rigid materials, stiffeners etc. The flexible materials are usually nonreinforced and thin. Detailed here are the most commonly used materials and processes for fabricating flexible printed wiring boards.

5.9.1.1 Single- and double-sided flexible circuit without plated-through holes (NPTH).

Material. The materials used to fabricate a single- or double sided-flex are

1. Copper-clad laminate (Table 5.7)

2. Adhesive-coated dielectric film (Tables 5.10 or 5.12)

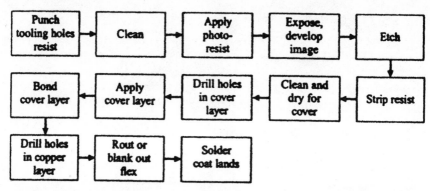

Figure 5.3 Process sequence for single- and double-sided flexible printed wiring boards without plated-through holes.

Processing. The most commonly used processing sequence for fabricating a flexible single- or double-sided flexible circuit without plated-through holes is shown in Fig. 5.3.

Fabrication options

1. The thin single-sided flexes may not need the conventional film material and may be constructed from thin (0.003-in) glass-reinforced copper laminates.

2. The imaging process for etching the conductive material may be performed using the silk-screen process as the etch resist. If the screen resist is left on after etching, it may act as a cover coat even though the sides of the conductors are not covered.

3. Cover-coat material is often applied without punching or drilling clearance holes for lands. The lands are later cleared of cover-coat material by skiving with a special drill. Care must be taken not to thin the copper on the lands.

4. Photoimageable solder masks can be used as a cover coat; however, brittleness at low temperatures or with repeated flexing can occur.

5. The lands are usually coated with solder or are treated with a protective coating in order to preserve the solderability of the copper lands.

6. Stiffeners may be added to the flex for support for either component installation or for connector engagement. These stiffeners are generally attached to the flex using one of the adhesive materials, as shown in Table 5.13.

5.9.1.2 Double-sided flexible printed wiring with plated-through holes (PTH)

Materials

1. Double-sided copper-clad flexible laminate (Table 5.7)
2. Cover-coat material (Tables 5.10 and 5.12)

Processing. The most common processing sequence for double-sided flexible circuits is shown in Figure 5.4.

Fabrication options

1. The use of carbon, graphite, or palladium conductive coatings that are later electroplated with copper can be substituted for the standard electroless copper process.

2. Copper conductors plated with electrodeposited copper by the pattern or panel plating process are not desirable for retaining flexibility. The elimination of electrodeposited copper overplate requires a double process, and only the plated-through holes and associated lands are plated by the process called *selective* or *button plating* (Fig. 5.5). After plating the drilled holes and the associated lands, the resist is stripped and the circuit pattern image is applied and used as an etch resist and the plated-through holes must be tented. Etching and resist stripping completes the circuit formation.

3. The cover-coat material is usually laminated to both sides of the double-sided flexible circuit. Since the cover-coat materials are very unstable and in order to maintain some type of registration accuracy between the cover-coat clearance holes and the drilled holes, you

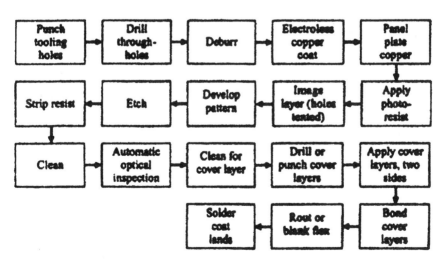

Figure 5.4 Process sequence for double-sided flexible printed wiring boards with plated-through holes.

should either secure the package with tape or heat tack the cover coat to the base laminate with a soldering iron prior to laminating the cover coat to the clad laminate.

5.9.1.3 Multilayer flexible printed wiring board.

The multilayer flexible printed wiring board consists of three or more layers with plated-through holes connecting the various layers. The multilayer is made up of several double-sided printed wiring laminates laminated together with their electrical circuits interconnected by plated-through holes.

Materials

1. Flexible copper-clad laminate (Table 5.7 or 5.11)
2. Adhesive-coated dielectric film (Table 5.10 or 5.12)
3. Bonding film (Table 5.12)
4. Adhesive (Table 5.13)

Processing. The most common fabrication process for fabricating multilayer flexible circuits is shown in Figure 5.6.

Fabrication options

1. The high z-axis expansion can cause cracked plated-through hole barrels when subjected to thermal cycling. One of the methods used to decrease z-axis expansion is to eliminate the cover coat in the multilayer section. Another is to use glass-reinforced prepreg as the bond sheet.

2. The multitude of materials that are available to fabricate multilayer flex circuits provides a wide range of options. One of the earliest materials, polyimide film with acrylic adhesives, is still widely used. An early major disadvantage of using this structure was drill smear. Plasma desmearing was developed to overcome this problem. However, this process was, and is still, inconsistent as the acrylic adhesive etches at a faster rate than the polyimide film (approximately 2 times as fast). Polyimide or epoxy adhesives could be used but, if possible, adhesiveless laminates are the preferred material.

3. Copper pattern plating of the outer layers is the preferred process followed by solder plating and fusing (See Figure 5.5.) Hot-air leveling does not work well.

4. Vacuum-assisted lamination with conformal press pads works best in removing air entrapment and volatiles.

5. Solder mask or cover coats are often applied to the outer layers of the multilayer flex circuit (See Figure 5.6.) The covercoat (same material as the core) is the most common, but new solder mask materials will probably replace the cover coat on new designs.

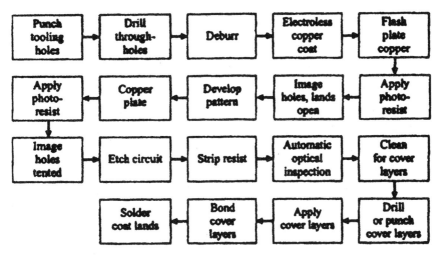

Figure 5.5 Process sequence for pattern, button, and selective copper plating.

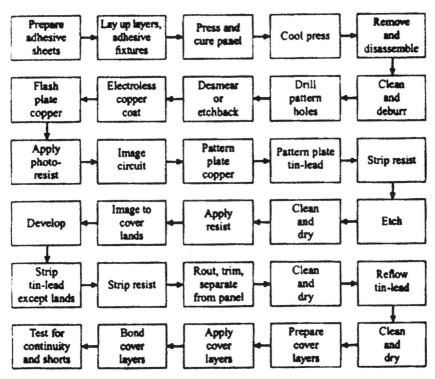

Figure 5.6 Process sequence for multilayer flexible printed wiring boards with plated-through holes.

Acknowledgments

I wish to express my sincere thanks to David Bergman and John Riley of the IPC for the generous support in providing technical and specification information for this chapter.

Bibliography

Coombs, Clyde Jr.: *Printed Circuits Handbook*, 3d ed., McGraw-Hill, New York, 1988.

Dean, Steven: "Covercoating Techniques," Institute for Interconnecting and Packaging Electronic Circuits Technical Review, 1995, San Diego, Calif.

Dean, Steven: "Flexible Circuits Covercoating Techniques," IPC Technical Meeting, 1995, San Diego, Calif.

Fjelstad, Joseph: "Expanding the Role of Flexible Circuitry in Electronics," IPC National Conference on Flex, 1996, Bedford, Mass.

Fjelstad, Joseph: "Flexible Circuit Technology Overview," Printed Circuit Design Conference, 1996.

Fjelstad, Joseph: "Key to Flexible Circuits Design Principles," Printed Circuit Design Conference, 1996.

Forman, G.: "Flex Base for MCM," Institute for Interconnecting and Packaging Electronic Circuits Flex Circuits Meeting, 1995, Providence, RI.

Gilleo, Ken: "Flex Based Packaging," *Flexible Circuits Engineering*, January–February 1997.

Ginsberg, Gerald: *Printed Circuit Designs*, McGraw-Hill, New York, 1990.

Institute for Interconnecting and Packaging Electronic Circuits Documents: *Proceedings of the First IPC National Conference on Flex Circuits*, Minneapolis, Minnesota, May 16–17, 1995; *Proceedings of the Second IPC National Conference on Flex Circuits*, Bedford, Massachusetts, April 22–23, 1996; *Proceedings of the Third IPC National Conference on Flex Circuits*, Phoenix, Arizona, May 19–20, 1997; *Proceedings of the Fourth IPC National Conference on Flex Circuits*, San Jose, California, March 18–19, 1998.

Institute for Interconnecting and Packaging Electronic Circuits Specification Document referenced in this section: "Metal Foil for Printed Wiring Applications," IPC-MF-150; "Flexible Base Dielectrics for Use in Flexible Printed Wiring," IPC-FC-231; "Adhesive Coated Dielectric Films for Use as Cover Sheets for Flexible Printed Wiring and Flexible Bonding Films," IPC-FC-232; "Flexible Adhesive Bonding Films," IPC-FC-233; Flexible Metal Clad Dielectrics for Use in Fabrication of Flexible Printed Wiring," IPC-FC-241.*

Jawitz, Marty: *Printed Wiring Board Materials Handbook*, McGraw-Hill, New York, 1997.

Kanakarajan, K.: "New Adhesiveless Substrate for Flexible Printed Circuits and Multichip Modules, E. I. Du pont de Nemours, IPC National Conference on Flex, 1995, Minneapolis, Minn.

Pollack, H., and R. C. Jacques: "Adhesiveless Laminates Improve Flex Performance," *Electronic Packaging and Production*, May 1992.

Scarlett, C.: *An Introduction to Printed Circuit Board Technology*, 1st ed., Electrochemical Publications Limited, Ayr, Scotland, 1984.

Stearns, Tom: "Dielectrics Influence Circuit Performance and Laminate Performance," *Electronic Packaging and Production*, May 1992.

Stearns, Tom: *Flexible Printed Circuits*, McGraw-Hill, New York, 1995.

Wallig, Lyle: "Materials," *Electronic Packaging and Production*, May 1992.

---

*IPC, 2215 Sanders Road, Northbrook, IL 60062-6135.

6

Design of
High Performance
Circuit Boards

Sam R. Shaw

AMP / TYCO Electronics
Harrisburg, Pennsylvania

Alonzo S. Martinez, Jr.

Dell Computer
Round Rock, Texas

6.1   Introduction and PCB Design Methodology

Major advances in personal computers, data communications, networking, and consumer electronics, as well as the proliferation of electronics everywhere, has been facilitated by improvements in printed circuit board (PCB) technology. Printed circuit boards typically represent 15 to 25% of the price of electronic assemblies.

The PCB must be considered as just one part of the total interconnect path. Edge rates below 100 ps and ever-increasing data throughput require that the total interconnect system be simulated. Impedance, cross-talk, ground bounce, reflections, and frequency-dependent skin effects are just some of the key factors that have to be considered during the design and layout of printed circuit boards. This chapter summarizes major aspects of the printed circuit board design process.

6.1.1 PCB design flow

This section describes the typical flow (Fig. 6.1) that is made when the engineer formulates the PCB design. The engineer *must* maintain close contact with the mechanical and layout groups from system inception through production to ensure that producibility, reliability, and performance are "designed in." The criticality of impedance, signal layout, grounding, etc., demands that plots be reviewed as layout pro-

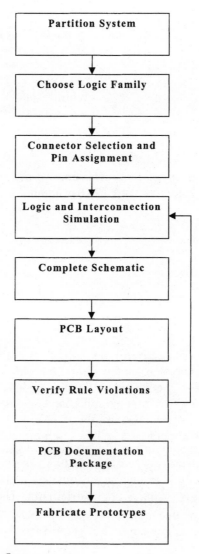

Figure 6.1 PCB design flow.

gresses and that the impact of situations which do not meet desired design rules are assessed.

Each block shown in Fig. 6.1 is explained in detail in the text that follows.

6.1.1.1 Partition system. The total system must be partitioned into functional elements, usually circuit board modules. This partitioning always results in design compromises because the introduction of connectors, backplanes, and other interconnects is a potential source of unwanted noise, delays, and reliability problems. Large PCBs minimize this problem. The requirements of shock, vibration, and thermal management, along with maintainability and producibility, however, limit the size of large cards.

6.1.1.2 Choose logic family. The interconnection requirements of the specific logic families chosen will have a significant impact on system cost/producibility and overall performance, including electromagnetic interference (EMI). Table 6.1 is a comparison chart of standard logic families. As a basic rule-of-thumb, choose the *slowest* logic family that can adequately meet worst-case performance and timing goals. This will ensure fewer cross-talk and EMI problems and fewer stringent layout guidelines.

6.1.1.3 Connector selection and pin assignment. It is at this point in the design that the selection of the connector system is made. Each system design will have its own unique requirements, but preassigning pins to minimize layers should be considered. The number of signals required at each interface is known along with the edge speeds associated with the particular logic families. The signal-to-ground ratio required to meet noise margins will be determined by this noise analysis.

6.1.1.4 Logic and interconnection simulation. This step ensures that the design is both *logically* sound, that is, for any input, there is the correct output, and the interconnection does not affect system performance. The analysis will determine the optimum system impedance, net topologies, and placement and layout rules. Section 6.3.1 describes how impedance interrelates with all other system parameters, and Sec. 6.3.1.5 recommends impedance ranges as a starting point based on logic family and packaging requirements. Use the simulation results to design a stackup based on the layer estimate that meets desired impedance with producible line widths and spacing. Remember that the number of power or ground planes needs to be adequate for optimum power distribution. It is very important to review

TABLE 6.1 Standard Logic Family Comparison

Parameter	CMOS 4000	HCMOS	CMOS ACL	LS TTL	FAST TTL	ECL 10K	ECL 10H	ECL 100K	ECL ECLinPS
Gate delay maximum (ns)	300	23	9.5	15	6	3.3	1.4	0.95	0.5
Flip-flop toggle (MHz)	2	24	125	25	100	125	250	400	800
Gate power maximum (MW)	0.0094	0.010	0.50	5.5	13	38	38	72	38
Edge speed maximum (ns)	200	19				3.7	2.3	1.1	0.8
Logic swing typical (V)	5	5	5	3.1	3.1	800	800	800	800
Noise margin (mV)	1450	1400	1400	300	300	125	150	130	150
Fan-out	—*	—*	—*	20	33	—*	—*	—*	—*
Input current maximum (µA)	1	1		400	600	265	265	350	150
Output drive maximum (µA)	0.36	4	24	8	20	22	22	22	22
Power supply (V)	3–18	2–6	2–6	5	5	−5.2	−5.2	−4.5	−4.5–5.2
Package pins	14	14	14	14	14	16	16	24	28
Cost (relative)	Low	Low	Medium	Low	Low	Medium	Medium	High	High

*AC limited.

stackup design with mechanical and producibility engineers. Another method of estimating the layer count is called the *equivalent integrated circuit count (EIC) technique.* In this technique, the 14-pin dual inline package (DIP) is selected as the reference device and defined as a unit EIC. All the devices mounted on the PCB are then compared to the 14-pin DIP to determine the correlation factors.

A popular method of determining the correlation factors for DIPs is to calculate the ratio of the number of leads between the DIP under consideration and the 14-pin DIP. For example, a 64-pin DIP has a correlation factor of 64/14 = 4.5.

Correlation factors for other devices are:

149-pin pin grid array (PGA), 12

Single inline package (SIP), 0.5

Card edge connector. Number of pins/64

High density connector. Number of pins/14

The correlation factors are used for calculating the total number of EICs and the board density. The number of signal layers is estimated as a function of the board density. The typical package area of a 14-pin DIP is $(0.8 \times 0.4$ in $= 0.32$ in^2).

The following chart shows the total layer count using the EIC method:

Density (in^2/14-pin DIP)	Number of layers
>1.0	2
0.8–1.0	2
0.6–0.8	4
0.42–0.6	6
0.35–0.42	8
0.2–0.35	10
0–0.2	12+

As an example, consider a board with a total area of 140 in^2 with 120 in^2 of routable area. From the device configurations and the correlation factors given here, the EIC for this board was determined to be 240 and the corresponding density is 0.5 (120 in^2/240 = 0.5). This means that six layers are the estimate for this design.

Keep in mind that designs that require many different voltages may require additional voltage and ground planes. Additionally, the cross-sectional geometry must be balanced in copper weight and layer distribution to reduce warping.

6.1.1.5 Complete schematic. As the design progresses, keep in mind the partitioning of the circuit into functional units that will be allocated to the various daughter cards. As soon as the circuit design is reasonably complete, give rough drafts of the schematics to layout personnel to begin estimating signal layer requirements.

Given the area available and the rules for populating the area, the engineer must break up the system schematic and enter it into the design capture system. The raw schematic is input into the computer-aided design (CAD) design capture system. This system contains the library of parts and check algorithms to prevent the errors of multiple connections, unconnected signals, etc. The parts to be used are called up and placed on the workstation screen. The engineer enters the schematic signal by signal. The signal names and pin assignments are specifically entered. All inputs and outputs (I/Os) and test points are designated.

6.1.1.6 PCB layout.   The design is downloaded into the routing system in the form of a net list and parts library. The net list contains the point-to-point connect instructions for each specified signal. The library of parts contains the pin assignments of each part. It is important that the design capture system and the router use the same library. If not, mistakes are made because of revision differences or obsolete parts.

Most high density designs do not route completely under autoroute. The designer then analyzes the routing failures and attempts to complete the route interactively with the CAD system. Sometimes it is necessary to move parts or swap gates. The engineer must approve any changes to the placement or gate assignment. Many CAD routers do not consider stubs or branching. The engineer must be aware of the impact of interconnect on system performance. Newer routers are incorporating more and more high speed design practice.

6.1.1.7 Verify rule violations.   A review of the fully routed board is the next step in the layout process. Wire length and rule violation reports are used during this checkout. Critical net performance can now be analyzed based on actual layout. Continue the interactive process with layout until all rule violations and critical nets are verified.

6.1.1.8 PCB documentation package.   After final approval of the layout, the CAD engineer prepares the documentation package for board fabrication, assembly, and test. This will include the Gerber data, drill data, silk screen, surface-mount masks, solder masks, and aperture lists. The data are usually transmitted electronically to the PCB fabrication shop.

6.1.1.9 Fabricate prototypes. During this phase, producibility and functional problems are uncovered. The success of the prototype is largely dependent on the early simulation, analysis, and attention to detail.

6.1.2 PCB design tools

Table 6.2 lists some popular tools for computer-aided engineering (CAE) and CAD. This list is meant to be used only as a representative sample and not an exhaustive compilation. Also, the list only describes those products related to PCB CAE and CAD. Section 6.7.1 describes some of the advances in CAD and CAE tools.

6.2 Mechanical Design and PCB Materials

Printed circuit board costs are driven, to a large extent, by board size and shape, panel utilization, laminate material, and the number of layers.

6.2.1 Printed circuit board sizes

The PCB size is dictated by the mechanical design constraints of the system packaging. There are, however, industry standards and practices for rack and panel equipment.

The 19-in (482.6-mm) rack and panel equipment practice standard defines backplane and daughter card sizes based on the "U" which is 1.750 in (44.45 mm). Backplane board sizes, taken from International

TABLE 6.2   PCB Design Tools

Company	Product	Function	Platform
Cadence Design Systems	Allegro	PCD CAD	Unix, NT
	Specctra	Rules-driven autorouter	Unix, NT
	Specctraquest	Signal integrity analyzer	Unix, NT
	Concept	Schematic capture	Unix
Mentor Graphics	Board Station	PCB CAD, schematic capture, autorouter	Unix
OrCAD, Inc	OrCAD	Schematic capture	NT
	OrCAD Layout	PCB CAD	NT
VeriBest	Ascent	PCB CAD (links to Mentor Graphics tools)	NT
	Pinnacle	Autorouter	NT
	Signal Analyzer	Signal integrity analyzer	NT
Zuken-Redac	CADstar	PCB CAD, schematic capture, autorouter	NT

Electrotechnical Commission (IEC) IEC 297-3, are shown in Table 6.3 with the most common heights being 3U, 6U, and 9U.

Standard daughter card sizes are shown in Table 6.4, taken from IEC 297-3, IEC 297-4, Institute of Electrical and Electronics Engineers (IEEE) IEEE 1101.1, and Institute for Interconnecting and Packaging Electronic Circuits (IPC) IPC-PD-335. The depth increases by increments of 60 mm (2.362 in).

Standard rear daughter card sizes for a midplane application are shown in Table 6.5, taken from IEEE 1101.11. The depth increases by increments of 20 mm (0.787 in) for boards under 160 mm (6.300 in), and by 60 mm (2.362 in) for cards over 160 mm (6.300 in).

6.2.2   Board thickness

Typical daughter cards in electronic equipment have a thickness of 0.062 to 0.093 in (1.6 to 2.36 mm). Backplane thickness range is from 0.093 in (2.36 mm) to more than 0.300 in (7.62 mm). The particular packaging standard being utilized and whether the application is a backplane or a midplane can limit board thickness.

6.2.3   Panel sizes

Panel sizes can be divided into three groups: full-size, primary, and secondary panel sizes. The full-size panel is the largest manufactured size. Table 6.6 shows the common full-size panels. For the most part, equipment, material, and human engineering cannot handle full-size panels; therefore, the primary panel size is used. The primary panel size is the largest submultiple of a full-size panel. The most common size primary panel is 457 × 610 mm (18 × 24 in).

The secondary size panels are again submultiples of the full-size panels, not of the primary panel. Table 6.7 shows common secondary panel sizes. Table 6.8 shows board size recommendations as a function of panel size. The boards have a 25.4-mm (1-in) border around the boards and test coupons.

6.2.4   Mechanical partitioning

There are slightly different issues to consider when designing backplanes and daughter cards. The mechanical partitioning of the backplane should take into consideration the following factors:

- Space requirements for components, including connectors, devices, heat sinks, power supplies, I/O devices, etc.

TABLE 6.3 Standard Backplane Sizes [mm (in)]

	2U	3U	4U	5U	6U	7U	8U	9U	10U	11U	12U
Height	84.25 (3.317)	128.70 (5.067)	173.15 (6.817)	217.60 (8.567)	262.05 (10.317)	306.50 (12.067)	350.95 (13.817)	395.40 (15.567)	439.85 (17.317)	484.30 (19.067)	528.75 (20.817)
Width						≤84 × 5.08(0.2)					

TABLE 6.4 Standard Daughter Card Sizes [mm (in)]

	2U	3U	4U	5U	6U	7U	8U	9U	10U	11U	12U
Height	55.55 (2.187)	100.00 (3.937)	144.45 (5.687)	188.90 (7.347)	233.35 (9.187)	277.80 (10.937)	322.25 (12.687)	366.70 (14.437)	411.15 (16.187)	455.60 (17.937)	5 00.05 (19.687)
Depth						100.00 (3.937) 160.00 (6.300) 220.00 (8.661) 280.00 (11.024) 340.00 (13.386) 400.00 (15.748)					

TABLE 6.5 Standard Rear Daughter Card Sizes [mm (in)]

	2U	3U	4U	5U	6U	7U	8U	9U	10U	11U	12U
Height	55.55 (2.187)	100.00 (3.937)	144.45 (5.687)	188.90 (7.347)	233.35 (9.187)	277.80 (10.937)	322.25 (12.687)	366.70 (14.437)	411.15 (16.187)	455.60 (17.937)	500.05 (19.687)
Depth						60.00 (2.362) 80.00 (3.150) 100.00 (3.937) 120.00 (4.724) 140.00 (5.512) 160.00 (6.300)					

TABLE 6.6   Common Full-Size Panels—Domestic and Foreign [mm (in)]

Length	Width
914 (36)	1219 (48)
1219 (48)	1829 (72)
1219 (48)	3658 (144)
1219 (48)	1067 (42)
914 (36)	1067 (42)
1000 (39.4) [European, Far East]	1000 (39.4)
1000 (39.4) [European, Far East]	1200 (47.2)
1055 (41.5) [European, Far East]	1165 (45.9 0)

TABLE 6.7   Common Secondary Panel Sizes [mm (in)]

Length	Width
457 (18)	533 (21)
406 (16)	457 (18)
356 (14)	457 (18)
305 (12)	457 (18)

- Keep-out areas for routing and components due to mounting hardware, including mounting rails, stiffening bars or rails, screws, washers

- Guidance features for daughter cards

- Space required for cable routing, through the board or on/off of the board

- Space around slot connectors for EMI daughter card boots

The mechanical partitioning of the daughter cards should take into consideration the following factors:

- Space required for components (including height, for board-to-board spacing) such as connectors, devices, heat sinks, power supplies, I/O devices, etc.

- 2.5-mm (0.100-in) no routing or component keep-out area spacing on top and bottom of card for card guide or electrostatic discharge (ESD) strips (IEC 297-3, IEEE 1101.1)

- Space on front edge of card for injector/extractor handles or PCB holders for front faceplates

- Component keep-out area for stiffening bars for large cards (usually greater than 6U high)
- Standoff hardware for mezzanine cards

TABLE 6.8   Recommended Board Sizes Based on Panel Size

Panel size [mm (in)]	Quantity of boards per panel	Board size [mm (in)]
457 × 610 (18 × 24)	1	406 × 559 (16 × 22)
457 × 533 (18 × 21)	1	406 × 483 (16 × 19)
457 × 610 (18 × 24)	2	267 × 406 (10.5 × 16)
457 × 610 (18 × 24)	3	178 × 406 (7 × 16)
457 × 533 (18 × 21)	2	234 × 406 (9.2 × 16)
406 × 457 (16 × 18)	1	356 × 406 (14 × 16)
457 × 533 (18 × 21)	3	152 × 406 (6 × 16)
406 × 457 (16 × 18)	2	196 × 356 (7.7 × 14)
356 × 457 (14 × 18)	1	305 × 406 (12 × 16)
356 × 457 (14 × 18)	2	196 × 305 (7.7 × 12)
305 × 457 (12 × 18)	1	254 × 406 (10 × 16)
406 × 457 (16 × 18)	3	127 × 356 (5 × 14)
305 × 406 (12 × 16)	1	254 × 356 (10 × 14)

6.2.5 Mechanical support

The mechanical support of a backplane is crucial for reliable, noninterrupted mating of the connector during daughter card insertion, normal operating conditions, and for disaster conditions. Backplanes should be supported with enough mounting hardware, rails, screws, and stiffening bars to allow proper mating of connectors without bowing. Daughter cards should have enough float in the card guides to account for board bow, warpage, board drilling tolerances, and connector tolerances to mate, but not enough to allow large vertical or horizontal motion that could cause intermittent failure or catastrophic failure if a card jumps out of its guides. The most rigid design criteria is required for earthquake-proof designs, network equipment building system (NEBS 4 testing), where a design should remain working through and after an earthquake.

Vibration analysis of a chassis system can be performed to improve on a design. Analysis consists of determining the first five harmonics and seeing if any of the harmonics could occur in everyday situations. If there is a problem, extra stiffening measures can be taken, such as changing or thickening the chassis, or adding mounting rails, gussets, or support beams.

6.2.6 Thermal analysis

The placement of components on the boards can drastically affect the thermal profile and efficiency of the system. High power components should be placed in free-flowing air, close to the air inlets or fans, to allow unhindered airflow with the coolest temperatures. Do not place such high power components behind tall ones or heat sinks. Do not line up a series of high power components with the airflow direction. Tall components create vacuums and eddy, and do not allow for good airflow or mixing. Heat sinks are the worst of both worlds, heating the air and also drastically changing the flow of the air. Lining up high power components in the airflow direction creates a scenario of ever-increasing air temperature flowing over the next one in line.

On the other hand, noncritical, or low to no power producing components, can be placed in areas that have little or no airflow, or near the air outlets, where the air temperature will be the highest. Flow analysis will show locations of low airflow, eddies, and obstructions. The failure criteria of the system should be well defined to determine whether the system will fail or pass. Some failure criteria are

- Maximum air temperature increase
- Maximum critical component temperature
- Maximum component junction temperature

Each type of failure criteria is a more detailed analysis from an overall system analysis to a macroscopic analysis between component and board. The minimum width and thickness of conductors is determined on the basis of the current-carrying capacity required, and the maximum conductor temperature rise. Figure 6.2 shows the conductor thickness and width for internal and external printed circuit board layers for safe operating temperatures. Figure 6.3 shows the temperature rise versus current for a 1-oz copper thickness.

6.2.7 Mechanical ICD

The interface control drawing (ICD) relates the mechanical chassis constraints to the board connector constraints. ICDs ensure proper

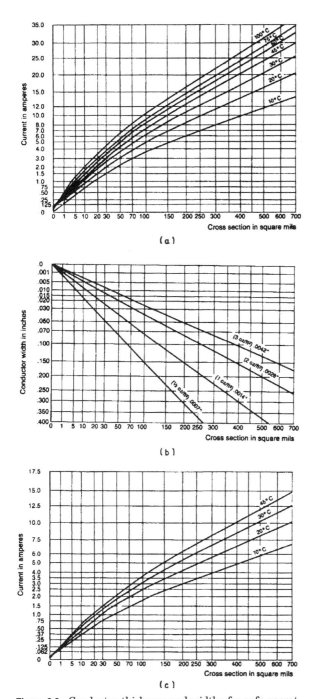

Figure 6.2 Conductor thickness and widths for safe operating temperatures (a) and (b) for external layers, (c) for internal layers.

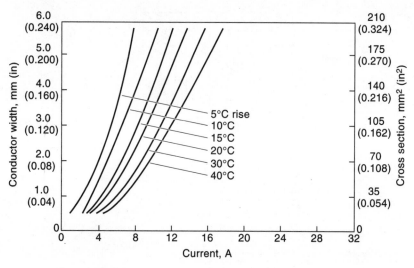

Figure 6.3 Temperature rise versus current for 1-oz Cu.

slot spacing, daughter card alignment (both horizontal and vertical), and component placement.

An ICD shows a fully dimensioned layout for the board designer, including overall board dimensions, mounting holes, component and routing keep-out areas, cutouts, and essential components. The essential components are any items that must be placed in a defined location for mechanical alignment and compatibility such as, but not limited to, slot connectors, guidance features, power connectors, fan connectors, etc. Other components, such as resistors, capacitors, I/O connectors, etc., whose location is not critical to proper mechanical alignment of the overall system, are not shown, but rather are left to the board designer to place.

Figure 6.4 shows an example of an ICD for a 6U × 84HP backplane, and Fig. 6.5 shows an example of an ICD for a 6U × 160-mm daughter card. The 0,0 location can be either a corner of the board or from a mounting hole, depending on the board designer or the program being used for routing. Pin 1 locations are used for dimensioning components. Keep-outs are labeled and dimensioned. Component outlines are placed to show the slot spacing and available space for placement.

6.2.8 PCB materials

The proper laminate and prepreg for a specific application must take into account the electrical, mechanical, chemical, and thermal properties of the material.

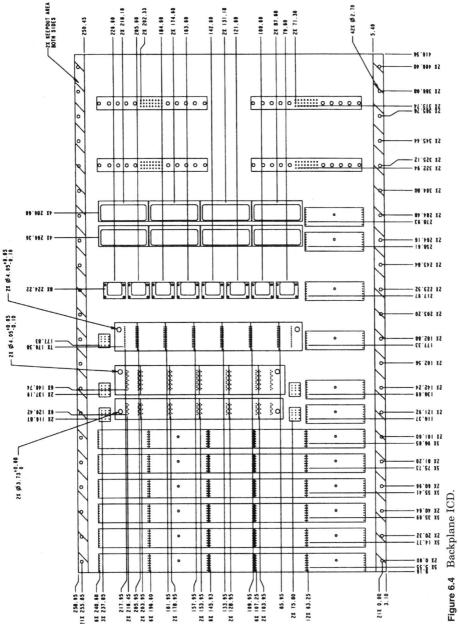

Figure 6.4 Backplane ICD.

6.15

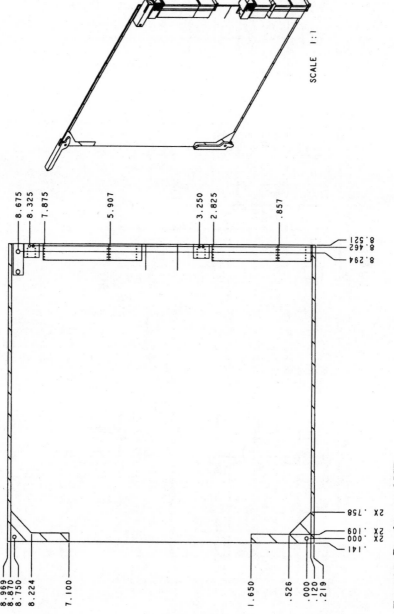

Figure 6.5 Daughter card ICD.

6.2.8.1 Electrical properties. The increasing speeds in digital circuits and microwave applications have raised concerns about lowering the dielectric constant and dissipation factor. The dielectric constant, ε_r is defined as the ratio of the material capacitance, C_x, to the capacitance of an air dielectric, C_v, for a given construction. It should be noted that ε_r is frequency dependent, and will vary according to the frequency at which the capacitance is measured.

The dissipation factor at high frequencies is defined by the ratio of the effective series resistance of a capacitor to the capacitive reactance. Table 6.9 summarizes the electrical properties of epoxy-glass (E-glass) laminates. Other electrical properties that give an indication of the electrical integrity of the laminate are electric field strength and dielectric breakdown.

6.2.8.2 Mechanical properties. The important mechanical properties of copper-clad laminates are the degree of copper adhesion to the laminate and flexural strength aging. The flexural strength test measures the time it takes for the laminate at elevated temperatures to lose 50% of its original flexural strength. The copper adhesion at ambient conditions for 1-oz copper foil is 140 to 180 kg/m (8 to 10 lb/in) for many of the resin systems (see Table 6.10).

6.2.8.3 Chemical properties. The laminate chemical resistance is typically measured by a methylene chloride test and a water absorption test (see Table 6.11). The sensitivity of a material to water is a good

TABLE 6.9   Typical E-Glass Electrical Properties

Material	Dielectric constant at 1 MHz*'†	Dissipation factor at 1 MHz*'†	Electric field strength‡ kV/mm	V/mil	Dielectric breakdown‡ (kV)
FR-4 epoxy	4.10–4.20	0.028–0.030	48–56	1200–1400	70–75
Polyfunctional FR-4	4.10–4.20	0.028–0.030	48–56	1200–1400	70–75
High-temperature, one-component epoxy system	4.45–4.55	0.020–0.022	36–44	900–1100	70–75
Bismaleimide triazine epoxy	3.85–3.95	0.011–0.013	48–56	1200–1400	70–75
Polyimide epoxy	4.00–4.10	0.011–0.013	48–56	1200–1400	70–75
Cyanate ester	3.50–3.60	0.0045–0.0065	32–40	800–1000	65–70
Polyimide	3.95–4.05	0.008–0.010	48–56	1200–1400	70–75
PTFE	2.45–2.55§	0.001–0.003§	32–40	800–1000	40–45

*Two-ply 1080 laminate at 57% resin content.
†Conditioning with humidity, for 24 h, at 23°C (73°F).
‡Conditioning with immersion in distilled water, for 48 h, at 50°C.
§Laminate at 73% resin content.

TABLE 6.10   **Typical E-Glass Laminate Mechanical Properties**

Material	1-oz copper adhesion, ambient*		1-oz copper adhesion*'†		Flexural aging‡
	kg/m	lb/in	kg/m	lb/in	
FR-4 epoxy	180–195	10.0–11.0	150–170	8.5–9.5	24–48
Polyfunctional FR-4	140–160	8.0–9.0	107–125	6.0–7.0	48–72
High-temperature, one-component epoxy system	160–180	9.0–10.0	140–160	8.0–9.0	168–192
Bismaleimide triazine epoxy	90–107	5.0–6.0	80–100	4.5–5.5	168–192
Polyimide epoxy	150–170	8.5–9.5	125–140	7.0–8.0	Not available
Cyanate ester	90–107	5.0–6.0	80–100	4.5–5.5	288–312
Polyimide	140–160	8.0–9.0	140–160	8.0–9 .0	>6000
PTFE	180–215	10.0–12.0	140–180	8.0–10.0	Not available

*Tested on multilayer laminates.
†Conditioning at temperature for 1 h at 125°C (255°F).
‡Hours at 200°C (390°F) to lose 50% of original flexural strength.

indicator of the degree of control required to produce consistent high quality laminates with good thermal intensity.

6.2.8.4 Thermal properties.   The three laminate thermal properties are the glass transition temperature T_g, thermal resistance, and the UL flammability rating (see Table 6.12). The glass transition temperature, T_g, is the point at which an amorphous polymer changes from a hard

TABLE 6.11   **Typical E-Glass Laminate Chemical Properties**

Material	Methylene chloride absorption,* % weight gain	Water absorption†,‡
FR-4 epoxy	7.0–8.0	1.10–1.20
Polyfunctional FR-4	3.0–4.0	1.10–1.20
High-temperature, one-component epoxy system	5.0–6.0	0.40–0.60
Bismaleimide triazine epoxy	1.0–2.0	0.80–0.90
Polyimide epoxy	0.10–0.30	1.15–1.25
Cyanate ester	0.40–0.60	0.60–0.70
Polyimide	0.05–0.25	1.40–1.50
PTFE	Not available	0.20–0.30

*0.13-mm (0.005-in) laminate made with two-ply 1080 style.
†1.6-mm (0.062-in) laminate made with 7628 style.
‡Conditioning with immersion in distilled water, for 24 h at 100°C (212°F).

and relatively brittle condition to a viscous condition. In this temperature region, many physical properties, such as hardness, brittleness, and thermal expansion, undergo significant and rapid changes.

The length of time it takes for copper-clad laminates to blister on 290°C (550°F) molten solder is a measure of thermal resistance. FR-4 laminates blister in about 175 to 200 s. Most resin systems have a UL flammability rating of V-0.

6.3 Electrical Design Considerations

This section examines impedance, cross-talk, and signal distribution as they relate to the PCB. Trade-offs between the three must be considered for a successful design.

6.3.1 Impedance

Impedance relates to all system parameters including delay, noise, and noise tolerance. Advantages and disadvantages of microstrip and strip-line configurations are analyzed and relationships of physical parameters to impedance are discussed.

6.3.1.1 Characteristic impedance. Characteristic impedance Z_0 is the single most important interconnection system parameter. The physical geometry of the circuit card and the system signal distribution properties are directly related to impedance. Some important relationships are listed here:

Line width (w). As line width increases, Z_0 decreases.

Dielectric spacing (h). As a line gets farther from the ground plane, Z_0 increases.

TABLE 6.12   Typical E-Glass Laminate Thermal Properties

Material	Glass transition temperature* °C	Glass transition temperature* °F	Solder blister(s) at 290°C (550°F)*	Flammability*
FR-4 epoxy	125–135	255–275	175–200	V-0
Polyfunctional FR-4	140–150	285–300	225–250	V-0
High-temperature, one-component epoxy system	170–180	340–355	300–325	V-0
Bismaleimide triazine epoxy	180–190	355–375	300–325	V-0
Polyimide epoxy	250–260	480–500	300–325	V-0
Cyanate ester	240–250	465–480	750–800	V-0
Polyimide	>260	>500	>1200	V-1
PTFE, melting point	327	620	>1200	V-0

*1.6-mm (0.062-in) laminate made with style 7628 at 40% resin content.

Dielectric constant (ε_r). As ε_r increases, Z_0 decreases.

Propagation delay (t_{pd}). Propagation delay for an unloaded line (no gate loads connected) is a function of ε_r. As ε_r increases, t_{pd} also increases (that is, signals propagate more slowly). t_{pd} does increase as distributed gate loads are added and the amount of increase is a function of impedance. The lower the Z_0, the less t_{pd} is affected by gate loading. This is a very important effect.

Intrinsic capacitance (C_0). As impedance decreases, C_0 increases. A 100-Ω system typically exhibits approximately 1.5 pF/in, whereas a 50-Ω system exhibits approximately 3 pF/in. In lumped nets, higher capacitance can mean longer delays due to the RC charging effect.

Cross-talk. As impedance decreases, cross-talk noise also decreases. This is due to the close proximity of the signal traces to the reference planes in a low impedance construction.

Switching noise. Switching noise is a direct function of the amount of transient current switched when a gate changes logic states along with inductances inherent in the power distribution system. Since a lower impedance system draws more transient current, switching noise increases as impedance decreases. In situations where connector noise is marginal, higher impedance may solve the problem. There are producibility issues that also need to be considered (for example, board thickness and line width constraints).

Noise tolerance. As impedance increases, system noise tolerance tends to get better but cross-talk gets worse.

As can be seen from the preceding relationships, there are many interdependent factors and trade-offs involved in choosing a system Z_0. For example, lower impedance yields a quieter system with respect to cross-talk at the expense of increased loading and delay due to the inherently higher intrinsic capacitance. These trade-offs are analyzed in the next section.

6.3.1.2 Impedance effects. The effects of impedance on delay, noise, and noise tolerance are covered in this section.

Delay. Predicting delay is one of the most important aspects of characterizing a particular net. Impedance effects contribute delay adders in four distinct ways depending on net configuration:

1. Voltage divider effect of gate output impedance versus line impedance adds delay.
2. Lumped load capacitance RC charging delay.
3. Discrete transmission line delay due to reflections at a discontinuity.

4. Distributed transmission line delay due to distributed capacitance lowering Z_0.

These delay effects are explained in this section to gain a good understanding of their importance.

Driver loading delay.   Every driver has finite output impedance. The voltage amplitude initially put into the line is a function of the voltage divider consisting of Z_0 and Z_D (see Fig. 6.6):

$$V_{\text{line}} = V_S \left(\frac{Z_0}{Z_0 + Z_D} \right)$$

For any driver output impedance, Z_D, the input voltage to the line is reduced as Z_0 gets smaller. Note that V_S is not necessarily equal to V_{CC} due to resistive drops, but these are typically minimal. Since the rise time of the signal, t_r, remains the same, the amount of time required for the line voltage to reach the receiver threshold increases as the line voltage decreases. Thus, a delay factor is added, as shown in Fig. 6.6. This factor becomes increasingly important as lines get electrically longer and take on transmission-line characteristics.

Lumped load capacitance charging delay.   When lines are electrically short, the driver essentially sees a single capacitive load composed of gate load capacitance, trace capacitance, and connector capacitance. The charging time of this "capacitor" is the lumped delay factor.

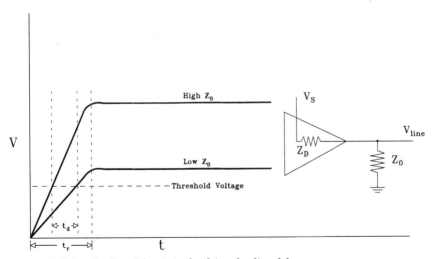

Figure 6.6   Driver loading delay. t_d is the driver loading delay.

Lumped load delay is essentially independent of impedance. This is because the lines are short. The delay is really more a function of the driver's ability to drive a capacitive load.

Discrete transmission line delay. When gates are spaced far apart $[(t_{pd} \times 1) > t_r/2]$ on an electrically long line, reflections are generated at each gate input which represents a capacitive discontinuity. A delay is created since a portion of the signal is reflected back, decreasing its amplitude and increasing the amount of time required to reach the threshold of subsequent gates. This delay is expressed as $T_D = (C_D \times Z_0)/2$ where C_D = the capacitance of the discontinuity.

Note that as impedance increases and/or the gate capacitance increases, the delay becomes longer. This can be understood by realizing that as Z_0 increases, the intrinsic capacitance, C_0, of the line goes down so the same gate load capacitance represents a larger relative discontinuity, causing a larger reflection and more delay.

Distributed transmission-line delay. When gate loads are spaced on short segments $[(t_{pd} \times 1) < t_r/2]$ on an electrically long line, the reflections from the gate loads tend to merge together. The gate load capacitance, C_L, and the line capacitance, C_0, add together and appear to the driver to be uniformly distributed on the line. *The capacitance/unitlength of the line increases and acts to lower the line impedance, thereby increasing propagation delay.* This is the primary delay adder in buses.

This new "effective" impedance

$$Z_{\text{eff}} = \frac{Z_0}{\sqrt{1 + C_L/C_0}}$$

The additional delay factor

$$T_L = t_{pd} \sqrt{1 + C_L/C_0}$$

where Z_{eff} = new effective impedance (Ω)
  Z_0 = unloaded impedance (Ω)
  C_L = total load capacitance per unit length (pF/in)
  C_0 = unloaded intrinsic capacitance (pF/in)
  t_{pd} = unloaded prop delay (ns/in)
  T_L = the new prop delay as a function of loading (ns/in)

Note that the more loads distributed on the line, the lower the effective impedance and the greater the delay.

A low impedance system is less sensitive to the effects of gate loading. Its impedance and delay change a smaller percentage than a high impedance system with equivalent loading. This is because the gate loads represent a smaller percentage of the total net capacitance in a low impedance system in which C_0 is relatively high. For example,

10-pF loads spaced 2 in apart (5 pF/in) on a 100-Ω line drop the effective impedance to 48 Ω (52% drop), whereas the same loading on a 50-Ω line drops Z_{eff} to only 31 Ω (38% drop).

Noise. Three types of noise are of concern to the designer: reflection noise, switching noise, and cross-talk.

Reflection noise. Reflections occur when signals propagating down a line encounter an abrupt change in impedance (see Sec. 6.3.3.2 for more detail on reflections). This will occur between daughter cards and backplanes having dissimilar impedances. Energy is reflected back from the point of mismatch and becomes noise propagating throughout the system. For very fast edges ($\leq$250 ps), even right-angle trace bends represent impedance discontinuities and create reflection noise.

These noise pulses can cause data errors if the magnitude of the reflection is sufficient to reach the gate threshold regions. Large reflections can also add substantial delays, as described in the previous section.

Gate loads also represent discontinuities causing reflected noise. The magnitude of the discontinuity must be controlled to prevent excessive noise and the possibility of false switching.

Two important design guidelines will help to minimize the effects of reflected noise:

1. All system PCBs and interconnecting cables should have matched Z_0 within 10 to 15%. If daughter board stubs are electrically short, they're seen as essentially capacitive and impedance matching to the backplane is not necessary.
2. Limits must be placed on maximum gate loading at any given point on a line.

Switching noise. Switching noise is caused by inductances inherent in the power distribution system and the rate at which the drivers switch signal-line currents. As impedance gets lower, more current is drawn from the driver. As it switches, the di/dt across these inductances creates spikes. Increasing driver currents cause this "switching noise" to increase.

Unfortunately, impedance has the opposite effect on switching noise than cross-talk. Lowering Z_0 in an effort to reduce cross-talk may just shift the problem to one of switching noise. This is further complicated by the fact that cross-talk and switching noise can be additive.

Two important design guidelines will help to minimize the effects of switching noise:

1. Minimize the power distribution inductances.
2. Effectively decouple switching gates as close as possible to the V_{CC} and ground pins. This will supply the necessary transient currents through the shortest possible inductive path.

Section 6.4.2 outlines the most effective means of accomplishing these guidelines.

Cross-talk.   Cross-talk is discussed in detail in Sec. 6.3.2.

Noise tolerance. *Noise tolerance* is the degree to which a gate is impervious to noise spikes at its input. This tolerance is usually expressed as a direct current (DC) noise margin. *Noise margin* is defined as the difference between the worstcase input logic level (V_{IH}min or V_{IL}max) and the guaranteed worst-case output (V_{OH}min or V_{OL}max) specified to drive these inputs (see Fig. 6.7).

Impedance directly affects V_{OH}min by changing the initial voltage step driven into the line. This is due to the voltage divider effect previously discussed in Sec. 6.3.1.2. Lower impedance reduces the initial V_{OH}min while not affecting V_{IH}min and, therefore, "shrinks" the high level noise margin as shown in Fig. 6.7. Higher impedances (100-Ω range) allow the driver to produce a larger initial voltage and thus preserve the noise tolerance.

Another consideration is that, at higher impedances, a larger reflection takes place at each discontinuity for a given loading. Thus, the remainder of the wave traveling past this discontinuity is of smaller amplitude, which reduces V_{OH} and lowers the noise margin of subsequent gates.

6.3.1.3 Microstrip configurations.

Microstrip is the name given to a signal line referenced above a single ground or power plane (Fig. 6.8). Although referencing to ground planes is generally discussed, from an alternating current (AC) standpoint a well-bypassed power plane is also at ground potential and makes an effective reference plane.

Height and line width has the most impact on microstrip impedance. Line width is usually chosen to enhance routability so height must be adjusted to provide the desired impedance.

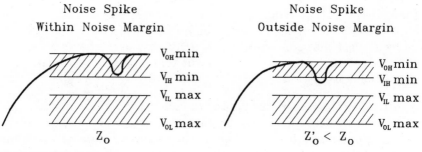

Noise Spike   Noise Spike
Within Noise Margin   Outside Noise Margin

V_{OH}min V_{OH}min
V_{IH} min V_{IH} min
V_{IL} max V_{IL} max
V_{OL}max V_{OL}max

Z_0 $Z'_0 < Z_0$

Figure 6.7 Noise margins.

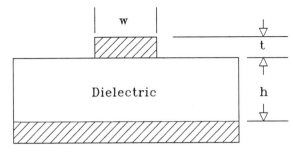

Reference Plane

Figure 6.8 Microstrip cross section.

Intrinsic capacitance C_0 is a function of impedance and propagation delay. The following equation can be used to determine capacitance per inch for any board geometry:

$$C_0 = \frac{t_{pd}}{Z_0} \times 1000$$

where C_0 is measured in picofarads per inch
Z_0 is measured in ohms
t_{pd} is measured in nanoseconds per inch

Microstrip configuration advantages are

1. Lines are on the surface and visible so they can be cut and jumpered easily for rework. Keep in mind that the jumpers introduce impedance discontinuities that can impact performance of high speed nets.
2. Higher impedances (>70 Ω) can be easily designed by adjusting height h from the reference plane.
3. It can be easily fabricated in a producible manner.
4. Empirical testing has indicated that solder mask or conformal coatings have negligible effect on microstrip impedances.

Microstrip configuration disadvantages are as follows:

1. Surface lines are subject to physical damage (nicks, cuts, and solder splashes).
2. Radiated emissions are about 15 dB higher than from strip line. They are also about 15 dB more susceptible to radiated noise in the 30-MHz to 1-GHz range. Thus, coupling onto nearby low level signals can be increased up to 30 dB.
3. Parallel traces in a microstrip configuration will exhibit more coupling and cross-talk.

Embedded microstrip.   In many cases, two consecutive signal layers are stacked on top of a reference plane. At least one signal layer in each pair is now completely embedded in the dielectric material at a specific distance below the surface (see Fig. 6.9). Surface mounting generally requires use of "pads only" outer layers, making embedded microstrip a common configuration.

The impedance of microstrip signal layers completely buried in dielectric material (surrounded by a homogeneous dielectric) is lower compared to a surface microstrip spaced the same distance from the reference plane. The degree of impedance lowering is a function of the distance below the surface.

As a consequence of the impedance-lowering effect in embedded microstrip, it is hard to closely match impedances in a pair of signal layers where one is embedded and one is not. Therefore, a "pads only" outer layer is preferred so as to embed both signal layers in dielectric. Line widths can be adjusted between surface and embedded microstrip layers to better match impedances if "pads only" layers are not used.

Embedded microstrip advantages are that it

1. Allows inclusion of a second signal layer in addition to surface microstrip.
2. Helps when many signal layers are required in a constrained board thickness.

Embedded microstrip disadvantages are:

1. Radiated emissions and susceptibility are about 15 dB higher than in strip line.
2. Signal propagation is slower than in surface microstrip.

Strip-line configurations.   *Strip line* is the name given to a signal line referenced between two ground planes at a defined spacing (Fig. 6.10).

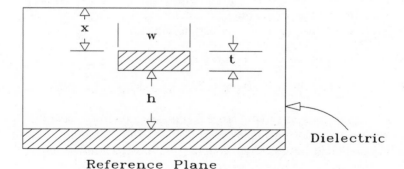

Reference Plane

Figure 6.9 Embedded microstrip cross section.

Although referencing to ground planes is generally discussed, from an AC standpoint a well-bypassed power plane is also at ground potential and makes an effective reference plane.

Spacing h and line width w have the most impact on strip-line impedance. Line width is usually chosen to maximize routability so an appropriate spacing will provide the desired impedance. However, higher impedances ($>70\ \Omega$) require large spacings which are not usually practical due to board thickness constraints. Thus, used strip-line configurations are usually used in systems with dense interconnection requirements, where low impedance (50 to 60 Ω) is unavoidable.

Strip-line configuration advantages are

1. Signals are very quiet and exhibit less cross-talk than microstrip.
2. Printed circuit process allows very accurate line definition for better impedance control.

Strip-line configuration disadvantages are

1. Signals are buried so they are difficult to repair or rework during test.
2. Signal propagation time is somewhat slower than in surface microstrip.

Dual strip line. It is common practice to embed two perpendicular signal lines on adjacent planes between reference planes to create a dual strip line (Fig. 6.11). This arrangement maximizes interconnection density. The layers are routed perpendicularly to reduce cross-talk noise. Generally, some parallelism is required to accomplish routing and a rule-of-thumb is to allow about a 0.25- to 0.5-in (0.64 to 1.27 cm) maximum layer-to-layer parallelism.

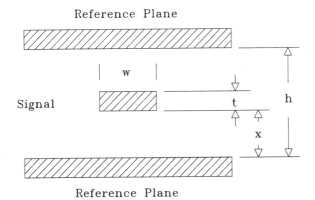

Figure 6.10 Strip-line cross section.

Reference Plane

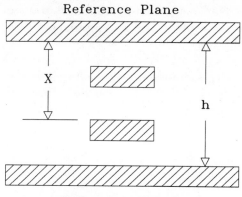

Reference Plane

Figure 6.11 Dual strip-line cross section.

Most bused backplanes use single strip lines since the signal layers use horizontal routing. Dual strip lines would result in very high layer-to-layer coupling.

As can be seen in Fig. 6.11, this configuration will create strip lines that are not centered with respect to the reference planes. Because the signal trace is closer to one reference plane than the other, its impedance will be somewhat lower than if it was centered.

6.3.1.4 System impedance.
From a practical standpoint, both mechanically and electrically, impedances in the range of 50 to 75 Ω are the most effective for high-speed digital systems.

Impedances higher than 75 Ω are typically difficult to implement on multiple signal layers in boards with 0.062- or 0.093-in thickness constraints. Additionally, these higher impedances are more susceptible to cross-talk effects, so more considerations need to be given to controlling parallelism. These parallelism restrictions can sometimes affect the routability of dense layouts.

As the number of logic devices per board increases and interconnect density becomes very high, mechanical constraints on overall board thickness warrant a strip-line approach and impedances in the 50-Ω range become more common.

6.3.2 Cross-talk

This section outlines the basic nature of cross-talk and explains its relationship to impedance, board geometry, and risetime. Specific guidelines for controlling worst-case system cross-talk are presented.

6.3.2.1 The nature of cross-talk.
Cross-talk is defined as the undesired energy imparted into a transmission line due to signals in adjacent

lines. Cross-talk magnitude is dependent on risetime, signal-line geometry, and net configuration (type of terminations, etc.).

Figure 6.12 indicates two signal lines in close proximity that are capacitively coupled (C_M) and inductively coupled (L_M). Both lines have the same characteristic impedance, Z_0, and are fully terminated to avoid complications due to reflections. One line is active and transmits a digital pulse while the other is passive.

Due to the mutual capacitive coupling and the fact that an equal capacitance is "seen" looking either way into the passive line, two equal and opposite currents, I_C, are induced into the passive line. At the same time, an inductive current due to L_M is also induced but only in one direction. It travels in the opposite direction of the active line source pulse.

At the source end (point B) of the passive line, I_C and I_L are additive. These summed currents produce a voltage drop in the *same polarity* as the source voltage. This voltage drop is termed *backward cross-talk* since it travels in the opposite direction than the source pulse.

At the far end (point F) of the passive line, I_C and I_L are of opposite polarity and thus subtractive. These subtractive currents, $(I_L - I_C)$, produce a voltage drop termed *forward cross-talk* since they travel in the same direction as the source pulse.

Normally, $I_L > I_C$, so the forward cross-talk is opposite in polarity from the source voltage. This is the case in microstrip configurations whose signal lines are in two different (nonhomogeneous) dielectric materials: epoxy-glass and air (or solder mask). If both lines are

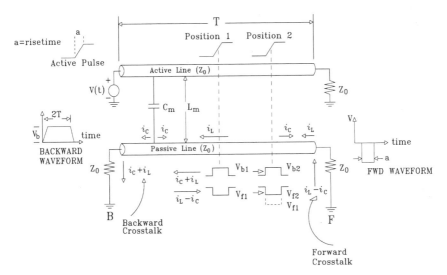

Figure 6.12 Cross-talk between two parallel lines.

completely surrounded by the same dielectric material (that is, they are homogeneous), I_L and I_C have equal magnitudes and cancel at the far end. This is why strip-line configurations where lines are completely embedded in dielectric exhibit no forward cross-talk. This is also true of embedded microstrip when the signals are buried under more than 20 mils (0.5 cm) of dielectric.

The ground plane and its proximity to the signal lines have a very definite effect on cross-talk magnitudes. The closer the signal line is to the ground plane (that is, the lower the impedance), the lower the cross-talk. This is because the shielding effect of the ground mass tends to reduce the mutual coupling terms, C_M and L_M.

Forward cross-talk. At any instant on the line, the active pulse induces two voltages into the passive line. One is the backward cross-talk voltage (produced by $I_C + I_L$) and the other is the forward cross-talk voltage (produced by $I_L - I_C$).

As the active pulse moves from position 1 to position 2 (Fig. 6.12), the forward pulse induced at position 1 (V_{f1}) also moves to position 2. The new forward pulse induced at position 2 (V_{f2}) is added to the one produced at position 1. These forward pulses continue to add directly on the passive line as the source wavefront travels down the active line. *Therefore, when the end of the line is reached (point F), the forward waveform has an amplitude directly proportional to the length of the line and in opposite polarity to the signal voltage. Its pulse width is approximately equal to the source voltage risetime.*

Backward cross-talk. As the source voltage travels toward point F, it also continues to induce backward cross-talk pulses (V_b). It generates the first pulse at the beginning of the line and the last pulse at the end of the line during a total time, T, which is the line delay. Thus, the string of backward pulses follows each other sequentially down the line toward point B where they add together. *The last backward pulse induced at time T at the end of the line takes another time T to travel back to the beginning so the total width of the summed backward pulses at point B is 2T, twice the total line delay.*

The amplitude, however, is independent of line length (unlike the forward cross-talk) as long as the source pulse risetime is less than the total line delay. (This is the familiar line length relation, $t_r/2$, seen in Sec. 6.3.3, "Signal Distribution"). For lines that are "electrically short," that is, $(t_{pd} \times 1) < t_r/2$, backward cross-talk amplitude is directly a function of line length.

Cross-talk effects

1. Mutual capacitive and inductive coupling between signal lines produces cross-talk.

2. Moving a ground plane closer to the signal line (that is, lowering Z_0 and increasing C_0) reduces mutual coupling and, therefore, reduces cross-talk amplitudes.
3. Induced currents of opposite polarity produce forward cross-talk: $I_L - I_C$. It occurs at the far end (furthest from the driver) of the passive line in nonhomogeneous constructions such as microstrip.

   Amplitude. Directly proportional to line length.
   Polarity. Normally opposite of signal source.
   Pulse width. Approximately equal to active source risetime.
4. Backward cross-talk is produced by induced currents of the same polarity: $I_C + I_L$. It occurs at the near end (driver end) of the passive line.

   Amplitude. Independent of line length and t_r when the line is "electrically long." Directly proportional to line length and t_r when the line is "electrically short."
   Polarity. Same as signal source.
   Pulse width. Twice the line delay.

6.3.2.2 Backward coupling coefficient. A *coupling coefficient* represents the percentage of the source voltage that is induced into the passive line, K_B for backward cross-talk and K_F for forward cross-talk. These coupling coefficients include both the inductive and capacitive components. They are, therefore, a function of line geometry and proximity to reference planes.

In practical digital systems, backward cross-talk is the worst case and, therefore, the limiting factor to parallelism. Also, keep in mind that in strip lines, forward cross-talk is zero. For this reason, only the backward coupling coefficient, K_B, is considered here and forward cross-talk is not considered a problem.

6.3.2.3 Worst-case cross-talk. As mentioned previously, backward cross-talk is considered the worst case. K_B is larger than K_F for most geometries since backward cross-talk is the *sum* of currents whereas forward cross-talk is the *difference* between the same two currents. Thus, for worst-case design, if the backward cross-talk is controlled, forward cross-talk need not be considered.

The situation depicted in Fig. 6.12 has both lines fully terminated to avoid reflections. In practice, many lines are only partially terminated or essentially open-ended as when connected to high impedance gate inputs. Reflections can increase cross-talk amplitudes and create worst-case situations. When lines are fully terminated ($R_T = Z_{\text{eff}}$), use K_B in calculating cross-talk. When lines are unterminated or partially terminated, use $2K_B$ to be conservative.

The worst-case configuration of parallel lines is when they are driven in opposite directions. In this configuration, percentage coupling can reach $2K_B$. In practical systems, the parallelism restriction should be formulated assuming the worst-case coupling, or $2K_B$.

6.3.2.4 Cross-talk control. *First and foremost: keep lines short!* The engineer must oversee the layout as it progresses to keep an eye on critical nets:

1. Design assuming a worst-case coupling coefficient of $2K_B$. Analyze critical individual nets to determine if $2K_B$ or just K_B is warranted. Termination will reduce cross-talk amplitudes.

2. Make sure that traces are not routed over clearances in reference planes. This will let traces couple noise even though isolated by planes.

3. Use the maximum edge-to-edge spacing between lines that the layout density will allow. Look for areas in the layout where lines are parallel when they do not have to be.

4. Use fast risetime devices only if necessary for performance in specific nets. Short risetime equals increased cross-talk. This will also reduce EMI emissions.

5. For adjacent signal layers (that is, dual strip line), signals must be routed orthogonally. Limit maximum adjacent layer parallelism to 0.75 in (2 cm) for transistor-transistor logic (TTL), 0.5 in (1.2 cm) for complementary MOS (CMOS), 0.25 in (0.7 cm) for emitter-coupled logic (ECL), and 0.1 in (0.3 cm) for ECLinPS to avoid cross-talk problems.

6. Avoid parallel runs of high speed logic parallel to low level analog circuits. If analog and digital circuits must share the same card, they must be completely segregated. Sandwich analog signals between analog power distribution layers and, if at all possible, place an earth guard ring around analog circuits on every layer where there is analog track.

7. Terminate if possible. Even partial termination will reduce reflections, resulting in decreased cross-talk.

8. Keep data buses separate from address buses. Keep all buses physically separate from clocks and strobes, especially through card edge connectors. "Separate" is a term that depends on risetime.

9. The 3-W rule that was first defined by EMC Expert W. Michael King states that the distance separation between critical traces must be three times the width of the traces measured from centerline to centerline. As an example, assume that a clock trace is 5 mils (0.13 mm) wide. No other trace can exist within a minimum of 2×5 mils (2×0.13 mm) of this trace, or 10 mils (0.26 mm), edge to edge.

10. The use of thin traces will always reduce cross-talk. This is because finer lines allow the ground plane to be closer while still maintaining the same impedance. Additionally, fine lines yield a greater edge-to-edge spacing than thicker lines on the same grid.

11. To reduce cross-talk in large backplanes use the 3-W rule when traces are routed on the same layer, or orthogonal (90°) when routed on adjacent signal planes (horizontal versus vertical routing). Another technique is to separate traces at 2 mils/in of trace length.

6.3.3 Signal distribution

Proper signal distribution is mandatory to ensure predictable performance and timing. *In fact, net layout is the single most important factor in ensuring good signal integrity.* This section analyzes both lumped and transmission line nets. Gate loading and its effects on delay and impedance is also discussed. Clock and strobe distribution, reflections, termination, and differential signals are given special emphasis.

6.3.3.1 Signal interconnections. A well-designed interconnect system begins at the system level. The following guidelines are recommended during the initial design phase:

1. Cross-talk needs to be carefully considered in designs with extensive bus structures due to the high degree of parallelism.
2. Preassign I/O pins of the daughter cards to aid in the bus structuring. Avoid pin assignments that would cause signal traces to cross each other between connectors.
3. Simulate and analyze the connector pinouts to ensure acceptable levels of cross-talk and common mode noise.
4. Partition logically complete functions so that they do not grow beyond the boundary of a card.
5. Match characteristic impedance Z_0 of the daughter cards and backplane as closely as possible, within 10%. In a bused system, daughter card stubs that are short relative to the signal risetime make impedance matching to the backplane unnecessary.
6. Allocate card slots on the backplane to minimize interconnection length to other cards. The shorter the length, the better!

Matched impedances. Matching impedances and minimizing interconnect lengths improves system performance. Reflections from mismatches can add propagation delay. These mismatches can also be caused by large variations in gate loading. If the mismatch is large enough, the reflection produced can cause false switching in previous gates on the line. Generally, impedance varies on a PC board by 15 to

20% due to manufacturing tolerances and the nonideal ground plane effects. Critical applications may require 10% impedance control. Reflections are usually no problem here since the percent reflection is about half the percent change in impedance. *By matching impedances, the largest amount of energy is delivered with the smallest reflection.*

Branching. When a line of a specific Z_0 branches into two lines, each equal to Z_0, the impedance at the point of branching effectively becomes $Z_0/2$. See Fig. 6.13.

If the branched lines are very short stubs (for example, about 2 in for ABT), they will appear more like lumped loads and not be a problem. Longer branched lines cause a large impedance mismatch and can be responsible for added delays and the possibility of false switching due to shelves or steps in the transition region.

Most CAD layout designers route from a purely "DC" standpoint and do a lot of branching. The designer must be aware of this and analyze critical signal nets and line lengths to avoid this problem. The following two design guidelines are recommended:

1. Prioritize the signals for the layout so critical signals are routed first when they can be effectively handled.
2. *Oversee the layout and verify the routing of these signals.* It is important that this be an ongoing process so that desired routing changes can be implemented with the least amount of impact on surrounding traces.

6.3.3.2 Reflections. This section presents the basic concepts concerning reflections on transmission lines as they apply to good interconnection design. Although reflections are generally considered as sources of noise and delay, they can be used to advantage by increasing line voltage on a bus.

Reflections need to be considered when lines become electrically long. In short lines, reflections overlap during the initial risetime of the incident signal and tend to mask each other out.

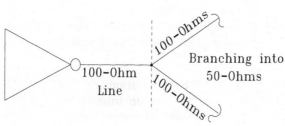

Figure 6.13 Branching effects.

Causes of reflections. When a voltage step is traveling down a constant impedance transmission line and encounters an abrupt change in impedance, a portion of the energy is reflected back. This abrupt change may be due to a capacitive discontinuity, such as a connector or gate input, or a transmission line of different characteristic impedance. The amount of energy reflected back is a function of the degree of mismatch and is expressed as a reflection coefficient, ρ, where $\rho = (Z_1 - Z_0) / (Z_1 + Z_0)$.

The amount of voltage reflected back to the source equals $\rho \times$ incident voltage. The amount of voltage which continues down the line away from the source equals $(1 + \rho) \times V_{incident}$. Thus, ρ represents the percentage of the incident voltage that is reflected:

$$V_{reflected} = V_{incident} \times \rho$$

or

$$V_{ref} = V_{inc} \times \left(\frac{Z_1 - Z_0}{Z_1 + Z_0} \right)$$

Controlling reflections

1. As a rule-of-thumb, limiting reflections to 10 to 15% ($|\rho| = 0.1$ to 0.15) will prevent severe noise, ringing, and the possibility of false switching. Using terminations in addition to good layout practice can control this.

2. Sometimes, the initial voltage step into a line may not be enough to reach the threshold of gates distributed on it. A controlled positive reflection from the end of the line ($\rho > 0$) can be used to raise the amplitude of the line voltage. In this case the gates will transition sequentially from the end of the line toward the source. This is typical of VME bus (Versa Module Europe) and CompactPCI buses.

3. Proper net routing with minimal branching and balanced gate loading will help prevent impedance mismatches and resulting reflections or shelves.

6.3.3.3 Clock and strobe distribution. Clocks, strobes, and control signals require special attention in layout to minimize skew and preserve edges. The following design guidelines will help avoid potential problems.

1. Preferrably use point-to-point clock configurations.
2. Centrally locate the clock drivers to shorten interconnection distances and allow easier length matching to control skew.
3. Avoid branching which causes Z_0 mismatch problems, additional delay, and unwanted excursions into the transition region.

4. Make clocks, strobes, and control signals first priority in the layout so they can be effectively handled. Review the layout and make the necessary changes before the rest of the layout is begun.
5. Do not use the same buffer to drive on and off the card simultaneously. Buffers with symmetrical drive characteristics for high and low transitions are preferred.
6. Do not assign clocks of different frequencies to the same clock driver package. Asymmetrical waveforms or other interference effects may result.

6.3.3.4 Terminations. *Termination* serves to reduce or eliminate reflections from discontinuities in a transmission line. Reducing reflections can prevent additional delay, false switching, ringing, undershoot, and can also decrease cross-talk (see Sec. 6.3.2, "Cross-talk").

Termination is required when nets are long enough to be transmissive. It is typically used in clock and strobe distribution, on long lines and long bus configurations. The added expense of parts required for terminating can be avoided if the layout of critical signals is accomplished with electrically short interconnects. With many of today's devices being subnanosecond, lines that are physically short can still be electrically long. *Termination and ground currents.* One important issue to keep in mind that is not usually discussed is ground currents. In a 64-bit bus with each data line terminated, a simultaneous switching of all logic levels can create tremendous ground currents. For example, an ACT bus with an AC termination of 70 pF and 56 Ω on each end will do an excellent job of limiting reflections. But if all 64 bits simultaneously change state, almost 180 mA of ground current will be generated for each data line or a total of 11 A.

Even with good grounding practices, these small inductances in the ground system can translate into large glitches due to these *di/dt* effects and the large currents involved. Doubling the resistor value in this case will still adequately dampen reflections for acceptable bus settling time while halving the ground currents. Also, the grounding scheme must be low inductance as described in Sec. 6.4.1. Series termination, parallel terminations, AC termination, and Schottky diode termination techniques are discussed.

Series termination. Using a damping resistor at the output of the driver is called *series termination* (see Fig. 6.14). The voltage divider action between the net series resistance and the line impedance causes an incident wave of half amplitude to start down the line. When the signal arrives at the unterminated high impedance end of the line, it doubles ($\rho = 1$) and is thus restored to full amplitude. Any reflections returning to the source are absorbed since the line and source impedance match.

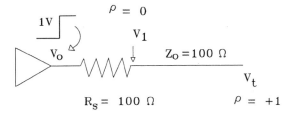

(Ideal zero impedance source)

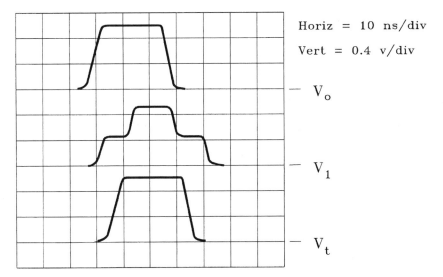

Horiz = 10 ns/div

Vert = 0.4 v/div

— V_o

— V_1

— V_t

Figure 6.14 Series termination.

Choosing series resistors. The value of the series termination resistor should be approximately $Z_0 - Z_D$, where Z_D is the output impedance of the driver. For ECL, the driver is typically 8 to 12 Ω in both logic states. For unsymmetrical drive bipolar devices, the impedance is dynamic and changes from about 3 to 5 Ω in the low state to about 30 to 50 Ω in the high state. Since low state reflections are more critical than high state reflections due to smaller low state noise margin, and because of DC voltage drop, the series resistor should be selected assuming a low value of Z_D.

Advantage of series termination. Series termination is ideal when using point-to-point interconnects. Also, series termination does not increase power dissipation.

Disadvantage of series termination. Gate loads cannot be distributed along the line but must be lumped very close to the end of the line.

Otherwise, false switching could occur due to the half-amplitude pulse traveling down the line. Multiple loads can be driven with series-terminated lines as long as monotonic edges are not required.

Parallel terminations. Parallel DC termination uses a resistor or a resistor network at the receiving end of the line. Buses generally use a parallel termination (see Fig. 6.15). This serves four purposes:

1. Reduces reflections from the ends of the bus. The reflections are not eliminated entirely if the actual loaded line impedance does not match the termination.

2. Provides a high state pull-up for open-collector drivers.

3. Provides a discharge path for the line when three-state devices are disabled. Tri-state buses should be properly terminated and not allowed to float.

4. Since the termination appears in parallel with the driver output impedance, it modifies the DC quiescent operating point of the driver that allows it to put a larger initial voltage step into the line on positive transitions. This helps to overcome the voltage divider delay effect previously discussed.

For 50- and 75-Ω buses, a 220/330-Ω value is recommended as a starting point to provide a compromise between items 1 and 4 in the preceding list. The address and data setup times should account for the fact that the drivers may not be able to drive a loaded bus and provide first incident switching. A high level threshold may not be reached until a reflection is received from the end of the bus. This adds to the propagation delay of the bus system since the gates will transition sequentially starting at the ends of the bus.

If the termination matches the loaded line impedance, Z_{eff}, no reflections take place. The bus will never reach the threshold level, however,

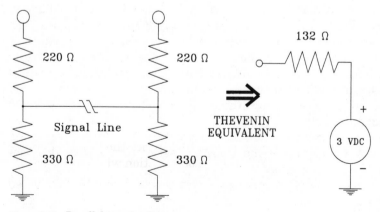

Figure 6.15 Parallel termination.

if the divider formed by R_T and Z_D limits the final line voltage to a value less than the logic threshold. Also, if the reflections were not partially reduced with the termination, false triggering and noise could result. The low state quiescent point is also raised enough by the termination to put a larger initial step into the line, but not enough for the low level noise margin to become too small.

Important guidelines

1. The Thévenin equivalent impedance requires that +5 V be well decoupled to ground. If not well bypassed, it behaves as a higher resistance causing more undershoot. Bypass caps, typically 0.1 μF, should be placed as close as possible to the V_{CC} pin of the termination package.

2. Beware of single-inline packages (SIP). Many types are subject to cross-talk within the package due to high inductance of the common lead. It is better to choose types that give access to the individual resistor leads since they afford better grounding. Intermixing clock and data signals in the same SIP also causes interference.

Advantages of parallel termination

1. Used for lines with distributed loads (data buses, etc.) since signals travel relatively undistorted along the full length of the line.
2. Parallel-terminated lines have the advantage when speed is the main factor since the loading delay is about half that of series terminated lines.
3. Parallel termination helps to offset the driver versus line voltage divider delay by increasing the initial voltage amplitude into the line.

Disadvantages of parallel termination

1. Higher power consumption than series termination.
2. Somewhat higher cross-talk levels since more signal energy is sent down the line versus less signal energy for series-terminated lines.
3. Since the operating point of the driver is changed, V_{OL} is increased. This is what enables the larger initial input step. However, this larger V_{OL} *decreases* the low level noise margin of V_{IL} min − V_{OL}max since V_{IL}min is unchanged.

AC termination. Another termination technique uses an RC network at the end of a data line, as shown in Fig. 6.16. Since the line is now terminated at the receiver end, the degree of reflection will depend on the relative values of the effective line impedance, Z_{EFF}, and the resistance value, R. The advantage of the AC network is that, in the quiescent state, very little power is consumed. This is especially important since

DATA OR BUS LINE

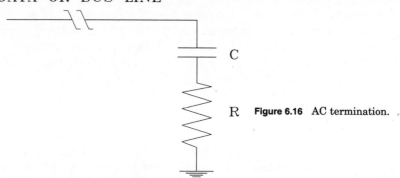

C

R **Figure 6.16** AC termination.

the value of the resistor can be quite low so as to match Z_{EFF}. It is very important to make sure the time constant of the RC network is less than the pulse rate on the bus. Otherwise, the RC network may prevent logic levels from being reached.

AC termination values are a function of signal propagation delay, which is, in turn, a function of loading and the transmission line configuration. For unloaded lines use the standard t_{pd} of the particular layer in the board. For loaded buses, calculate the *effective* propagation delay using

$$T_{pd} = t_{pd} \sqrt{1 + C_L/(C_0 L)}$$

where T_{pd} = the effective propagation delay in the net (ns/in) ; multiply this value by the total net length to get total propagation delay (T_{DELAY})

   t_{pd} = the unloaded propagation delay on the particular layer of the multilayer board (ns/in)

   C_L = the capacitance of the gate, connector, and stub (pF)

   C_0 = the distributed capacitance of the particular layer in the multilayer board (pF/in)

   L = the average distance between the loads (in)

Once the effective propagation delay is calculated, use the following formula for determining the capacitance value:

$$C = \left[\left(\frac{2\pi}{4T_{\text{Delay}}} \right) (0.2Z_0) \right]^{-1} \text{nF}$$

The resistance value, R, will be the characteristic impedance, Z_0, of the signal trace.

Example

- Signal trace is 12 in long
- $Z_0 = 75 \ \Omega$ @ 2.5 pF/in

- Signal layer is strip line on epoxy-glass with t_{pd} = 0.175 ns/in
- Gate loads @ 4 pF are spaced an average length (L) of 3 in apart

Therefore

$$T_{PD} = 0.175 \sqrt{1 + \left[\frac{4}{(2.5 \times 3)} \right]} = 0.217 \text{ ns/in}$$

Total delay in the net, T_{DELAY} = (12 in)(0.217 ns/in) = 2.6 ns

The RC values for the AC termination network are

$$C = 0.110 \text{ nF} \quad \text{or} \quad 110 \text{ pF}$$

$$R = 75 \ \Omega$$

The nearest decade values are acceptable. These values should be considered as starting points only. For critical nets, simulation is mandatory to confirm signal integrity and termination values. The time constant of the network is 8.25 ns. This must be less than the pulse rate on the bus.

Since the capacitor must be charged and discharged through the line, the data rate is reduced and some degradation of the rise and fall times can be expected.

Schottky diode termination. Schottky diodes can also be used as line terminators, although they handle the problem of undershoot with more of a "brute force" approach. For example, if a forward-biased diode were used, as shown in Fig. 6.17, it would start conducting at −0.3 V and would damp all undershoots greater than that value. But because of the parasitic inductances and capacitances, undershoots of about −1 V will occur. Even though undershoot is relatively large, it will now be limited to a known value and can be dealt with. Choosing diodes with fast turn-on times will help alleviate this problem.

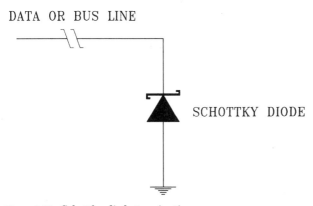

DATA OR BUS LINE

SCHOTTKY DIODE

Figure 6.17 Schottky diode termination.

A Schottky diode approach is advantageous for "cleaning up" reflections during initial breadboarding since only one part is needed. Since the diodes don't actually provide damping or slow edges, they tend to allow for the fastest signal propagation.

6.3.3.5 Differential signals.

Many systems require balanced or "differential" transmission lines with a specific differential impedance. *Differential impedance* is the impedance seen between a pair of lines as opposed to *characteristic impedance* which is referenced to a ground or power plane. In multilayer circuit cards, the close proximity of the reference planes has a definite influence on differential impedance, as shown by the inclusion of Z_0 in the differential impedance expression. The approximate differential impedance can be calculated from the following expression:

$$Z_{\text{diff}} \approx 2Z_0 \, (1 - 2K_B)$$

where Z_{diff} = the differential impedance
Z_0 = the characteristic impedance
K_B = the coupling coefficient between the lines

Even- and odd-mode coupling. To maximize signal integrity, discontinuities need to be minimized. The entire interconnect path needs to be examined. Consider a differential pair in a typical system, as shown in Fig. 6.18. The interconnection path from driver to receiver includes a driver, daughter card, connector, backplane, connector, daughter card, and receiver.

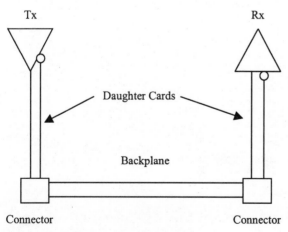

Figure 6.18 Differential system topology.

In most digital applications what is referred to as "differential signals" are in reality "complementary signals." As differential signals are launched from silicon devices, the gates that drive these signals are both referenced to ground, resulting in even-mode coupling. In some applications the differential signals are driven from a transformer, which is not referenced to ground. In this situation the signals are launched with primarily odd-mode coupling. Similarly, the receiver may be a silicon device or another transformer. Thus, the methods by which signals are driven and received determines the type of coupling that should be used throughout the interconnect path.

This leaves the user to choose the mode of coupling for the traces on the backplane and daughter cards, which will be determined by the trace geometry. To minimize discontinuities in the signal path, the engineer should choose a geometry that will match the mode of coupling in the driver, receiver, and connector(s) chosen. It should be realized, however, that tight odd-mode coupling cannot be achieved throughout the entire backplane and daughter cards, as the signals will have to split when entering into the signal pins.

For a typical digital application, based on interfacing with integrated circuits whose gates are ground referenced, even-mode coupling should be the primary mode of coupling throughout the entire interconnect path.

Differential signal topologies. Differential signals have traditionally been routed in a coplanar fashion, as shown in Fig. 6.19. Both traces couple equally to the reference planes that are adjacent on either side, and, therefore, are balanced. Varying the space between the two

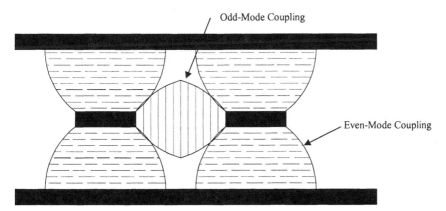

Figure 6.19 Coplanar differential structure.

traces controls the coupling between the two traces. This configuration maximizes even-mode coupling, while allowing easy control of odd-mode coupling. Furthermore, common mode rejection is inherent to this configuration. Any difference in voltage between the two ground planes is seen equally by both traces. The differential nature of the signals negates the difference in potential between the ground planes.

The suggested configuration of routing differential signals parallel to each other on adjacent layers in a broadside configuration is shown in Fig. 6.20. The primary method of coupling will depend on the configuration. As the traces move closer to each other and farther away from the ground plane, odd-mode coupling becomes more dominant. As the traces move further away from each other and closer to the ground plane, even-mode coupling becomes more dominant. The coupling between the two traces is controlled by varying the dielectric between the traces or by varying the width of the traces.

With some connectors, a perceived advantage to routing with the broadside approach is that a differential pair of traces can be routed through a connector pinfield with wider traces than a coplanar geometry. This advantage disappears however with the introduction of other issues into the system, namely, losses due to mode conversion, dielectric thickness required to achieve the desired impedance, and introduction of common-mode noise.

When using a predominantly even-mode coupled structure, a given trace will be within close proximity to one reference plane. Thus, the traces are not balanced with respect to the reference planes. Any difference in potential between the ground planes will contribute to increased common-mode noise.

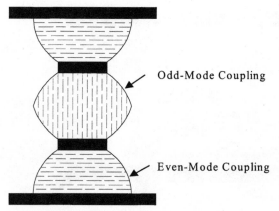

Figure 6.20 Broadside differential structure.

Before deciding to route differential pairs in a broadside configuration, one should consider

1. How even-mode coupling and odd-mode coupling will be affected.
2. The thickness impact on the PCB stackup.
3. The reduction in operating margins due to signal attenuation from skin effect losses.
4. The interaction between common-mode noise and how the differential pairs are routed.

Differential signals stackup. The impact of the geometry chosen on the overall thickness of a backplane or daughter cards is examined. The following parameters are used as an example to permit a fair comparison between the two proposed geometries:

1. Differential impedance equals approximately 100 Ω.
2. The trace width was fixed to 8 mils.
3. The coupling coefficient ($2K_B$) was fixed to 5%.
4. One-ounce copper plating was used for all layers.

The corresponding geometries that meet these conditions are shown in Fig. 6.21 for coplanar routing and in Fig. 6.22 for broadside coupling. The coplanar geometry results in a thickness of 0.026 in. The broadside geometry results in an overall thickness of 0.060 in. It would be possible to fit two of the coplanar geometries in the same amount of board thickness as one broadside geometry. Furthermore, it would be nearly impossible to implement the broadside geometry on both the backplane and daughter cards, since a typical board thickness for a daughter card is 0.062 in.

In the broadside coupling mode, by allowing $2K_B$ to increase to 0.10 and 0.15, the revised geometries are shown in Figs. 6.23 and 6.24. The overall thicknesses for these two geometries were, respectively, 0.047

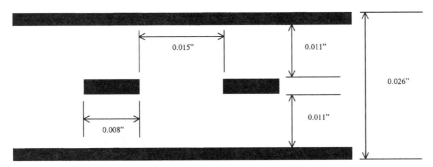

Figure 6.21 Coplanar geometry.

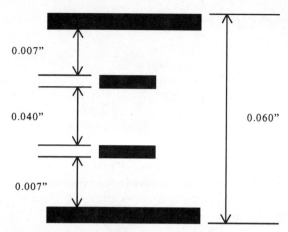

Figure 6.22 Broadside geometry $2K_B = 0.05$.

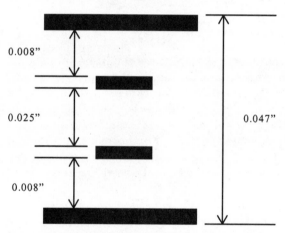

Figure 6.23 Broadside geometry $2K_B = 0.10$.

and 0.044 in. These thicknesses are still much more than the 0.026 in that the coplanar geometry permitted.

Common-mode noise. When working with a system where strong even-mode coupling needs to be implemented, common-mode noise is critical. When strong odd-mode coupling is implemented, this is not an issue. The signals in the differential pair are primarily coupled to each other, and should be as far as possible from a reference.

Today's digital applications typically need to be implemented with strong even-mode coupling; therefore, control of common-mode noise is critical. It is easy to see that, in the coplanar approach, any difference

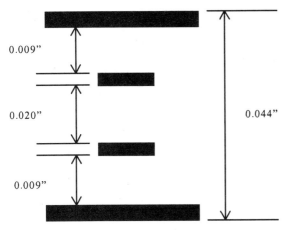

Figure 6.24   Broadside geometry $2K_B = 0.15$.

in potential between the two reference planes will be seen equally by both traces, so the configuration itself inherently rejects common-mode noise. The same cannot be said for the broadside approach. The two traces will couple primarily with the reference plane they are near. Thus, any difference in potential between the two reference planes will result in common-mode noise.

Assume that the broadside approach is implemented and that the signals are referenced to two ground planes. Typically, an ideal ground is assumed. The reality of the situation is that there is a physical aspect to the ground planes, which results in some potential difference between the ground planes. Thus, the broadside will inherently introduce common-mode noise.

To reduce the difference in reference potentials, vias could be used between the reference planes. However, this will use up valuable routing real estate.

Differential signal guidelines. Implementation of a system is always dependent on the needs of a system, and each of these needs must be examined to develop the best approach to optimize the design. For those systems where strong even-mode coupling is needed, the coplanar approach to routing differential pairs should be used over a broadside approach for the following reasons:

1. Less dielectric is required to achieve desired impedance.

2. Odd-mode coupling is easier to control with the strip-line approach than broadside approach. Reducing odd-mode coupling with the broadside approach requires varying the dielectric thickness between the two signals, which can result in a thick PCB.

3. The broadside configuration yielded approximately 8% more in signal attenuation due to skin effect losses over the 500-MHz to 2.5-GHz frequency range than the strip-line configuration.

4. It is anticipated that dielectric losses will be higher with the broadside configuration than the coplanar configuration, due to inherently higher field concentrations from the stackup required to achieve the desired impedance and coupling.

5. Common-mode noise is rejected in the coplanar approach, while it is inherently introduced with the broadside configuration.

6. Control of common-mode noise in the broadside configuration requires the use of vias, resulting in the use of valuable board real estate.

Understanding the application is key to determining the best way to approach its design. While a single benefit may point to one approach, the system as a whole needs to be considered.

Differential signal terminations.   Termination is very important in differential transmission. For a dedicated single driver, locate the termination resistor at the receiver end, as shown in Fig. 6.25. It should equal the differential line impedance. For multiple drivers, terminate at both ends as in Fig. 6.26. For party line systems, terminate at the extreme ends only as in Fig. 6.27. Extra terminations will only adversely load the line.

Figure 6.26 was terminated "line to line." This is often desirable since it requires minimum power from the driver and only one resistor. However, in high noise environments, using line-to-ground termination, as shown in Figs. 6.27 and 6.28 results in a lower impedance path to ground for externally generated noise currents and, therefore, a quieter system. This method is preferred for long line applications even though driver power is increased.

6.3.4   Net layout

The specific layout of critical nets is key in terms of maintaining signal integrity, that is., preserving edges, maximizing noise tolerance,

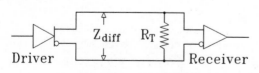

Twisted Pair Line
$R_T = Z_{diff}$

Figure 6.25 Termination at receiver.

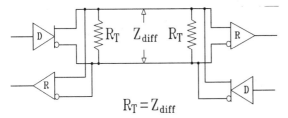

$$R_T = Z_{diff}$$

Figure 6.26   Termination at both ends.

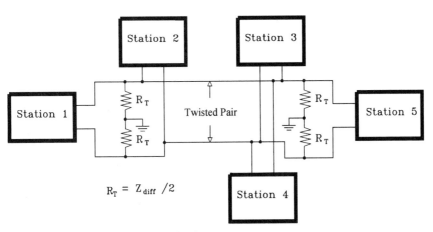

$$R_T = Z_{diff}/2$$

Figure 6.27   Party line system terminations.

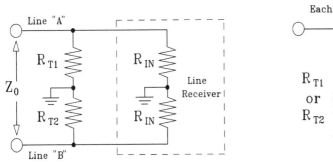
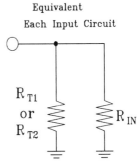

Figure 6.28   Line-to-ground terminations.

and minimizing delay. The edge rates associated with today's logic demand close attention to a transmission line environment.

This section will focus on those issues that need to be addressed in the layout of high speed nets. A set of practical design rules is discussed to aid the designer in translating a valid schematic into optimally performing hardware.

6.3.4.1 Impedance discontinuities. Maintaining a constant characteristic impedance, Z_0, along a signal path will provide maximum energy transfer with minimal attenuation or loss of waveform integrity. When signal lines are electrically "short" ($<t_{rise}/2$), reflections merge together and damp out during the risetime so that overall signal integrity is basically unaffected. These nets are referred to as "lumped" since they can be approximated by a driver with a single lumped capacitance on its output equivalent to the loading of the circuit traces, gate inputs, connectors, etc.

In the transmission line case, however, line lengths are long enough such that reflections do not overlap and damp out during the risetime. Instead, each reflection from a discontinuity is seen as a distortion of the waveform. The amplitude of this reflective distortion is a function of the degree of impedance mismatch at the point of the discontinuity.

The reflection pulse width is directly proportional to the *length* of the impedance discontinuity. For example, a via will have very little effect on ABT waveforms because the reflection pulse width generated is very small with respect to the device risetime. However, for considerably faster ECLinPS devices, the same reflection pulse width is a greater percentage of the risetime and thus will have a greater impact on overall signal integrity.

In past designs, risetimes were sufficiently long such that lumped characteristics prevailed. However, today's ever-quickening edge rates make more stringent control of the transmission line environment a necessity.

6.3.4.2 Net layout guidelines. The following guidelines are intended to provide a practical approach to proper net layout.

1. Make critical signals first priority in the layout so they can be most effectively handled. Provide the layout department with a list of critical signals (clocks, strobes, etc.) along with layout rules to route first.

2. Keep signal lines short whenever possible. As described previously, lumped nets avoid many of the problems associated with transmission lines. Most impedance discontinuities can be ignored.

3. Use the *slowest* logic family that will adequately meet timing requirements. Using longer risetimes will allow longer lines to still

appear lumped, thereby easing layout constraints. Lower EMI emission is an additional benefit.

4. Terminate transmission line nets to reduce reflections. There are many ways in which to accomplish this, including parallel split resistors, AC termination, series termination, and Schottky diodes. Each has its advantages and disadvantages. See Sec. 6.3.3.4 for detailed information on these termination methods. The physical placement of the terminations must be considered during the layout process.

5. Keep loading along distributed nets equal and well balanced. This particular aspect of layout has a tremendous effect on the resulting signal integrity. When loads are evenly distributed on a transmission line, their input capacitance and stub capacitance combine to essentially *increase* the distributed capacitance of the line. This approximates a lowering of the effective impedance of the line. When the loading is well balanced, the lowered effective impedance is relatively constant so reflections are minimized. However, when loading is unbalanced, the various effective impedances of the different sections create multiple reflections. Unfortunately, this is very common in CAD layouts. If the layout cannot be changed due to routability restrictions, varying the trace width of different sections to change Z_0 so that the various loading densities create matched effective impedances will significantly reduce reflections.

6. Avoid branching in the layout of critical signals. This is a very common pitfall of CAD layout systems. If the signal branches, the impedance at the point of the branch is essentially $Z_0/2$. This discontinuity creates significant reflections. If routability restrictions require branching, try to keep each branch short enough to meet lumped criteria of the specific logic family. If this cannot be accomplished, consider a separate buffer for each leg of the branch.

7. Avoid driving on and off card simultaneously with the same driver. This is another way of stating that each signal leaving and entering the card should be separately buffered.

8. Keep stubs off of transmission lines <2 in for TTL/ABT/CMOS/ECL 10K and <1 in for ECL100K and GTL. Since stubs are essentially branches, it is important to keep them clearly lumped. On backplane/daughter card systems, this means keeping all drivers and receivers as close as possible to the card edge connector.

9. Proper grounding practices are mandatory to maintain signal integrity. If proper return current paths are not provided, even the best layout will not perform satisfactorily. See Sec. 6.4.1, "Grounding Considerations."

10. When using logic devices with risetimes of 1 ns or less, use 45° as opposed to right angles in the circuit trace path. Also, keep the use

of vias to a minimum. Right angles and vias create impedance discontinuities. With very fast edge rates, even these small reflections can significantly affect signal integrity. Routing within these constraints can severely limit routability, which is another reason to use slower devices if timing allows.

11. Avoid long test point circuit traces. Attaching a long unterminated circuit trace path for a test point will create numerous reflections. Keep test point traces very short or use a separate buffer if at all possible.

12. Testability should also be considered during the entire PCB design process. The requirements for in-circuit testing (ICT) must be reviewed with the test group.

6.4   Grounding, Power Distribution, and EMC Design

This section discusses the relationship between noise and grounding. Design guidelines are presented to avoid the adverse effects of long ground returns. Power distribution techniques are analyzed, with emphasis on minimizing voltage drops and noise margin effects. Guidelines are presented to increase power distribution effectiveness, choose the number of power/ground layers, and improve decoupling. Finally, EMI guidelines are presented to control emissions.

6.4.1   Grounding considerations

The relationship between system performance (especially noise) and grounding is fully discussed in this section. Design guidelines are presented to avoid the adverse effects of long ground returns. Additional guidelines concerning ground pin distribution within connectors are outlined. Refer to Sec. 6.4.3, "Minimizing EMI," since grounding and EMI considerations go hand in hand.

6.4.1.1   System grounding.   The importance of good system grounding cannot be overemphasized in high speed digital systems. Many unexpected and hard-to-trace noise problems, including some attributed to cross-talk, are in reality due to poor grounding.

The basic requirement for a good system ground is that it has low impedance and low inductance. The use of multilayer PCBs with dedicated power/ground planes is definitely a step in the right direction, but the grounding problem is complicated at the daughter card/backplane interface. The connector must be relied on to adequately maintain a low impedance/low inductance path. Two specific problem areas need to be considered in system grounding: (1) the loss of low level DC noise margin due to ground shift and (2) the effect of long ground returns.

6.4.1.2 Low level DC noise margins. Any difference in ground potential decreases noise margins when two circuits are interconnected in a single ended mode, as shown in Fig. 6.29. Thus.

$$V'_{OL} = V_{OL} + I_g R_g$$

$$I_g R_g = (V'_{OL} - V_{OL}) = \text{noise margin decrease}$$

The return current flowing in the system ground produces a voltage drop, $I_g R_g$, across the total ground resistance between the two devices. This voltage drop adds directly to the V_{OL} of the driver, thus creating the higher V'_{OL} seen by the receiver. Since the low level DC noise margin equals $V_{IL}\text{max} - V_{OL}\text{max}$, the net effect is shrinkage. In a back-panel/daughter card system, the total resistance, R_g, includes the incremental resistance of the ground distribution plane from card to card. The ground current, I_g, is successively reduced by the contribution of each card, as shown in Fig. 6.30.

Note that current flow through the incremental resistance closest to the ground input is the largest since it is the summation of all the subsequent ground currents. This means that the ground shift between slots 1 and 3 is greater than the shift between slots 13 and 15 since the currents are larger. Thus, card location on the backplane

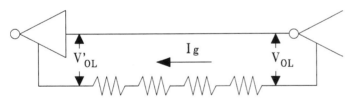

Figure 6.29 Ground potential. Total resistance $= R_g$.

Ground Input

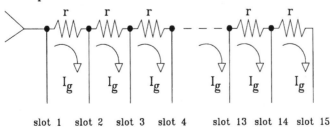

slot 1   slot 2   slot 3   slot 4   slot 13   slot 14   slot 15

Figure 6.30 Successive reduction of ground current. $r =$ incremental bus resistance between slots. $I_g =$ aboveground current per card.

will definitely affect the ground shift seen by these cards with respect to each other.

Reducing ground shift

1. Do not use thermal relief "butterflies" on compliant pin components since this increases resistance and inductance from the ground plane to pin. Thermal relief is not required as no wave soldering takes place on the board. When thermal relief is used, keep the connection "fingers" as thick as possible to minimize inductance and IR drops, yet thin enough to wave solder efficiently.

2. Use the appropriate connector. The pin assignments and use of grounds are critical. Use as many ground pins per connector as possible and keep them well distributed throughout the connector. This also enhances signal distribution by decreasing long ground return paths.

3. Logic cards that represent the heaviest current drain should be located nearest the end where the ground enters the backplane. This will help reduce ground currents for the remaining slots, and, in turn, reduce ground shifts between other cards.

4. Cards with the most single-ended logic interconnects between them should be assigned to slots as close together as possible. This also improves signal distribution characteristics by reducing interconnect lengths.

5. If the ground shift between two card slots proves unacceptable with respect to loss of noise margin, consider differential driving, which eliminates ground shift as a noise margin factor.

6.4.1.3 Long ground-return effects. The most often overlooked culprit of poor system performance is the adverse effect of long ground-return paths. For every current flowing in a signal line, an equal return current flows in the transmission line ground. If the signal and return currents begin and end close to the driving and receiving devices, there is no problem. However, if the ground return is longer than the signal line, the return current is out of phase and sees the added length as a higher inductance path. This creates a noise spike in the ground system. As signal rise and fall times grow short with respect to the delay length of the ground return path, spike amplitude increases.

This situation is very common, as shown in Fig. 6.31, which indicates that the ground pin is far from the signal pins on the edge connector. Figure 6.32 shows a schematic representation of the situation in Fig. 6.31.

When gate 1 switches, ground current i flows, as indicated by the arrows. Because the ground current is forced to travel a considerably greater distance than the signal current, it develops a voltage spike across the additional series inductance, L. If gate 2 is at logic low, the

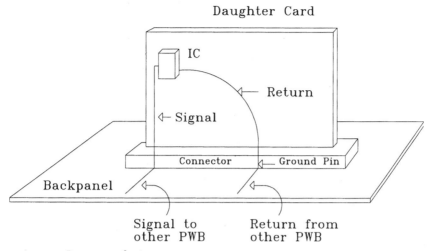

Figure 6.31 Long ground returns.

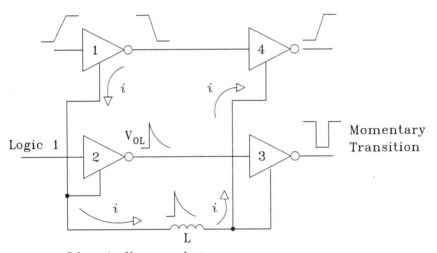

Figure 6.32 Schematic of long ground return.

spike on its ground will also appear at its output since its V_{OL} is referenced with respect to ground. Once the quiescent V_{OL} of gate 2 plus the spike amplitude are large enough, gate 3 will transition momentarily and propagate the spike through the system. Because the spike appears on the output of gate 2 when gate 1 switches, it misleadingly looks like cross-talk is the culprit.

Even if the long return path is not bad enough to cause false switching, the increased ground shift results in a decrease in available noise margin. Unfortunately, this may not be observed until worst-case testing is performed or, worse still, in production, when variations will occur in the tolerances of device and board manufacturing lots.

Note that the larger the return current, the larger the potential inductive spike in the ground circuit. Thus, lower impedances in the 50-Ω range, which require larger source and sink currents, may aggravate the problem.

In some systems, the problem may show up as being software dependent. Different combinations of 1's and 0's on data and address busses create varying return currents. These in turn may cause false switching of a sensitive strobe when certain instructions are executed or memory locations accessed.

Clearly, grounding issues need to be addressed early in the system design phase to prevent these erratic and hard-to-trace problems that always seem to chew up schedules and budgets.

Handling long ground returns

1. Choose the appropriate connector for the interface. Include as many grounds as possible in the connector. They must be well distributed. The importance of this guideline cannot be overemphasized.
2. Isolate critical signals, such as clocks and strobes, with ground pins to avoid long ground returns and to reduce impedance mismatch at the connector.
3. Fast edge rate device will require more connector ground pins for adequate performance.
4. Try to partition board logic in functional units to minimize the number of inputs and outputs.
5. Don't break up ground planes to run additional voltages or signals. This can cause ground loops and create indirect return paths. If planes must be split, make sure that critical signals do not traverse the split.
6. Use the slowest logic family that will adequately perform under worst-case conditions. Faster rise/fall times yield larger di/dt and increase potential risk of long ground-return problems.

6.4.2   Power distribution

In this section, power distribution techniques are analyzed, with emphasis on minimizing voltage drops and noise margin effects. Guidelines are presented to increase distribution effectiveness, choosing the number of power/ground layers, and improving decoupling characteristics.

6.4.2.1   Impedance and inductance.   The key characteristics of a good power distribution system are low impedance and low inductance. Since impedance equals L/C, this means maximizing C and minimiz-

ing L. The impedance of the distribution at the circuit terminals needs to be kept small compared to the impedance of the circuit load to avoid large potential voltage drops. These drops can reduce high level noise margin in the same way that ground shifts can reduce low level noise margin. See Sec. 6.4.1, "Grounding Considerations." V_{OH} tracks almost directly with V_{CC}. Therefore, if V_{CC} drops from 5.0 to 4.8 V, V_{OH} will drop a corresponding 200 mV. This, in turn, shrinks the high level noise margin, $V_{OH}\text{min} - V_{IH}\text{min}$ because V_{IH} remains constant.

Multilayer couplets. The use of a multilayer configuration utilizing power and ground "couplets" is the most effective method of distributing power. A *couplet* refers to placing power and ground planes closely together on adjacent layers in the multilayer stackup (Fig. 6.33). The distributed mutual capacitance between the parallel planes augments the high frequency bypass characteristics of the distribution system. The optimum efficiency of the couplet applies only when the power planes are closely spaced, preferably 5 mils for high speed applications.

Multilayer power distribution

1. Use a multilayer configuration with power/ground couplets. Keep plane spacing as close as possible (within producibility constraints) to maximize mutual capacitance between them. In systems with multiple planes, alternate power and ground planes to maintain couplets. They are effective even if there are signal layers in between.

2. Multiple 1-oz planes arranged in couplets are recommended for distributing backplane power. The number required will be a function of current density. However, many backplane layer stackups do not have enough overall thickness to accommodate them. In this case, 2-oz planes are acceptable. At least one power/ground couplet with 1-oz copper is recommended for 5-V daughter cards.

3. Distribute card power over multiple connector pins. The number of amperes per pin will be a function of the particular connector.

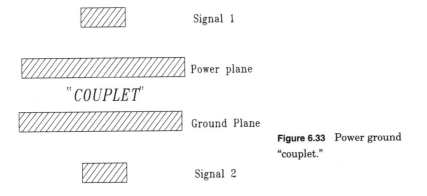

Signal 1

Power plane

"*COUPLET*"

Ground Plane

Figure 6.33 Power ground "couplet."

Signal 2

Remember that many backplane connectors are right angles so that the DC resistance of the pins will vary with the particular row. Shorter pins will have less resistance and will, therefore, tend to "current hog." This must be considered when assigning pins.

4. Stagger power/ground inputs on the backplane so as to distribute power more evenly into the plane (Fig. 6.34). It is much better to distribute along the *length* of the backplane to minimize the distance from power input to cards.

5. The point on the backplane where the low frequency bypass capacitors are placed is a good place to sense the feedback voltage for controlling the regulator. Here the power is well filtered and the voltage drops in the supply connecting cables will be taken into consideration.

6. Keep the power supply as close as possible to the backplane to minimize voltage drops in connecting cables. Use thick (<12 gauge) multiple cables to decrease inductance and voltage drops. Twist or tie together the power and ground cables to minimize their inductance.

7. As a rule-of-thumb, a 2% maximum voltage drop from the supplies to any circuit device is recommended. This also includes ground return losses.

8. Use additional planes to distribute auxiliary power in the backplane. Don't break up existing planes. Another option is to use standard bus bars. Both of these methods have much lower inductance than trying to run long meandering traces on signal planes.

9. In very high speed circuits RF currents exist on the edges of power planes due to magnetic flux linkage. This interplane coupling is called *fringing*. When using high speed logic circuits, power planes can couple RF currents to each other and thus radiate RF energy into free space. To minimize this coupling effect, all power planes must be physically smaller than the closest ground plane per the "20-*H*" rule developed by W. Michael King. To implement this rule, determine the spacing between the power plane and its nearest ground plane. For example, assuming a separation of 6 mils between planes, $20 \times H = 120$ mils. Make the power plane 120 mils smaller than the ground

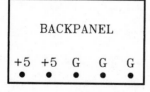

POOR   BETTER

Figure 6.34 Staggering power inputs.

plane. Should a power pin to a component be located inside this iso-lated area, the power plane may be jogged to provide power to this pin. Also, careful attention must be paid to ensure that there are no traces on the adjacent signal layer located over the absence of copper area.

6.4.2.2 Bypassing considerations.

Proper bypassing is extremely important in high speed digital systems. Not only does bypassing sup-ply transient currents required for switching surges but also provides a low impedance AC path between power and ground planes, enabling them to serve as references for impedance-controlled signals.

Effective bypassing. Bypassing should be accomplished at three distinct levels; these are basic guidelines and many special chips have their own requirements (for example, Pentium processors):

1. *Low frequency (<1 MHz) on the backplane.* This helps isolate the power supply and connecting cable noise from the backplane. Tantalum capacitors from 33 to 100 µF are recommended. Use two to three capacitors close to the backplane power input terminals and dis-tribute four to six lower value capacitors around the backplane. This paralleling will lower the ESR (equivalent series resistance) of the capacitors and will enhance their low frequency bypass characteris-tics. Use chip capacitors if possible to minimize lead inductance. If axi-al leaded capacitors are used, keep lead length minimal.

2. *Midfrequency (1 to 20 MHz) bypassing on daughter cards.* This helps to isolate backplane noise from each card and also prevents card-generated noise from reaching the backplane. Use three to six 25- to 47-µF solid tantalum capacitors mounted as close as possible to the power input on the edge connector. Memory cards require the use of five to eight capacitors due to the increased switching and refresh cur-rents.

3. *High frequency (>20 MHz) on daughter cards.* This level of bypassing is intended to supply transient surge currents to individual ICs. Use 0.01- to 0.22-µF RF type ceramic SMT capacitors as close as possible to the device they are intended to bypass. Tests indicate that the larger capacitance values tend to provide better decoupling, espe-cially if the smaller 0.010-µF values are interspersed.

Although power and ground plane couplets do contribute to high fre-quency decoupling, they are not recommended as a substitute for dis-crete decoupling capacitors. The number of holes and clearances in the planes make calculation of their AC effects difficult to determine.

Capacitor packages. The chart in Fig. 6.35 displays the relative suitabil-ity of various capacitor dielectric materials as a function of frequency. High quality ceramic (NPO), glass, and polystyrene offer the best high frequency performance. Polystyrene is not volume efficient, however. Multilayer ceramic chip capacitors are available with three different dielectrics:

NPO. NPO is the most popular formulation of the "temperature-compensating" materials. It is one of the most stable dielectrics available. Capacitance change with temperature is in the order of ±0.3% from −55 to +125°C. Typical capacitance change with life is <±0.1%. NPO is typically difficult to obtain in high capacitance values.

X7R. X7R formulations are called "temperature-stable" ceramics. Its temperature variation of capacitance is within ±15% from −55 to +125°C. The capacitance change is nonlinear. Change in capacitance over time is about 5% in 10 years. X7R is preferred for most designs.

Z5U. Z5U formulations are "general-purpose" ceramics and are meant primarily for use in limited temperature applications where small size and cost are important. They provide the highest capacitance possible in a given size compared to NPO and XR7. They also

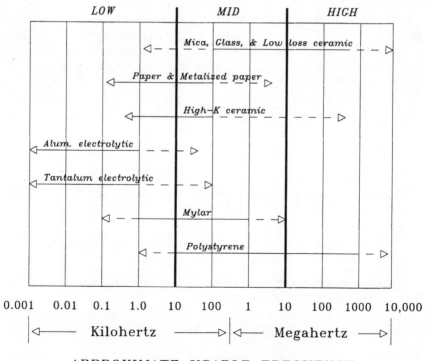

APPROXIMATE   USABLE   FREQUENCY

Figure 6.35 Bypass capacitor dielectrics.

show wide variations in capacitance under influence of environmental and electrical operating conditions. Capacitance can drop 25% in 10 years. Despite this instability, Z5U is popular due to small size, low ESL, low ESR, and excellent frequency response.

6.4.3 Minimizing EMI

FCC restrictions, along with European requirements, are becoming more and more stringent with respect to controlling emissions and susceptibility of electrical noise in systems. This section outlines some specific guidelines to help solve these problems.

Generally, if emissions can be controlled, susceptibility is likewise reduced and vice-versa. This is known as the *principle of reciprocity*.

The use of multilayer PCBs with their inherently low inductance power/ground planes goes a long way in reducing EMI problems. However, incorporating proper system EMI design practices initially will avoid costly redesigns.

The subject of EMI control is very broad, and this section presents a practical set of design guidelines.

6.4.3.1 The basics of interference.
EMI interference contains three essentials: an EMI source, a receptor, and a coupling path between the two. Sources include transmitters, radar, local oscillators in receivers, power lines, fluorescent lights, atmospheric noise, lightning, etc. EMI receptors include receivers, digital circuits, and sensitive analog circuitry. Coupling paths include spacing between conductors/shielding; filters and cables, ground connections, and power lines.

Coupling between a source and receptor takes place by radiation, induction, and conduction. Radiation is applicable when the source is at least several wavelengths away from the receptor—a far-field situation. Induction relates to a near-field configuration, in which the source and receptor are only a fraction of a wavelength away. Conduction occurs when hard wires or metal parts connect the source and receptor.

Most typical EMI problems are the result of radiation effects from cables, circuit traces, and logic devices.

6.4.3.2 EMI layout guidelines

1. Use multilayer boards with dedicated power and ground planes. This reduces the size of the current loops and hence the radiation area.

2. Use of strip-line is preferred over microstrip to reduce radiated emissions by up to 15 dB. This, in turn, can lead to reduced box shielding requirements.

3. When auxiliary power or ground traces are required, keep them as wide as possible to lower inductance and DC resistance. Arranging

power and ground traces in couplets on adjacent layers helps lower inductance even further. See Sec. 6.4.2, "Power Distribution."

4. Adequate decoupling reduces current in inductive traces and thus helps to lower emissions. See Sec. 6.4.2.2, "Bypassing Considerations."

5. Keep highest speed logic closer to the card connector to keep high transient current loops minimal in length.

6. Keep signal traces as short as possible! Not only will delays be shorter, but also the length of current loops will be minimized, resulting in less emission.

7. If a noisy card is anticipated, use a strip-line approach with power and ground on outer layers to act as shields and restrict emissions within the card as well as to and from external devices.

6.5   Electronic Components

This section provides information on designing with both through-hole and surface-mount components.

6.5.1   Through-hole components

The varieties of circuit components are extensive and vary in type, size, and shape. Even common components, like resistors and capacitors, are available in a wide range of configurations. Both through-hole and surface-mount components are available. Discrete components with leads for through-hole mounting are of two basic types, axial-lead and radial-lead. Resistors, diodes, and capacitors are available in both these lead configurations.

Transistors typically have three radial leads and some common packages are designated TO-3, TO-5, TO-107, etc. Integrated circuits are available in both dual-in-line (DIP) and quad-inline (QUIP) packages. The DIP's in-line pitch is 0.100 in (2.54 mm), with the inline-to-inline spacing 0.3 to 0.6 in (7.62 to 15.24 mm), depending on I/O count. The lead count varies from 14 to 50 pins.

Single-inline package (SIP) components typically have resistor and or capacitor combinations with a standard 0.1-in (2.54-mm) spacing between leads.

6.5.2   Surface-mount components

Surface-mount technology (SMT) has made it possible to shrink the size, weight, and cost of printed circuit board assemblies. The first SMT applications for the consumer market were the small-outline (SO) or small-outline integrated circuit (SOIC) packages. Plastic leaded chip carriers (PLCC) have leads on all four sides, with each lead bent down and under like a "J." These SMT components are sometimes referred to as *J-lead devices.*

Another version called the *leadless ceramic chip carrier* (LCCC) has gold-plated, groove-shaped indentations known as *castellations* that provide shorter signal paths. The most common pitch is the 50-mil (1.27-mm) family.

The most common plastic packages used in commercial electronics are the SOICs with gull-wing leads, PLCCs with J leads, and fine pitch devices with gull-wing leads that are known as *plastic quad flat packs* (PQFPs). Fine pitch packages are available in 0.8-, 0.65-, and 0.5-mm (33-, 25-, and 20-mil) pitches.

6.5.3 SMT land patterns

SMT land patterns are also referred to as *footprints* and define the sites at which components are to be soldered to a printed circuit board. The IPC-SM-782 document provides good land pattern guidelines and must be used for designs with SMT devices. Conductors connecting to a land can act as a solder thief by drawing solder away from the land and down the conductor. Routing conductors in a necked-down config-uration can prevent component swim.

Some basic parameters that must be considered to achieve reliable SMT land patterns are

1. Solderable lead and termination width and height
2. Lead contact points/areas on PCB
3. Extension of pads beyond contact points

The basic variables in the land pattern design guidelines are as follows:

1. The center-to-center distance between adjacent pads should be equal to the center-to-center distance between adjacent component leads.
2. The pad width should equal the component lead or termination width plus or minus a constant.
3. The height and width of solderable terminations or leads and the contact points or areas of components determine the pad length.
4. The gap between opposite pads of the land pattern depends on the component body width and dimensional tolerances and is most crit-ical in properly locating the components on the pads.

6.5.4 SMT CAD layout

SMT designs generally require narrower trace widths and spacing (<5-mil line/space). The location of via holes that are required for interconnection and testing is critical with SMT components. To preserve routing channels, the number of vias must also be minimized and traces routed from surface pads. Newer auto routers can handle this situation.

Since many dense designs require placement of SMT components on both sides of the circuit board, blind and buried vias may be needed. Newer designs incorporate microvia technology. Most CAD tools can handle blind, buried, and microvias. Typically, SMT designs require extra postprocessing for creating solder paste artwork and generating test maps for assembly testing.

6.6   Useful Design Reference Data

This section is a compilation of various figures and tables to aid the CAD engineer during the design and layout process.

Table 6.13 shows the conductor spacing requirements at different voltages for commercial applications. Figure 6.36 depicts millivolt drop versus length of trace and trace width. Table 6.14 summarizes the propagation delay for both microstrip and strip-line configurations. Figures 6.37 and 6.38 show both the U.S. and European conducted and radiated emission limits. Table 6.15 shows examples of stackup assignments for PCB's from two to 10 layers.

6.7   Future Trends

This section gives an overview of advances in CAD tools, and new design concepts to help with the design of complex PCBs.

6.7.1   CAD tools

As complex designs move from concept to physical prototypes, timing requirements, thermal problems, manufacturing, and test issues all confound the problem of good PCB design. CAD tool vendors are attempting to help control the physical effects of PCB designs with analysis tools like timing, signal integrity, thermal management, and manufacturing and test control. The vendors are trying to embed human experience in their tools through rules-driven software that guides the engineer through the design process. Tools are becoming available whereby a design is decomposed into subcircuits so that a number of engineers can work simultaneously. This partitioning of the design can also function at the system level where backplanes, daughter cards, etc., can be designed separately and then combined for system analysis.

TABLE 6.13   Minimum Conductor Spacing, Commercial Specifications

| Circuit voltage | | Distance | |
Sinusoidal AC (rms) (V)	DC or AC plan mixed voltage (V)	(mm)	(in)
24	35	0.50	0.020
24–60	35–85	1.0	0.040
60–130	85–184	1.5	0.060
130–250	184–354	2.0	0.080
250–450	354–630	4.5	0.180
450–650	630–920	6.0	0.240
650–1000	920–1400	9.0	0.360
1000–1500	1400–2100	12.0	0.480
1500–2000	2100–2800	14.0	0.560
2000–2500	2800–3600	16.0	0.620

TABLE 6.14   Propagation Delays for Transmission Lines

Transmission line (FR-4)	Propagation delay (ns/ft)	Propagation delay (ps/in)	Propagation delay (ns/cm)
Microstrip	1.77	150	0.38
Embedded microstrip	1.72	143	0.36
Strip line	2.26	188	0.48

Shape-based routers like the Specctra from Cadence look at traces on PCBs as geometric shapes rather than lines. This approach gives the router flexibility to incorporate predefined routing rules. Analysis of EMI, cross-talk and timing has forced attention to interconnection component characteristics and placement. Newer routers are becoming adept at using these rules and are referred to as *rules-driven routers*. Intelligent routing technology is a must for high speed designs. Routing electrical signals with a minimum number of layers and vias continues to challenge the most sophisticated automation techniques.

6.7.2   EMC tools

Newer advances in EMC analysis tools are discussed in this section. Electromagnetic radiation from electronic equipment has become a serious problem. To bring electromagnetic radiation under control it is important to move away from costly fixes at the end of the design and integrate EMC analysis in the PCB design cycle. The description that follows is an example of a typical simulation tool from Quantic EMC.

Computer simulation consists of four steps. In the first step, the transmission line simulation, all selected nets are divided into several

TABLE 6.15   Example of Stackup Assignments

	Layer number										Comments
	1	2	3	4	5	6	7	8	9	10	
Two layers	S1	S2									Lower speed designs.
	G	P									
Four layers (two routing)	S1	G	P	S2							Difficult to maintain high signal impedance *and* low power impedance.
Six layers (four routing)	S1	G	S2	S3	P	S4					Lower speed design, poor power high signal impedance.
Six layers (four routing)	S1	S2	G	P	S3	S4					Default critical signals to S2 only.
Six layers (three routing)	S1	G	S2	P	G	S3					Default lower speed signals to S2–S3.
Eight layers (six routing)	S1	S2	G	S3	S4	P	S5	S6			Default high speed signals to S2–S3. It has poor power impedance.
Eight layers (four routing)	S1	G	S2	G	P	S3	G	S4			Best for EMC.
Ten layers (six routing)	S1	G	S2	S3	G	P	S4	S5	G	S6	Best for EMC. S4 is susceptible to power noise.

Note: S = signal routing layer, P = power, G = ground.

trace segments. For each trace segment, current and voltage waveforms are calculated for both cross-talk and reflection. Based on these results, the fast Fourier transform (FFT) processes the current spectral analysis for each trace segment.

Using these results, the radiated magnetic field is computed for each spectral frequency. Each trace segment is taken into account to calculate the electromagnetic field at, for example, 5 mm (0.2 in) above the PCB. Regions of the PCB that radiate strongly are located and then the contributing net segments are identified.

A very useful approach is to produce a worst-case summary of the highest field amplitude at each frequency just above the PCB. This gives an excellent indication of which frequencies are the most significant contributors to EMC problems.

The advantage of the software screening is that it takes place within the PCB design process. Since the currents on the traces are responsible for the radiation, the critical nets for selected critical frequencies can be easily located. Once the nets have been isolated, the signal wave shapes should be investigated.

An example of a faulty net could be ringing due to transmission line lengths and mismatch. Edge rates account for a fair amount of energy at 100 MHZ.

Changing the line width to adjust the characteristic impedance or adding appropriate termination is a simple way to reduce ringing. Another approach may be to route high speed signals between two

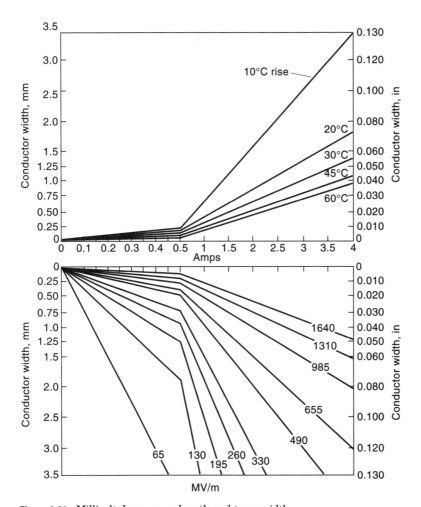

Figure 6.36 Millivolt drop versus length and trace width.

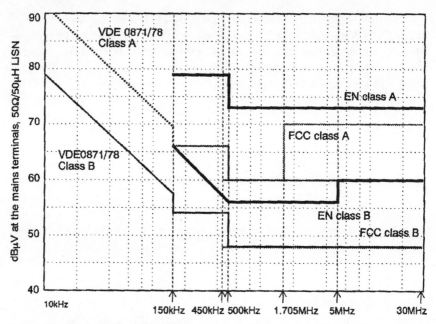

Figure 6.37 Conducted emission limits.

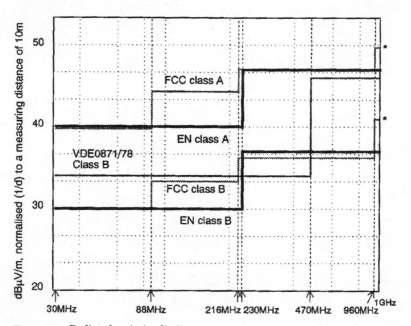

Figure 6.38 Radiated emission limits.

planes that effectively act as a shield. However, it should be kept in mind that vertical pins, which are not partly compensated by mirror-imaged tracks due to ground planes, could radiate strongly.

Radiation depends directly on the net location, current magnitude, and direction. To reduce radiation, the following measures can be taken:

1. Adding components like a resistor in series
2. Rerouting the trace and/or net
3. Moving trace to a different layer
4. Changing the placement of the component

The interactive nature of the layout with the simulation makes changes easy to implement and verify. Nets should always be simulated prior to layout. Signal integrity problems should be fixed first. EMC problems cannot be fixed without solving the signal integrity issues. EMC tools are also available from Cadence, Mentor, Ansoft, and others.

6.7.3 Final remarks

As designers continue to push technology and move to faster and faster clocks and edge rates, the challenges presented by high performance electronic systems will continue to grow. The CAD engineer must take full advantage of the CAD and CAE tools to build a product that works the first time to meet the ever-increasing demands of the marketplace. At the same time, relying solely on the tools is not enough; one must master the basics of the electromagnetic phenomena to be successful.

The authors wish to extend thanks to support from many individuals, in particular, Russ Moser, Steve Weller, Robert Jardon, and John D'Ambrosia.

6.8   Suggested Reading

1. Harper, C. A., *Electronic Packaging and Interconnection Handbook,* 3rd Edition, McGraw-Hill, New York, 2000.
2. Pecht, M., *Placement and Routing of Electronic Modules,* Dekker, New York, 1993.
3. Prasad, R. P., *Surface Mount Technology,* Van Nostrand, New York, 1989.
4. ASM International, *Electronic Materials Handbook,* vol. 1, *Packaging,* ASM International, 1989.
5. Ginsberg, G. L., *Printed Circuits Design,* McGraw-Hill, New York, 1990.
6. Blood, W. R., Jr., *Motorola MECL System Design Handbook,* 1988.
7. Bosshart, W. C., *Printed Circuit Boards Design and Technology,* McGraw-Hill, New York, 1983.
8. Montrose, M. I., *Printed Circuit Board Design Techniques for EMC Compliance,* IEEE Press, 1996.
9. Williams, T., *EMC for Product Designers,* BH Newnes, 1992.
10. Harris, S., and R. Cutler, *AMP Design Guide,* AMP Inc., Harrisburg, PA, 1997.

11. Cutler, R., "Interconnection Design Guide," Private Communication, AMP, Inc., Harrisburg, Pa, 1991.
12. D'Ambrosia, J., *Comparison of Differential Signal Routing Techniques,* AMP Inc., Harrisburg, PA 1998.
13. Clark, R. H., *Printed Circuit Engineering: Optimizing for Manufacturability,* Van Nostrand, New York, 1989.
14. "Tools and Teamwork Are Key to Successful PCB Design," *Computer Design,* June 1996.
15. Donlin, M., "CAE Vendors Strive to Improve Link between Design and Layout," *Computer Design,* May 1992.
16. Sullivan, B., "Meeting PCB Challenges with Team Design and IP Reuse," *Computer Design's Electronic Systems Technology & Design,* February 1999.
17. Kraemer, D., "Multi-board System-Level Analysis," *Printed Circuit Design,* May 1999.
18. Murugan, R. M., et al., "The Art of EMC," *Printed Circuit Design,* May 1999.
19. Davidson, E. E., "Electrical Design of a High Speed Computer Package," *IBM Journal,* vol. 26, no. #, May 1982.
20. Feller, A., H. R. Kaupp, and J. J. Digiacomo, "Crosstalk and Reflections in High Speed Digital Systems," *AFIPS Conference Proceedings,* Fall, Joint Computer Conference, 27, 1965.
21. DeFalco, J. A., "Predicting Crosstalk in Digital Systems," *Computer Design,* June 1973.
22. Catt, I., "Crosstalk (noise) in Digital Systems," *IEEE Transactions on Electronic Computers,* vol. EC-16, No. 6, December 1967.

6.9   Institute for Interconnecting and Packaging Electronic Circuits (IPC) Standards

1. "Controlled Impedance Circuit Boards and High Speed Logic Design," IPC-2141.
2. "Generic Standard on PWB Design," IPC-2221.
3. "Sectional Standard on Rigid PWB Design," IPC-2222
4. "Laminate/Prepreg Materials Standard for Printed Boards," IPC-4101.
5. "Generic Performance Specification for Printed Boards," IPC-6011.
6. "Qualification & Performance for Rigid Printed Boards," IPC-6012.
7. "Design & Application Guidelines—Surface Mount Connectors," IPC-C-406.
8. "Printed Board Dimensions & Tolerances," IPC-D-300.
9. "High Frequency Design Guide," IPC-D-316.
10. "Design Guidelines for Electronic Packaging Utilizing High-Speed Techniques," IPC-D-317.
11. "Guidelines for Selecting Printed Wiring Board Size," IPC-D-322.
12. "End Product Definition for Printed Boards," IPC-D-325.
13. "Printed Wiring Design Guide," IPC-D-330.
14. "Printed Board Description in Digital Form," IPC-D-350.
15. "Automated Design Guidelines," IPC-D-390.
16. "Design Standard for Rigid Mutilayer Printed Boards," IPC-D-949.
17. "Drilling Guidelines for Printed Boards," IPC-DR-572.
18. *Electronic Packaging Handbook,* IPC-PD-335
19. "Surface Mount Design and Land Pattern Standard," IPC-SM-782.

6.10   Institute of Electronics and Electrical Engineers (IEEE) Standards

1. "IEEE Standard for Mechanical Core Specifications for Microcomputers Using IEC 603-2 Connectors," IEEE Std. 1101.1-1991.

2. "IEEE Standard for Mechanical Rear Plug-In Units Specifications for Microcomputers Using IEEE 1101.1 and IEEE 1101.10 Equipment Practice," IEEE Std. 1101.11-1998.
3. "IEEE Standard for Additional Mechanical Specifications for Microcomputers Using the IEEE 1101.1-1991 Equipment Practice," IEEE Std. 1101.10-1996.

6.11   International Electrotechnical Commission (IEC) Standards

1. "Dimensions of Mechanical Structures of the 482.6 mm (19") Series," IEC 297-3.
2. "Mechanical Structures for Electronic Equipment—Dimensions of Mechanical Structures of the 482.6 mm (19") Series," IEC 297-4.

6.12   Bellcore Standards

1. "Equipment Physical Design for Reliability," FR-78.
2. "Components for Reliability," FR-357.
3. "Network Equipment-Building System (NEBS) Requirements Physical Protection," GR-63-CORE.
4. "Electromagnetic Compatibility and Electrical Safety Criteria for Network Telecommunications Equipment," GR-1089-CORE.
5. "Generic Requirements for Separable Electrical Connectors Used in Telecommunications Hardware," GR-1217-CORE.
6. "Generic Physical Design requirements for Telecommunications Product and Equipment," TR-NWT-000078.
7. "Generic Requirements for Assuring the Reliability of Components Used in Telecommunications Equipment," TR-NWT-000357.
8. "Reliability Assurance Practices for Optoelectronics Devices in Central Office Applications," TR-NWT-000468.
9. "Generic Requirements for Hybrid Microcircuits Used in Telecommunications Equipment," TR-NWT-000930.

6.13   Other Standards

1. "European Standard—Radiated/Conducted Emissions Requirements for Information Technology Equipment," EN 55022.
2. *Code of Federal Regulations,* Title 47, FCC CFR T47 C1 Parts 15, 18, & 68.
3. "Safety of Information Technology Equipment, Including Electrical Business Equipment," UL 1950.

7

Soldering and Cleaning of High Performance Circuit Board Assemblies

Dr. Charles G. Woychik

IBM Corporation
Endicott, NY

7.1 Introduction

The integration of all the components to the card requires a process in which excessive heat cannot be allowed to damage any of these individual elements, but it is also a reversible process. Ever since the beginning of the electronics industry, starting around the 1940s, solder has been a primary material used to attach components to a carrier. The forms of carriers have dramatically changed throughout the years; however, they have all been a composite structure comprising impregnated epoxy-cloth layers and copper circuitry layers. With time, the complexity of these cards increased, with high density circuits on each layer and also with an increasing number of layers. An example of a high performance assembly that consists of a high density circuit board along with fine-pitch–leaded surface-mount components is shown in Fig. 7.1. In Fig. 7.1 the heat sinks on two of the components are attached using an advanced low modulus thermally conductive adhesive to allow a very low thermal resistance of 0.8 C/W in the bond region of the chromated aluminum heat sink to the plastic overmolded package. Another example of an advanced type of electronic package is the tape ball-grid array (TBGA) package shown in Fig. 7.2. This package offers array surface-mount solutions for both flip-chip and tape automated bonding (TAB) peripheral die. By using ball grid array (BGA) surface-mount technology, the assembly yields are very high. Extensive reliability testing on the TBGA has demonstrated that the package design and materials choices effectively eliminate solder-joint fatigue as a failure mechanism. Another form of module is direct-chip attach (DCA) to a laminate carrier. Figure 7.3 shows a photograph of a direct-chip attach module (DCAM) developed at IBM. This early prototype DCA package showed that a flip-chip die could be successfully attached to a laminate substrate. The attachment of the chip to the substrate in the module is referred to as the *first-level interconnection*. Likewise, the interconnection of the module to the card is referred to as the *second-level interconnection*. In the following sections, a detailed description of both types of interconnection will be made.

7.1.1 First-level assembly using flip-chip attach

In this section emphasis will be placed on a growing form of first-level interconnection, flip-chip attach (FCA). IBM has been using this technology since the 1960s, primarily in high end applications. However, in recent years lower end applications have increased dramatically. The major difference between the FCA used by IBM in the early years versus that being used today is the melting temperature of the solder interconnect. When IBM used the FCA to attach the device to a ceramic

Figure 7.1 A printed circuit card used for card assembly. This card can accommodate surface-mount components as well as pin-in-hole (PIH) components.

Figure 7.2 Tape ball-grid array (TBGA)—an advanced type of electronic package.

Figure 7.3 A direct-chip attach module (DCAM).

substrate, a high melting Pb-rich Pb/Sn alloy was used, having a nominal melting temperature of about 320°C. Today, the applications for FCA require low cost laminate substrates that cannot withstand temperatures in excess of 240°C. As a result, the need to develop a low melt type of flip-chip interconnection became apparent. The low melt FCA demand is also extending to higher end applications, which is making the low melt a strategic technology solution for FCA.

The future demand for low melt FCA is expected to grow substantially provided the technology is capable of achieving the following requirements:

1. Industry standardization
2. Low cost method to bump die
3. Cost-competitive substrates
4. High yield assembly processes

There are many different methods to achieve a low melt flip-chip interconnection. However, there are two basic types: The first type uses a conventional Pb-rich bump and a eutectic Sn-Pb alloy to attach this high melting point bump to the mating pad on the substrate. The resulting duplex solder alloy interconnection is schematically shown in Fig. 7.4a. The other type of low melt flip chip interconnection can be achieved by using a low melt eutectic alloy for the bump and attaching this to the mating pad on the substrate. The resulting type of interconnection is shown in Fig. 7.4b. Both of these basic types of low melt FCA are used today. The benefits of each one will be elaborated on later in this chapter. As to what type will be pervasive in the future is yet to be determined.

The process in which a bumped die is attached to the substrate is critical to ensuring a reliable type of flip-chip interconnection. Included in the process is the proper materials selection that is compatible with the die and the laminate substrate. A generic process flow for FCA is outlined as follows:

1. Ensure bumped die is clean of major contaminants.
2. Ensure pads on laminate substrate are free of contaminants.
3. When solder is applied to substrate pads, ensure that the volume is within specifications and that the nominal composition corresponds to a eutectic Sn-Pb alloy.
4. Flux card pads.
5. Pick and place die on substrate pads.
6. Reflow.
7. Plasma clean the assembly.
8. Apply encapsulant.
9. Cure encapsulant.
10. Inspect integrity of underfill using sonoscan.

The critical materials selection focuses on the solder alloy, the flux types, and the encapsulant. When the flip-chip interconnection is of the duplex solder alloy type or the single low melt type, it is important to ensure that the low melt alloy corresponds with the 63Sn–37Pb eutectic composition. It should be mentioned that excessive contaminants in this alloy can affect the wettability of the sol-

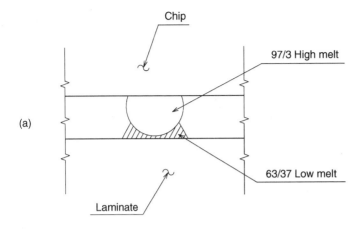

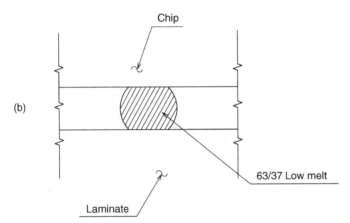

Figure 7.4 Illustration of two different methods to form a low melt flip-chip attach (FCA) type of interconnection. (*a*) In this configuration a high melt Pb-rich bump is attached to the substrate using low melt eutectic solder. (*b*) A completely low melt eutectic alloy is bumped on the die which in turn is attached to the mating pad. Here, either solder or no solder can be deposited on the mating substrate pad.

der. The next major material is the flux. Here it is desirable to have a high solids flux to ensure good wetting, especially when a Pb-rich bump is used. However, the high solids can inhibit the effective adhesion of the encapsulant in the next stage. Therefore, it is very desirable to use a low solids flux in order to minimize flux residues. This requires an inert atmosphere to prevent subsequent oxidation of the surfaces during the soldering operation. Finally, the encapsulant is a critical final material used to ensure that the chip is coupled to the laminate substrate.

When the preceding process is used to form a low melt flip-chip interconnection on a laminate substrate, the reliability is also a strong function of the die size. Basically, more problems arise with a larger die. Here intuition would have one believe that the larger die is less reliable than a smaller die. The predominant failure modes are either delamination of the encapsulant, most often at the die passivation interface, or chip cracking. Sometimes a combination of both can occur. The critical factor in minimizing or eliminating delamination is the flux type and the corresponding encapuslant material. It should be mentioned that defects related to the dicing of the die can also affect the overall tendency for die cracking. The success of low melt flip chip on a laminate requires very tight control of the semiconductor fabrication process as well as the laminate and assembly process used for the package.

7.1.2   Second-level assembly of ball-grid array packages

The next stage of assembly is to attach the component, or module, to the card. Here the use of a large solder ball between the laminate substrate on the module to the card pad is one method that has gained rapid acceptance over the past 10 years. This type of second-level interconnection is referred to as a ball-grid array (BGA). Today, most surface-mount packages are a BGA design. The two major types of packages that use BGA technology are classified into two groups: ceramic or organic.

In the ceramic BGA, more commonly referred to as CBGA, a Pb-rich solder ball is attached to the ceramic substrate using low melt eutectic Sn-Pb solder. Typically, a 35-mil 90Pb-10Sn solder ball is used. The module is then attached to the mating card pads using a eutectic Sn-Pb solder paste. The resulting type of second-level solder interconnection is shown in Fig. 7.5. This type of solder joint is known to produce reliable interconnections for substrate sizes up to 35 mm. For larger sizes, a solder column is used. This design is referred to as ceramic column-grid array (CCGA). Figure 7.6 shows a comparison of a CBGA package (top) with a CCGA type of package (bottom). The nominal standoff of the CBGA interconnection is 35 mils, corresponding to the Pb-rich ball diameter, whereas an 87-mil column height is typically used for a standard type of CCGA interconnection. Again, the solder column can be attached to the substrate using eutectic solder or the column can be directly attached to the substrate by forming the column and attaching to the pad in a single operation. This later design is preferred since a continuous high melting interconnection is formed that does not reflow at the top portion of the column during second-level assembly or upon subsequent removal during a rework process.

The organic BGA technology is rapidly growing as compared to CBGA/CCGA since the resulting reliability of the second-level interconnection is more reliable. Most often, the organic BGA packages use a full eutectic solder ball that completely collapses during reflow. Typically, an initial ball diameter of 35 mils is used. The coefficient of thermal expansion (CTE) mismatch between the mother card and the organic module is so low that the need to increase the standoff by using a high melt solder ball is not required for most applications. When the laminate substrate in the module is very thin, the effect of the die begins to dramatically reduce to an effective CTE of the module substrate to a point at which the module can behave more like a ceramic package than an organic one. In this case it is appropriate to use a high melting type of second-level interconnection such as that used in the CBGA package. Recent data at IBM have shown this to be true. Figure 7.7 shows an illustration of four different leading edge organic, or plastic ball-grid array (PBGA), types of organic packages.

The critical step in the second-level assembly process for a CBGA or an

Figure 7.5   A photograph of a Pb-rich type of BGA interconnection used on a ceramic ball-grid array (CBGA) type of module. A 35-mil 90Pb-10Sn solder ball is typically used to produce the BGA balls.

Figure 7.6 A photograph of two types of ceramic modules: A CBGA module having 35-mil Pb-rich solder balls (top) and a CCGA type module which has a 87-mil standoff (bottom).

organic low melt PBGA package is the control of the solder volume used to form the final interconnection. For a CBGA type of interconnection additional low melt solder is required to form the necessary connection between the Pb-rich ball and the card pad. However, in the case of the eutectic solder BGA ball, such as that used on the PBGA package, sufficient solder is present in the ball, and in certain cases only a minimal amount of solder paste is required, or just flux, to form the final interconnection. The fully collapsible feature of the eutectic ball can accommodate nonplanarity differences between the module substrate and the card. From this discussion it is apparent why the fully eutectic features offered in the organic BGA packages make this design preferable.

In addition to reliability, another factor in choosing a package is the cost. It is important to mention that typically when one refers to an organic package, it usually implies a comparison with a conventional wire bond type of low cost memory module. In the discussion thus far, most of the applications for a PBGA package with FCA are in the middle to high end range. This type of product requires a sophisticated laminate substrate design, which is, in general, the same price as that of a ceramic substrate, but in certain cases can be more expensive. Therefore, one major error in the justification for an organic package is lower cost. A major challenge the laminate supplier will have in the future is driving costs down, which I believe will drive even more applications toward use of PBGA packages.

The factors that drive the PBGA packaging solution are a high reliability BGA interconnection, elimination of the column type of interconnection as such on a CCGA package or even the Pb-rich solder balls used on the CBGA type of package, higher circuit densities that improve overall electrical performance, closer spacing between adjacent circuit and ground planes to reduce cross-talk, and easier assembly and reworking over comparable ceramic types of packages.

7.2   Solder alloys

In order to produce a reliable interconnection it is important to ensure that the assembly is not damaged by exposure to a high temperature. A low temperature soldering process has been used consisting of eutectic Sn-Pb solder with a rosin-based type of flux. Here, a low melting temperature solder alloy of 63 wt % Sn and 37 wt % Pb has been used to produce a reliable interconnection. A feature that makes this solder alloy attractive is that it is a reversible process, enabling the rework of an interconnection. Many of the different solder alloys used today are based on the binary Sn-Pb system. Therefore, by understanding the fundamentals of this alloy system, one can then extrapolate what will happen in other types of solder alloys.

7.2.1   Definition of a solder

A *solder* alloy consists of two or more elements that can wet to a surface, most often copper, and then react to form an adhesion layer, and, upon solidification of the alloy, produces an interconnection that has good

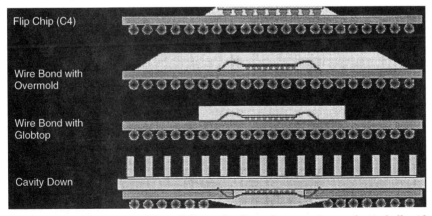

Figure 7.7 An illustration of four different leading-edge organic, or plastic ball-grid array (PBGA), types of organic packages. Flip-chip C4, wire bond with overmold, and wire bond with glob top are referred to as "chip up" package design, as compared with the "cavity down" design.

mechanical properties. Therefore, it is important first to understand the fundamentals of wetting and the experimental methods used to characterize wetting. Once the solder effectively wets the surface, it is important to understand the phase reactions that occur with the molten solder contacting the base metal, and then the subsequent solidification of the solder alloy. This discussion will emphasize the use of phase diagrams. Finally, the resulting metallurgical alloy composition and resulting microstructure is important in assessing the final mechanical reliability of the solder interconnection.

In a solder alloy one has to have an active element to produce the resulting adhesion layer. Again, in the case of the classic system in which a Sn-Pb solder alloy wets a copper base metal, the molten Sn in the solder reacts with the base copper metal to produce a Cu-Sn intermetallic phase layer. The Cu-Sn phase diagram, as shown in Fig. 7.8, can be used to predict the types of phases that must form between the solder alloy and the mating base Cu surface. Here the two phases that comprise this intermetallic phase layer, or generically the adhesion layer, consist of the Cu_3Sn and Cu_6Sn_5 phases. Again, when referring to the Cu-Sn phase diagram, in order to have a thermodynamically stable adhesion layer, this intermetallic phase layer must consist of these two phases. If both phases are not present, in a case where eutectic Sn-Pb alloy is contacting the copper metal, a nonstable condition will result which will cause reliability problems later on. For most alloys used today, the Sn is the active element that forms the intermetallic phase layer. However, in other alloys, indium acts in a similar way. However, it is important to emphasize that at least one of the elements in the molten solder alloy needs to react to form this adhesion layer.

Once the adhesion layer, or intermetallic phase layer forms, the net effect of this reaction on the alloy composition is to deplete the Sn content, resulting in the nominal alloy composition shifting to the Pb-rich portion of the phase diagram. This type of shift in nominal alloy composition will occur in any type of solder.

Some typical solder alloy systems that are commonly used are listed in Table 7.1. All of the compositions are in weight percentage, and most of these alloys are from Manko.[1]

In solder alloy selection it is desirable to use a binary system over a multicomponent, since the resulting phase structure can be predicted using binary phase diagrams. However, when a unique melting point range is required, or specific mechanical properties need to be achieved, then either a ternary or a multicomponent system may be required. A general rule of thumb is to keep the interconnection design as simple as possible. The fewer elements, the easier it is to troubleshoot if any problem arises.

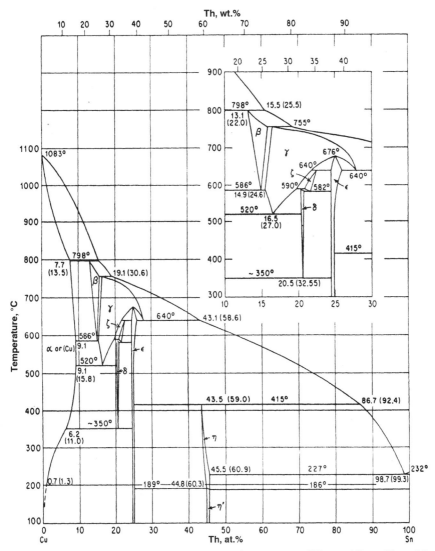

Figure 7.8   The Cu-Sn phase diagram shown in Constitution of Binary Alloys. (From M. Hansen and K. Anderko, *Constitution of Binary Alloys,* 2d ed., McGraw-Hill, NY, 1958.)

7.2.2   Solidification of binary solder alloys

To better understand the solidification of solder it is best to use a hypothetical eutectic system to explain eutectic and noneutectic solidification. Figure 7.9 illustrates a eutectic phase diagram, consisting of A and B elements. The composition X_e corresponds to the binary eutectic composition on the phase diagram. At this point the composition corresponding to X_e solidifies just like a pure metal, going from a liquid to a

TABLE 7.1   Typical Solder Alloy Systems

Binary alloys	Ternary alloys	Multicomponent alloys
Sn63Pb37	Sn37.5Pb37.5In25	Sn8.3Pb22.6Bi44.7In19.1Cd5.3
Sn42Bi58	Pb15In80Ag5	Sn12Pb18Bi49In21
Sn50In50	Sn70Pb18In12	Sn13.3Pb26.7Bi50Cd10
Sn70Pb30	Sn50Pb47Sb3	Sn12.5Pb25Bi50Cd12.5
Sn60Pb40	Pb90In5Ag5	Sn11.3Pb37.7Bi42.5Cd8.5
Sn50Pb50	Sn1Pb97.5Ag1.5	
Pb50In50	Sn62Pb36Ag2	
Sn96.5Ag3.5		
In90Ag10		
Pb75In25		
Pb60Sn40		
Sn95Cd5		
Sn95Ag5		
Sn20Pb80		
Pb97.5Ag2.5		
Pb95In5		
Sn3Pb97		

two-phase solid structure. When this alloy is at a temperature above the eutectic temperature, T_e, the alloy is a homogenous liquid consisting of A and B atoms. Upon cooling below the eutectic temperature, the liquid alloy segregates into an A-rich α phase and a B-rich β phase. The compositions of both the α and β correspond to the X_A and X_B compositions on the phase diagram (see Fig. 7.9). The unique property of a eutectic alloy is that it has a lower melting temperature than either of the pure element constituents, but it behaves like a pure metal in that it has a single invariant point at which solidification occurs.

The prior discussion is based on a pure eutectic alloy; however, in most cases, one must understand the solidification of a noneutectic solder alloy. In this situation, we assume an alloy composition of X_1, which corresponds to a B-rich alloy, as shown in Fig. 7.10. Here, B-rich refers to the fact that this composition is on the B side of the eutectic point. When the alloy is above the tèmperature T_1, a completely homo-

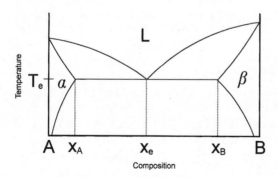

Figure 7.9 A hypothetical eutectic system comprised of A and B elements. The eutectic composition is denoted by X_e and the compositions of the A-rich α phase and a B-rich β phase at the eutectic temperature are X_A and X_B, respectively.

geneous liquid mixture exists. Upon cooling to the temperature of T_1, a small amount of precipitate begins to form of the β phase with an initial composition of X_β'. As the alloy begins to cool below T_1, more of the β phase forms, and the composition of this phase corresponds to the solidus line of the β-phase region on the phase diagram. The corresponding liquid that remains has a composition that also shifts. This changing liquidus composition corresponds to that of the liquidus curve on the phase diagram. When the alloy is at a temperature slightly above the eutectic temperature, two different phases exist, the solid β having a composition gradient from X_β' to X_B and the remaining liquid which has a composition corresponding to that of the eutectic composition. The solid β phase is commonly called a *proeutectic phase,* since it forms prior to the last solidifying eutectic. In most alloy systems, the shape of the proeutectic phase is that of a treelike structure which is called a *dendrite.* Figure 7.11 is a micrograph of a typical dendritic structure in a binary Sn-Pb solder alloy. When the alloy is cooled to below the eutectic temperature, the remaining liquid, which corresponds to the eutectic composition, totally converts to a duplex-phase structure as discussed in the previous paragraph.

In order to predict the resulting phase structure that forms upon solidification of a solder alloy, the use of phase diagrams is essential. In this discussion, again, we will reference the eutectic Sn-Pb system, assuming that the soldering reaction occurs on a base metal of copper. As was discussed previously, the solder must wet the copper surface, and then the intermetallic phase reaction occurs to produce the adhesion layer. However, upon completion of the intermetallic phase formation, the nominal solder alloy composition is shifted towards the Pb-rich side of the phase diagram. In the case of using a

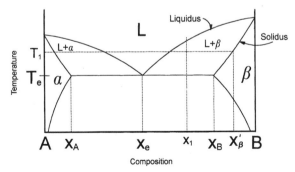

Figure 7.10 A hypothetical eutectic system comprised of A and B elements. Here a B-rich alloy, denoted by the composition of X_1 solidifies to form a proeutectic Pb-rich phase followed by the last remaining liquid of eutectic composition solidifying into the lamellar duplex phase structure.

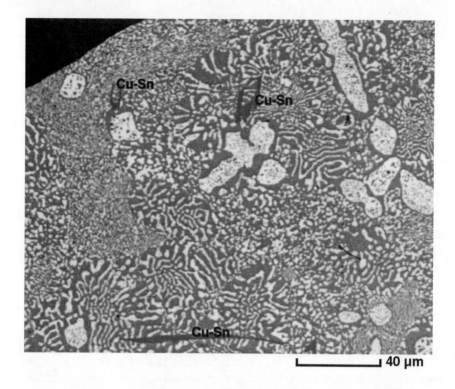

Figure 7.11 A classic micrograph of a noneutectic Sn-Pb, having a composition to the Pb-rich side of the eutectic point. Here, the presence of the Pb-rich dendrite along with the last solidifying eutectic structure can be seen.

eutectic Sn-Pb alloy initially, the resulting alloy that solidifies will no longer be the eutectic composition, but an alloy on the Pb-rich side of the phase diagram. This noneutectic alloy composition will result in the formation of a Pb-rich proeutectic dendrite, along with a two-phase eutectic which forms in a lamellar structure. Again, in Fig. 7.11, which shows a micrograph of a noneutectic Sn-Pb alloy after solidification, one can clearly see both the cored Pb-rich dendrites and the lamellar Sn-rich and Pb-rich phases that comprise the last solidified eutectic phase mixture.

7.2.3 Mechanical properties of solders

In this section an overview of the mechanical properties of solder will be given. However, one needs to realize that it isn't important to discuss only the mechanical properties of solder but also the overall properties of the specific type of solder interconnection. For example, Figure 7.12 illustrates three different types of solder interconnections, all of which use eutectic Sn-Pb solder and a copper pad or lead. Figure 7.12a shows a pin-in-hole (PIH) type of interconnection, Figure 7.12b shows a gull wing–leaded surface-mount lead, and Fig. 7.12c shows a full eutectic ball-grid array type of interconnection. It is well known that if everything on the package remains constant, but only the lead configuration changes, the final reliability of the assembled package will be different.

Another important consideration is the temperature at which the interconnection operates. For example, an important measure of susceptibility to creep is the homologous temperature, T_H, which is the operating temperature, in degrees Kelvin, divided by the melting temperature of the specific solder alloy, in degrees Kelvin. A general rule of thumb is that if the T_H is greater than 0.5, atomic diffusion can occur and, therefore, creep becomes a viable mechanism. Another dominant mode for solder-joint failure is fatigue caused by cyclic displacement–controlled strain, or strain-controlled fatigue. The combination

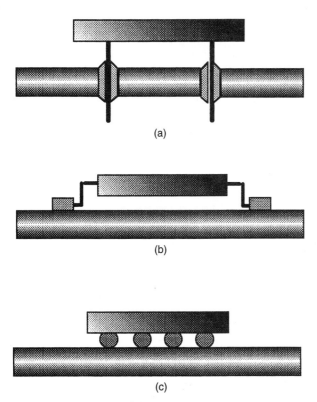

(a)

(b)

(c)

Figure 7.12 An illustration of three different types of solder interconnections: (a) standard type of pin-in-hole solder interconnection; (b) gull-wing type of surface-mount solder interconnection; (c) full eutectic ball-grid array (BGA) type of solder interconnection.

of both strain-controlled fatigue and creep contribute to the ultimate failure of a solder interconnection.

The first set of general bulk mechanical properties of solder are Young's modulus E, shear modulus γ tensile strength, and Poisson's ratio v. A plot of the normal stress-strain curve for a eutectic Sn-Pb alloy wire as a function of strain rate is shown in Figure 7.13. It is important to emphasize the strain-rate sensitivity of solder. In this specific example, the ultimate tensile strength is about 7800 lb/in^2 for a strain rate of 7.9E-3/s, whereas, for a much lower strain rate of 7.9E-6/s, the ultimate tensile strength is only 1000 lb/in^2. In addition, other mechanical properties, such as the elastic modulus, also change. When 2 wt % Ag is added to a eutectic composition, the 62%Sn-36%Pb-2%Ag alloy is known to have a higher strength. Roger Wild at IBM referred to this alloy as a creep-resistant alloy used for solder interconnections that are subjected to high strains. In Fig. 7.14, experimental data for a 2% Ag alloy wire was measured at a function of temperature. In Fig. 7.14 one can clearly see the dramatic influence of temperature on mechanical properties. The trends shown in both Figs. 7.13 and 7.14 apply to both the eutectic and 2% Ag alloys. The Poisson ratio can be calculated by using the formula

$$\gamma = \frac{E}{2\,(1 + v)}$$

where E = Young's modulus, v = Poisson's ratio, and γ = shear modulus. By experimentally measuring the shear and Young's modulus, one can calculate Poisson's ratio. The need for determining these parameters is for finite element modeling of a particular type of electronic package. Not only is it important to have the optimized mesh design for the model but also to use accurately measured material properties. Unless the model is done correctly and with precise measured data, the model will be meaningless. Good finite element modeling predictions are a result of expertise that is developed over years of experience. Figure 7.15 shows the results of using finite element modeling to predict high stress points, which can be used to highlight the reliability risk sites of this particular product.

In most electronic packages, the solder joint is the weakest element. Much emphasis has been placed on quality control specifications in order to ensure that the solder joints are optimized. In testing a soldered electronic package, the reliability test should reproduce the actual environment in which the product will live. Currently, there are numerous time-consuming reliability tests that have been used for a number of years—tests such as accelerated thermal cycling, power cycling, and shock and vibration—to name a few. These tests are very time consuming, and the electronics industry is looking for alternative

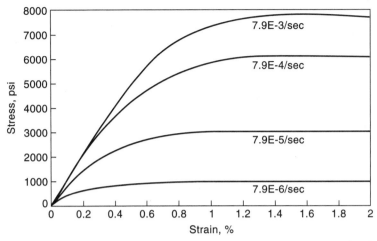

Figure 7.13 A plot of the normal stress-strain curve for a eutectic Sn-Pb solder alloy. 63Sn-37Pb wire pull tests at 20°C with four orders of strain rate.

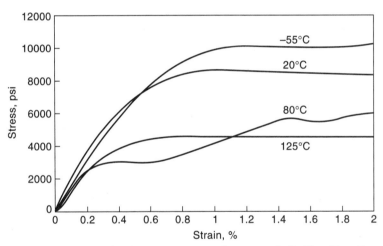

Figure 7.14 A plot of the shear stress-strain for a eutectic Sn-Pb solder alloy. 62Sn-36Pb-2Ag wire pull test; strain rate = 0.79%/s with temperature.

test methods that both accurately predict the product quality and shorten the qualification process. The test methodology to accomplish this objective is referred to as the *mechanical deflection system* (MDS) developed by Aleksander Zubelewicz at IBM.[16] The MDS method is intended to replace some of the traditional test techniques, such as the conventional time-consuming power cycling and accelerated thermal cycling test methods. In MDS, the cyclic out-of-plane deformation is imposed on an assembled printed circuit board. A portion of this type of deformation is then transferred to the solder joints, which in turn causes the solder joint to fail. Extensive research has shown that the failure mechanism in hardware tested using MDS is the same as that tested using conventional methods.[16]

A comparison of CBGA reliability was made by using both conventional thermal cycling and the MDS technique. Both types of tests used identical test vehicles. In all of the failed CBGA solder joints, the cracks initiated in the eutectic solder at the card side of the solder joint, which is typical for this type of interconnection. A micrograph of both types of solder-joint failure is shown in Fig. 7.16. Figure 7.16a shows the failed CBGA joint after thermal cycling at 20 to 55°C, with a frequency of 0.1 Hz, and Figure 7.16b shows a failed CBGA joint after MDS testing at ambient conditions, with an out-of-plane deflection of 0.6°/in of board length and a frequency of 0.1 Hz.

7.2.4 No-lead solder alloys

During the early 1990s, a great deal of attention was placed on replacing Pb in conventional solders with another appropriate element. One of the primary drivers in making this change was to ensure that the melting ranges for all of the solders used in the industry were not dramatically changed. The next major factor was the mechanical properties of

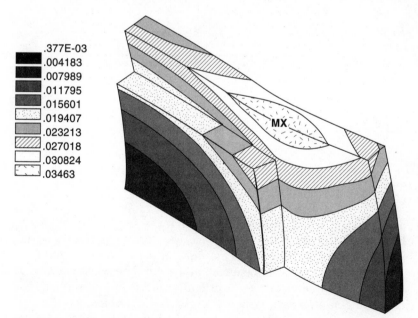

Figure 7.15 An illustration of finite element modeling analysis used to predict the high stress points in an electronic package.

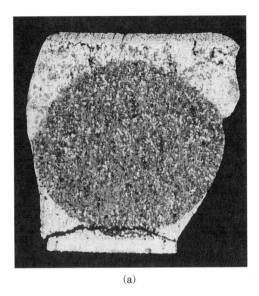

(a)

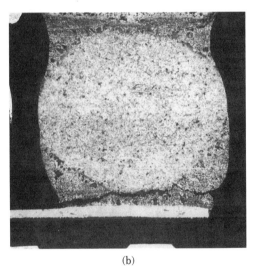

(b)

Figure 7.16 (a) CBGA solder-joint failure after conventional thermal cycling using the following conditions: 20 to 55°C and a frequency of 0.1 Hz. (b) CBGA solder-joint failure using MDS testing. The test conditions are: ambient temperature, out-of-plane deflection set at 0.6°/in of board length, and a frequency of 0.1 Hz.

the solder alloy, and the last concern was the wettability of the alloy. Many of the proposed alloys to replace Sn-Pb were based on the Sn-Ag binary system. For example, some of the proposed alloy systems to replace the Sn-Pb system were the Sn-Ag-Bi and Sn-Ag-In and Sn-Ag-Bi-In systems. The net conclusion from all of this work was that the current eutectic Sn-Pb system had the benefit of invariant solidification, due to the eutectic solidification, superior mechanical properties for both the eutectic and Pb-rich alloy compositions, and the best wetability.

Today, there is limited use of the Sn96.5Ag3.5 alloy, which melts at 221°C for second-level assembly. The Sn-Bi eutectic continues to be used for certain high end applications, mostly for immersion wave soldering a high aspect via in a large board. Also there is a no-Pb solder offered under the trade name of Castin by AIM Research, which is Sn-based.

The problem in replacing Pb-containing solders is that the most recent BGA and FCA technologies are all developed based on the Sn-Pb system. Even if a better alloy were identified, the cost to requalify all of the current products would be unacceptable. In addition, the quantity of solder used for Pb types of solders is so small compared with other uses, such as industrial paints, batteries, etc., that it would not make sense to put the focus on the electronics industry. Therefore, the urgency to change solders to a no-Pb alloy is not as critical today as earlier in the decade. The reliability database is so extensive on conventional Sn-Pb types of solders and their derivatives, along with the fundamental understanding of how these alloys wet and are processed for solder paste or in wave soldering applications, that a change would be extremely difficult, but not impossible. My view of this topic is that unless there is government legislation, the change to a no-Pb solder will not occur. The cost factors alone would be too great to justify any change, and the new alloys would require additional attributes in order to make any change at a global level possible.

7.3   Fluxes Used in the Soldering Operation

In the soldering operation, a Sn-based alloy is melted and makes contact with the suitable base metal, most often copper. In this description, it is assumed that there are no oxides or any other type of contamination that can impede the soldering reaction. In reality, one must be very concerned with the presence of any existing contamination or surface oxides on the solder or the copper base metal. The oxides can inhibit any flow of the molten solder and prevent any wetting. Therefore, it is imperative that all of these types of contaminates are eliminated during the soldering operation. Most often, the solder and base metal are contaminated to a point such that poor, or no, metallurgical reactions will take place. A flux is, therefore, essential to clean and prevent any subsequent oxidation.

7.3.1   Role of flux

A suitable flux will become activated prior to the onset of the melting of the solder alloy to ensure that the reduction of oxides has begun to occur. Also, the role of the flux is to encase the molten solder alloy in a protective enclosure to prevent any subsequent oxidation. The higher the solids content in the flux the greater the likelihood that it will be more effective at preventing any further oxidation during the soldering process. The classification method used for fluxes is presented in the following section, which is important to assess the degree of oxide and contamination removal by the flux.

In order to discuss the flux mechanism it is important to realize that the flux needs to fully encapsulate the solder and the base metal. The liquid flux needs to be activated at a temperature prior to reaching the melting temperature of the solder alloy, and the flux needs to maintain its capability to fully encapsulate the molten alloy during the reflow process. Not only does this prevent oxidation but it also helps to effectively transfer heat. Certain fluxes may contain a fugitive activator, such that after the necessary reduction of oxides, the harmful or active species will become volatile and can, therefore, escape into the atmosphere.

7.3.2 Classification of fluxes

Based on a recent article by Alvin F. Schneider from Alpha Metal, Inc.: "All soldering fluxes fall into 1 of 24 classifications.[2] The major focus of this article is to provide a basic understanding for the "Requirements for Soldering Fluxes," J-STD-004. This standard is used by both commercial and military electronics assembly contractors.

The J-STD-004 requirement divides all fluxes into one of four categories, and the category is determined by the basis of the composition. Each composition category is then subdivided into one of six flux activity levels, according to the corrosive or conductive properties of the flux and flux residues. The combination of the four composition categories and the six flux activity levels result in 24 flux classifications. The flux composition categories are defined as

Rosin (RO)

Resin (RE)

Organic (OR)

Inorganic (IN)

The flux activity levels are assessed by the combined results of copper mirror testing, corrosion testing, surface insulation resistance, and halide content. The three activity levels are

L—low flux/flux residue activity

M—moderate flux/flux residue activity

H—high flux/flux residue activity

Each of these three activity levels is categorized as either containing or not containing a halide in the flux. Therefore, the three flux activity levels now result in six flux types:

L0

L1

M0

M1

H0

H1

Of all of the different methods for classifying fluxes, I find the one just outlined to be the best. Over the years there has been much confusion on classification, and this method eliminates that confusion. Finally, for newer types of fluxes, such as low solids no clean and volatile organic compound (VOC) free, more work is required in properly classifying these materials using the J-STD-004 flux requirements.

7.4   Solder Pastes

A method in which both the flux and solder alloy are applied to the card to produce an interconnect is in the form of a solder paste or solder cream. Most often, a fine powder is suspended in the fluxing material and its vehicle. The ratio of solder power is typically 80 to 90 wt%, which corresponds to about 50 vol% . The amount of flux used is generally more than necessary for the interconnect formation, which is usually the result of the consistency requirements in the solder paste formulation. The paste can be applied by screening, extruding, brushing, or rolling onto the surface. Heat is applied, usually in the range of 220 to 230°C for a dwell above 183°C, which is the melting temperature of eutectic Sn-Pb, ranging from 1 to 1.5 min.

7.4.1   Description of solder paste

A *solder paste* is a homogeneous mixture of solder powder, flux, and a vehicle which is able to form a metallurgical reaction with the adjoining pads when using the appropriate soldering process to activate the flux and reflow the solder particles. The solder powder is the only component that permanently remains to form the interconnection. The flux aids in the wetting, and the vehicle is specifically designed to create the necessary rehological properties of the solder paste. The rheology of the solder paste is the critical element in how well the solder paste adapts to the manufacturing process. Some of the major challenges in developing new solder pastes for leading-edge processes are: (1) solder pastes must be able to withstand prolonged open exposure to temperature and humidity without undergoing changes that degrade its performance and (2) they must be able to withstand long delays or pauses in the process and resume printing with the print quality being equal to that before the pause.[3] A list of the solder paste performance criteria for the solder paste screening operation, the reflow soldering operation, and postsoldering process is outlined by J. S. Hwang in Table 7.2.[4] Solder pastes are well suited for high volume manufacturing processes used for attaching both surface-mount and also pin-in-hole components. The ratio of metal powder, flux, and vehicle is important in the optimization of the solder paste to meet aggressive demands of high volume screen printing as well as achieving the necessary tack for fine-pitch soldering applications. The details of the compositions are very highly guarded secrets of solder paste suppliers, and are never divulged to a customer.

7.4.2 Types of solder pastes

Today, most of the solder pastes can be classified into two basic groups: no-clean and water-soluble pastes. The no-clean chemistries consist of either rosin- or resin-based flux systems, along with a suitable vehicle system, that have sufficient activity to clean all the different types of contaminants during the soldering operation. An important feature of a no-clean type of solder paste is that the remaining residue does not pose any reliability concerns. The two major problems with any remaining flux residue is the potential for corrosion or migration. As the circuitization densities increase, the potential for one or both of these problems increase.

Sometimes there is an aesthetic concern over the remaining flux residue. When this is the case, it is recommended to use a lower solids flux in order to minimize the remaining flux residue. Water-soluble fluxes usually contain organic or inorganic acids in the flux. Again, the vehicle system is engineered to be compatible with the flux system. In general terms, these types of solder pastes react very well and produce good solder interconnections, due to their more active flux system as compared with that of a no-clean type of solder paste. In these systems, it is very important to ensure complete removal of the flux residue produced after soldering. Since ionic species remain in the flux residue, they can react with the surface of the card assembly and cause corrosion and/or migration problems. There is also a third type of solder paste, which is based on a rosin type of flux system. These systems require organic solvents to clean any remaining flux residues. They are very effective solder pastes that produce good interconnections; however, the rosin nature of both their flux and vehicle systems requires the sophisticated organic cleaning systems which are not environmentally friendly in today's manufacturing world that is emphasizing VOC-free types of card assembly processes. However, in certain instances, mostly high end processors, the reliability objectives can be met with only a rosin type of solder paste system. Today, the rosin types of solder pastes are used in sophisticated military and space electronics applications.

TABLE 7.2 Solder Paste Performance Criteria for Solder Paste Screening, Reflow Soldering, and Postsoldering

Before soldering	During soldering	Postsoldering
Physical appearance	Compatibility with surfaces	Residue quantity
Stability and shelf life	Flow properties prior to becoming molten	Residue cleanability
Cold slump	Flow properties when solder molten	Residue corrosivity
Dispensability through fine needle	Wettability	Electromigration
Screen/stencil printability	Dewetting phenomenon	Joint appearance
Tack time	Soldering balling phenomenon	Joint voids
Open time	Bridging phenomenon	Joint strength
Adhesion	Wicking phenomenon	Joint microstructure
Quality and consistency	Leaching phenomenon	Joint integrity versus mechanical fatigue
		Joint integrity versus thermal fatigue
		Joint integrity versus thermal expansion coefficient difference
		Joint integrity versus intrinsic thermal expansion anisotropy
		Joint integrity versus creep
		Joint integrity versus corrosion-enchanced fatigue
		Joint integrity versus interfacial intermetallics
		Joint integrity versus bulk intermetallics

7.4.3 Application methods
for solder pastes

For manufacturing card assemblies, the method most often used to dispense solder paste is stencil printing. In certain applications, a screen is used. Whether using a stencil or a screen metal mask, the forces contributing to the paste transfer are similar. A schematic illustration of a screening technique is shown in Fig. 7.17. Some of the important parameters during solder paste screening are summarized as follows:

Squeegee:

    Blade material

    Blade hardness

    Blade configuration

    Squeegee speed

    Down stop pressure

    Substrate

    Material type

    Flatness

    Size

SOLDER PASTE PRINTING

AN ADDITIVE PROCESS WHERE SOLDER PASTE IS PULLED ACROSS A TEMPLATE BY A SQUEEGEE, FILLING ITS APERTURES AND THEREBY TRANSFERRING THE SOLDER PASTE TO THE BOARD.

- MOST WIDELY USED SOLDER PASTE APPLICATION TECHNIQUE
- PARAMETER EFFECT & OPTIMUM SETTINGS VARY BY PRINTER
- EMPLOY SCREENS OR STENCILS TO DEFINE PRINT PATTERN

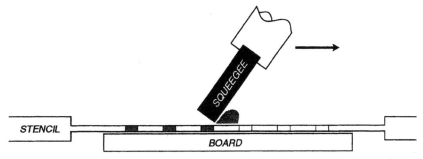

Figure 7.17 A schematic illustration of the screening operation.

Screening equipment:

Stencil/screen size

Pattern

Thickness

Material type

Snap off

Solder paste:

Paste rheology

Metal content

Alloy composition

Paste stability

Paste tackiness

Compatibility with screen and squeegee

The main operation parameters to monitor are[4]: squeegee speed, squeegee down stop, squeegee pressure, and snap-off. All four of these parameters are interrelated, and by changing one, all of the others will most definitely require some type of change. When the process is optimized, it is important to be able to repeat the process for a substantial number of cycles without any human intervention.

Of the printing, placement, and reflow operation in the assembly process, the most demanding operation is the solder paste printing. A survey of process engineers concluded that two-thirds of surface-mount defects were traced to the solder paste printing operation.[3] Another question that was asked in a survey of 350 surface-mount technology engineers was: "Which solder paste characteristics or properties contribute to the most printing defects?" There were four properties which accounted for about 85% of the response[5]:

1. *Stencil life.* The length of time that the solder paste can be worked on the stencil without drying out.
2. *Response to pause or to idle time.* This is the ability of the solder paste to be left idle on the stencil, and then recover, without kneading, to deliver acceptable prints.
3. *Insensitivity to changing temperature and humidity.* This relates to how the stencil life and response to pause change with changing temperatures and humidity.
4. *Solder paste release.* This relates to the ability of the solder paste to release cleanly from the stencil after multiple prints.

Items 2 through 4 are also strongly affected by the type of solder paste. The decision to use a no-clean or a water-soluble material can play a dramatic role in the performance of these three items. For example, water-soluble solder pastes are highly sensitive to humidity and temperature. In certain card assembly facilities where there is no humidity or temperature control, the performance of a water-soluble paste can be dramatically different in the winter than in the high humidity months in the summer (northern hemisphere). It is well known that these types of solder pastes exhibit poor screening definition in the summer, with slumping and poor tack. However, in the case of no-clean solder pastes, these materials are less sensitive to humidity and are known to perform much better in higher humidity conditions than water-soluble pastes.

7.4.4 Evaluation of solder pastes

Since the soldering process is the major source of assembly defects and the solder paste screening operation the major component, a quantitative test method for evaluating solder paste is necessary. Some of the key factors affecting solder paste performance are: the equipment and setup parameters, stencil fabrication method, pitch of components, lead density of components, operator skills, component and card solderability, and temperature and humidity. A suitable solder paste needs to be selected such that it can perform as best as possible in the particular application that is specified.[6] Here are some of the more common methods to characterize solder pastes:

Solder ball testing. Here a circular print with a 0.250-in-dia solder paste is made onto a ceramic substrate. The ceramic is used since the solder paste will not react with it, and, therefore, only the properties of the solder paste are being evaluated. Reflow using the recommended profile by the solder paste supplier, and characterize the amount of solder splash, or solder balling, using a scope with a 10× magnification. Characterize the number and size of the solder balls. The fewer, the better.

Residue. This applies to no-clean chemistries. Here, one needs to evaluate the amount and clarity of the remaining flux residue after reflow. It is important to ensure that there are no active ionic species, as discussed previously in Sec. 7.3. Also, in most cases a card assembly is probe-tested. Here it is important that the remaining residue can be penetrated by the probe, and also not gum up the probe. High residue fluxes are known to degrade probe life, and, therefore, it is recommended to use a lower solids flux to reduce the amount of flux residue.

Solder paste wetting. There are numerous tests to measure card and lead solderability. The best is to print a 0.250-in deposit on a larger coupon that is 1 in in diameter or greater. The surface of this coupon should represent the type of surface on the card that will be soldered. I recommend using the stencil thickness specified in the process. Subject the coupon to the reflow process and characterize the degree of solder spread on the coupon. Also, look for solder balling and the degree of flux residue migration. It is easy to characterize these aspects when using an organic copper coating, such as entek or a gold coating, since these surfaces provide a contrast with the solder. In the case of a hot air soldered leaded (HASL) surface, or an immersion tin surface, it is more difficult to differentiate the degree of solder paste wetting, particularly on the HASL surface.

Slump. Here one wants to characterize the ability of the solder paste to stay well defined after screening. The solder paste deposit can collapse, or spill over, which can cause bridging. It should be mentioned that the printing parameters or the stencil properties can be contributors; however, the rheology of the solder paste is a considered a prime contributor. The author recommends using a standard Institute for Interconnecting and Packaging Electronic Circuits IPC-21 or IPC-A-20.3 tests to characterize slump.

Tack. The solder paste must have a certain amount of adhesive strength to temporarily hold the component in place prior to the reflow operation. Tack is typically characterized as preplacement and postplacement. For preplacement, the solder paste is dispensed, and the tack is evaluated after certain time intervals. The maximum time interval should represent the maximum time the solder paste can be screened prior to component placement. In this test a simple passive component can be used along with a force gauge to characterize the tack force. Again, the same test methodology can be used to measure postplacement tack force.

Worklife. Here the concern is how long the solder paste can remain on the stencil before it can no longer print adequately. It is recommended to use the stencil representing the card to be screened. Finer pitches and higher component densities become more critical to evaluate. Screen a few times, and characterize the solder paste deposit looking for skips and bridges; wait 1 h and repeat the process. Leave the solder paste on the stencil.

7.5   Conductive Adhesives

In electronic packaging there are many different applications for conductive adhesives. The most common use is for heat-sink attach. In this application, a thermally conductive adhesive is used to conduct heat from the component to an extruded aluminum heat sink. Electrically conductive adhesives are now being used for many different types of low end applications. Toys, games, and calculators can be assembled using an electrically conductive adhesive. In this section I would like to focus on the use of electrically conductive adhesives in electronic packaging. There are two basic types of electrically conductive adhesives: anisotropic and isotropic. The aniosotripic materials are conductive in only one direction, and are sometimes referred to as z axis conductive adhesives, or z-axis films (ZAP). In the case of the isotropic adhesives, they behave in the same way as a solder, conducting in all three directions.

The anisotropic materials are becoming very well developed and are now being used in many different applications. In the case of the isotropic materials, they are less well developed, but they, too, are showing up in more and more applications each year. A new application for this technology is for producing a flip-chip type of interconnection using an isotropic electrically conductive adhesive.[7] There a three basic elements that make up the development of this technology: the materials development of the conductive adhesive, the assembly process development, and the development of an equipment set to form reliable flip-chip attach interconnections.

In the materials development, a polymer metal solvent paste (PSMP), which consists of a thermoplastic binder, conductive silver particles, and a solvent, was formulated to have the appropriate rheological properties to achieve fine-feature deposits. While the rheological properties are important for dispensing the PSMP material, it is important to ensure that the three-dimensional electrical conductivity is maintained by ensuring that the silver filler material density remains above the critical percolation volume fraction. As with any paste material, the PMSP was shown to have a long shelf life at room temperature, making it suitable for a manufacturing environment. Once the PSMP is deposited and the solvent is driven off, the resulting bump of material is referred to as a polymer metal composite (PMC). Once the PMC is formed, it is important to ensure that the adhesive of the bump is sufficient to yield a low contact resistance. This material was shown to exhibit low alpha particle emissions, which can minimize the occurrence of radiation-induced soft error rates in electronic devices.

In the assembly process development, the optimized paste formulation focused on paste deposition, chip bonding, encapsulation material selection and processing, and PMC reliability testing. The paste deposition process involved the development of a photo-bumping process to deposit PMSP on the wafer. Flat surface bumps ranging in height from 4 to 8 mils in diameter on a 20-mil pitch were achieved with a high yield. The chip bonding process was optimized for high speed assembly under controlled time, temperature, and pressure conditions. The optimized parameters had a pressure of 135 lb/in^2, a bonding temperature of

235°C, and a bonding time of 30 s. Of all of the encapsulants evaluated, the Matsushita CV5183S material was identified to be the best.

The Universal Instruments Corporation (UIC) was selected to develop the equipment set to do placement and bonding die with PMC bumps. As a part of this work, a computer-aided cost-estimation (CACE) software tool was developed to help understand the trade-off issues when comparing the PMC process to those using conventional solder-based flip-chip attach methods.[8] The set of optimal time, temperature, and pressure bonding parameters were found for the attachment of IBM-supplied PMC bumped flip chips to laminate substrates using the prototype bonder developed by UIC. With total bond cycles times of about 1 min, competitive assembly processes were achieved using the electrically conductive adhesive (ECA) material for flip-chip attach as compared with the conventional solder types of processes used for flip-chip attach. Figure 7.18 shows a photograph of a PMC bumped wafer, including some of the major technical challenges for this type of technology. This work was support by DARPA contract no. DE-FC-04094AL98817.

7.6   Assembly Processes

This section will be divided into two subsections. The first subsection will discuss the processes that are used to do first-level assembly, focusing on FCA. Different methods to bump a wafer with solder will be reviewed as well as methods using conductive adhesives. The second portion of this section will discuss the second-level assembly technologies. Again, the primary emphasis will be given to the state-of-the-art technology, that is, BGA technology. Both FCA and BGA technologies have great potential for packaging applications on organic substrates. There continues to be much interest in organic or plastic BGA (PBGA) packages, primarily because they can meet critical electrical requirements that are not achieved using conventional ceramic packages. Even with the demand migrating toward organic packages, a high demand still exists for ceramic BGA packages. Therefore, the primary discussion for BGA will focus on CGBA and CCGA technologies, much of which can also be applied toward PBGA technology.

7.6.1   First-level assembly—flip-chip attach

As cited previously there are many different approaches to form a FCA type of interconnection. Solder is used primarily, but today we are starting to see more applications using both isotropic and anisotropic electrically conductive adhesives. In this section I will discuss conventional solder FCA methods, focusing on only one technique that uses a

Goals, Objectives and Main Technical Approach	Major Technical Accomplishments
--Achieve Polymer-Metal-Solvent Paste Formulation that is screenable to form Polymer-Metal Composition bumps for Flip Chip Attachment --Deposit PMSP on 125mm Wafers with 0.010" diameter bumps, 0.004" high on 0.020" pitch --Identify key process variables for bond strength and fracture --Identify commercial underfill encapsulant compatible with PMC interconnection to organic cavities --Achieve high reliability bonding under standard stress conditions --Design and assemble manual and automated PMC bonding equipment --Establish optimal time-temperature-pressure bonding parameters --Model costs compared to conventional Flip Chip solder-based systems	--PMSP Shelf life of six months demonstrated --Bumping established for 0.008" diameter on 0.020" pitch with finer pitches feasible --Bump Reworkability demonstrated --Stability of resistivity under industry standard stress conditions demonstrated --Underfill material identified --Equipment test beds identify optimal time-temperature-pressure bonding parameters

Array of PMC bumps on a wafer for 0.008" x 0.004" on 0.020" pitch

Major Impact of Technology and Technology Transition Plan
--Competitive costs compared to solder processes --Technology extension to finer bump & pitch dimensions on large wafers --Contact Resistance Stability and High Reliability over wide temperature ranges. --Low alpha-particle emission reducing soft error rates --Flux and lead elimination --Licensing of PMSP and wafer bumping processes --Commercial availability of bond assembly equipment --Domestic industrial and IBM dual use application development for microelectronics products.

Figure 7.18 A summary of some of the goals, objectives, technical accomplishments, and illustration of a polymer metal composite (PMC) bump on a wafer. This work was support by DARPA, contract no. DE-FC-04094AL98817.

Pb-rich bump along with eutectic solder on the carrier.

In the solder-type FCA method, a wafer is bumped with either a low melt or a high melt solder and then diced. The next major step is that the corresponding substrate, either ceramic or organic, has to have a suitable pad metallization, such that the bumped die can then be placed on the pad and, upon reflow, a FCA interconnection will form. The type of pad metallization is a function of the melting temperature of the solder bump. In the case of a low melt solder bump, the mating pad on the substrate would be a copper pad protected with an organic type of protective coating such as entek. In some cases, a Ni/Au metallization is used; whereas, in the case of using a high melting temperature solder bump on the die, if a ceramic carrier is used, then usually only a Ni/Au pad metallization is used, and the solder will nicely reflow on this type of pad during a high temperature soldering reflow process. However, as mentioned previously, there is much interest in organic packages, and, therefore, when a high melting temperature bump is on the die, a lower melting solder needs to be deposited on the organic substrate. In this type of application, a copper entek pad on the organic carrier is usually covered with a low melt eutectic Sn-Pb solder deposit.

The most cost-effective method to deposit solder on the substrate is by solder paste screening. Upon screening the substrate with solder paste, the paste is reflowed and the resulting hemispherical bump creates some difficulty maintaining good alignment of the solder bump when placed on the solder-deposited pad. Therefore, the low melt deposit is coined to flatten the surface. One advantage of performing this operation is that the flattening operation makes it easier to visually examine the surface of the solder deposits to identify and major deviations from the nominal. In this type of flip-chip soldering operation, the high melt bump, usually a Pb-rich solder bump, is placed on the coined deposit of a eutectic Sn-Pb solder alloy. In most cases a very low solids type of flux is used in order to minimize flux residue. The previous discussion is a generic overview of soldering a high or low melt bump to either a ceramic or an organic substrate.

Since the focus of the industry is on PBGA, we will concentrate only on the organic carriers. Upon forming the FCA solder joint, the assembly is then subjected to a plasma treatment in order to remove any contaminants that are a by-product of the soldering operation. This explains why low solid fluxes are recommended. After the plasma, the FCA assembly is then encapsulated. Table 7.3 illustrates the many types of low melt FCA that can be accomplished either by low melt solder directly on the die or by using a high melting point bumped die and screening solder on the carrier. Another method to achieve low melt FCA with a high melt solder bump is to place some low melt solder on the high melt bump. This type of technology is referred to as *solder on chip* (*SOC*).

The methods outlined here are just a few of many that are being developed every year by the growing market for FCA. However, I believe that only a few of these methods will become the industry benchmarks, as they will become the optimized, lowest cost processes to bump a die, which is the major cost of FCA.

In order to best illustrate a low melt FCA soldering process, I will provide an overview of a standard type of FCA method, referred to in Table 7.3 as the solder paste screening process. In this operation, a Pb-rich bump is used along with screened solder paste on the carrier to provide the low temperature solder to form the interconnection. The process is outlined in Table 7.4.

In this process, both the wafer processing and laminate substrate

TABLE 7.3   Types of Low Melt FCA

	Low melt bump	High melt bump with low melt on carrier	Solder on chip (SOC)
Carrier surface	Ni/Au Entek	Ni/Au Entek	Ni/Au Entek
Solder deposition method	Not required	Solder paste screening Solder plating Solder jetting Solder injection	Not required

TABLE 7.4 Solder Paste Screening Process

Wafer processing	Laminate substrate processing	Module assembly
Deposit BLM (ball-limited metallurgy)	Coat carrier pads with protective layer	Flux coined solder deposits on laminate carrier (no clean)
Deposit high melt (97Pb-3Sn) bump	Screen solder paste (water soluble)	Place bumped die on carrier pads
Reflow	Reflow	Reflow
Probe testing	Water wash clean	Plasma clean
Inspect	Inspect hemispherical solder deposits	Encapsulate
Reflow	Coin	Sonoscan
	Inspect for volume specifications	Flux backside BGA pads with water-soluble flux
		Place eutectic Sn-Pb solder balls
		Reflow
		Water wash clean
		Inspect BGA balls
		Cap attach
		Inspect

processing are done independently and in parallel. For the wafer processing, the first step is to produce an adhesive layer between the base Al metallization that makes up the active circuitry of the semiconductor to the final solder. Here a ball-limiting metallurgy (BLM) layer that consists of Cr-Cu-Au is deposited. Next, the high melt Pb-Sn alloy is deposited on the BLM pad. In this step, the amount of solder deposited needs to be precisely controlled to ensure that the proper standoff will be produced in the final FCA interconnection. The bump is reflowed to ensure proper mixing of the alloy and to produce the necessary metallurgical reaction with the BLM to adhere the bump to the pad. The bump is then probe-tested, usually resulting in an indentation in the bump. The bumps along with the test data will identify the good and the bad die. It should be mentioned that, at this stage, it is very important to have known, good die (KGD). This has been a tremendous challenge facing the industry in the adoption of FCA. There are still many techniques being developed to address this issue, but few companies have been successful in adopting effective testing for KGD. In order to make FCA more acceptable, low cost and effective KGD methods will be essential.

In the laminate substrate screening process, the pads on the carrier are typically either an entek or a Ni/Au plating. Realize that the type of coating will not only be for the flip-chip pads but also the BGA pads. There have been recent articles published by the Universal BGA Consortium which highlights some of the reliability risks associated with Ni/Au BGA pads.[9] Solder paste is screened on the pads for flip-chip attach. In this operation, it is recommended to use a water-soluble paste, since any flux residue can be removed by water cleaning. After the screening, the paste is reflowed again to metallurgically attach the solder to the pad. The solder deposits are cleaned and then inspected to ensure that all pads are covered with solder and that no bridging has occurred. The next operation, which is the coining of the solder deposit, is very important to make this process successful at producing a high yield FCA interconnection. In this operation, the hemispherical solder deposit is flattened. This operation has been not only useful to maintain flatness of the coining flip-chip laminate pads, but also is a very effective way to inspect for eutectic solder volume control on the laminate carrier.

For the final module assembly, the first step is to flux the coined laminate carrier with no-clean flux. Here, most often, no-clean flux is used, since for most application, the gap between the die and the carrier is so small that it becomes very difficult to effectively remove flux residue when either rosin-based or water-soluble fluxes are used. The bumped die is then placed on the carrier, and then reflowed using a nitrogen atmosphere. After the solder joints are formed, even when a low solids, no-clean flux is used, a plasma cleaning operation is often recommended to remove flux residue and prepare the surface to effectively adhere to the encapsulant. After the plasma, the encapsulant is dispensed, usually on one side or on two adjacent sides. In this step, it is important to ensure that the encapsulant is dispensed without any voids or only a minimum number and size of voids. After the encapsulant is dispensed, it is then curved according to the manufacturer's specifications. In order to ensure that the encapsulant was effectively dispensed and that the voiding is within the specification, the encapsulated chip is then sonoscaned. A typical sonsocan showing excessive voiding in the underfill is shown in Fig. 7.19.

The next operational steps are directed at the BGA balling operation. Here, the back side of the laminate is fluxed, again recommending a water-soluble flux; the eutectic Sn-Pb solder balls are placed on the pads, usually held in place with a template, and then reflowed in a nitrogen atmosphere furnace. The module is then water washed to remove the harmful water-soluble flux residue, and then inspected. Typical defects are bridging, or nonwets, usually due to pad contamination. Both of these defects are readily apparent during visual inspection. In most flip-chip modules, a cover plate is attached to the top of the die, in order to protect this surface from handling damage. In this operation, a low modulus silicon adhesive is used to attach either a copper or a stainless-steel cap.

A final remark concerning this type of package. Since the die which has a CTE in the range of 2 to 3 ppm/°C is anchored to the laminate

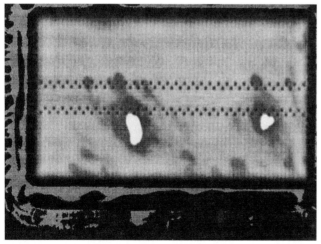

Figure 7.19 A sonoscan showing excessive voiding in the underfill. This type of condition can greatly impact the reliability of the first-level interconnections.

substrate which has a CTE of 18 to 20 ppm/°C using the encapsulant, the amount of stress produced in this package is extremely high. This type of module can be very susceptible to either die cracking or encapsulation delamination. Two papers have been published on these two types of failure mechanism in PBGA packages that have large die attached using FCA.[10,11] In addition, it is well known that defects due to the dicing operation are known to greatly impact the amount of module fallout due to die cracking. Very little fundamental work has been done to better understand the relationship between various defects produced during the dicing operation and how they impact die cracking.

7.6.2   Assembly processes—second-level assembly of BGA modules

The use of BGA modules has gained substantially over the past decade. In general, BGA assembly encompasses many different types of packages. The ceramic ball-grid array (CBGA) modules were the first type of BGA package in the industry, followed by ceramic column-grid array (CCGA) packages, which were able to increase the reliability of the interconnect by increasing the height of the interconnection. The organic BGA packages also became available in a number of different types, and typically, these packages are referred to as plastic ball-grid array (PBGA), with most of these package types being wire bond chip up and cavity designs that are either protected with overmold or glob top. A unique type of PBGA package is the tape ball-grid array (TBGA) design, which can accommodate either a perimeter die that is thermocompression-bonded to the package or a flip-chip version of the package. Both types of packages are known to be highly reliable and offer superior electrical performance. One of the distinct differences between the ceramic BGA technologies and the organic BGA is the type of solder ball. All of the ceramic types use a Pb-rich solder that is attached to the module and the motherboard using low melt eutectic solder. Most of the organic BGA packages, except for TBGA, use a low melt eutectic solder BGA ball. From an assembly and rework concern, the CBGA technology can be considered to be more challenging, especially during rework.

BGA modules cover a range of performances—from low end that use plastic types of packages to CBGA/CCGA modules that are high density, high performance surface-mount packages. Since the array of balls cannot be inspected once the module is attached to the card, it is very important to optimize the package design and the assembly process in order to achieve both high yields and reliability of the package. A major reason for the rapid acceptance of BGA technology is that these types of packages achieve very high yields, typically in the range of 1 to 3 ppm/lead, and form the required interconnection configuration to ensure that the reliability is met.

7.6.2.1 Ceramic ball-grid array assembly. CBGA packages are very robust in design. Lidless CBGA packages are not moisture sensitive, and lidded packages are either JEDEC level 2 or 3 moisture classification. They have a very good shelf life and maintain BGA coplanarity very well during shipment due to the ball type of BGA configuration. Figure 7.20 illustrates a CBGA type of solder interconnection using a 0.89-mm (0.035-in)-dia 10/90 Sn-Pb solder ball that is joined to the ceramic substrate and the printed wiring card with eutectic Sn-Pb solder. The high melt solder ball does not reflow during card assembly and, therefore, creates a predetermined standoff of 0.89 mm (0.035 in). The CBGA module is mounted on a card using a "dog bone" pad design.[12] A solder dam between the landing pad and the via is required to prevent loss of solder paste to the via. A nonsolder mask–defined pad is used, which has been shown to produce a high reliability in the BGA interconnection. The nominal pad diameter of 0.72 mm (0.0285 in) at the top of the pad is required in order to ensure joint reliability. Warping of the card is also considered to be a major contributor to the z axis tolerances for all surface mount technology (SMT) devices, and especially the BGA modules. Generally, by using the optimized paste printing processes, the card warpage is overcome during the reflow operation. It is recommended to consider the following design pointed

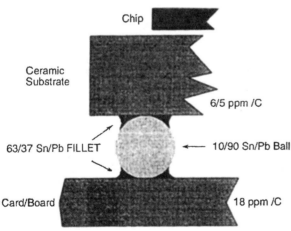

Figure 7.20 An illustration of a CBGA type of second-level solder interconnection. An 0.89-mm (0.035-in)-dia Pb-rich solder ball is joined to both the ceramic substrate and the organic card using eutectic Sn-Pb solder.

to prevent local warpage that can occur during assembly and rework processes[13]:.

Symmetrical card cross sections required to minimize warping of the card, especially during the reflow operations.

Minimize local card warpage created by adjacent components that "anchor" the printed circuit board (PCB) such as large PIH connectors.

Maximize uniformity of assembly thermal mass across the PCB.

Consider the form factor of the card. Large thin cards are more likely to warp during processing and may require fixtures.

Establish a reliable PCB supplier that delivers quality product and not "potato chip" cards. It is very difficult to specify and enforce a card flatness specification other than the IPC standard.[14]

A flatness requirement of 0.025 to 0.076 mm (0.001 to 0.003 in) average is common for a 32-mm (1.26-in) CBGA site on the card. There are other considerations that need to be considered in the design of the card. When fine-pitch components are integrated on the card, and adjacent to a CBGA module, enough clearance between the two modules sites is required for a step-down stencil. When tented or plugged vias are used on the card, they can become entrapped with contaminants, such as flux residue, which can be very difficult, if not impossible, to remove.

CBGA packages are assembled to printed circuit boards using standard SMT tool and processes.[13] A single-sided process flow is as follows:

1. Apply solder paste on card.
2. Solder paste verification.
3. Place components.
4. Reflow solder.
5. Clean, if required.
6. Test.
7. Rework component, if required.
8. Attach heat sinks.

CBGA packages are also capable of being attached in a double-sided card configuration. An important consideration here is the ability of the inverted reflow position to maintain the module on the card. The maximum weight per lead for a backside module in reflow has been experimentally determined to be 0.08 g/lead.[13] Although the solder-joint structure of the backside versus the front side modules is not that different, the reliability is reduced when modules are back to back with a shared via in card configuration.[13] A double-sided assembly processes is as follows:

1. Solder paste application on back side.
2. Solder paste verification.
3. Place components.
4. Reflow solder.
5. Clean, if required.
6. Solder paste application on front side.
7. Solder paste verification.
8. Place components.
9. Clean, if required.
10. Test.
11. Rework, if required.
12. Attach heat sinks.

In both of these assembly processes, the solder paste screening operation is the critical process step. Most often, either a type 3 or 4 solder paste is recommended, depending on whether additional fine-pitch components are used on the card. The solder alloy is a eutectic Sn-Pb, which is 50% vol of the paste mixture, or 90wt%. CBGA assembly is known to work very well for both no-clean and water-soluble types of solder pastes. For CBGA assembly, a minimum paste volume of 0.089 mm^3 (4800 mils^3) is recommended to meet reliability objectives for this type of package. A minimum z axis print height of 0.18 mm (0.007 in) is required in order to prevent opens and to facilitate high yields. It should be noted that this minimum height is more critical as the package size increases: a 21-mm (0.83-in) module may be able to tolerate a shorter print height than a 32-mm (1.26-in) module. A maximum volume of 0.16 mm^3 (10,000 mils^3) is recommended, since exceeding this quantity will result in bridging.

After screen printing, the paste deposit needs to be inspected, especially for the high reliability requirements for ceramic types of BGA packages. Here, the paste volume is measured on a fully automated tool, such as a synthetic vision systems (SVS) tool. After screening, the module is placed on the solder paste deposit. The CBGA type of module is very forgiving during placement. As long as the solder balls contact the paste, they will self-align during reflow. Both the top eutectic on the substrate side of the BGA ball, and the solder paste reflow, allows the high melt BGA ball to float and to equilibrate between both the substrate and card pads. This ability to float better enables the module to accommodate warpage in the card.

During the solder reflow operation, it is important to maintain the solder profile recommended by the paste supplier. A recommended reflow profile for Kester 244 no-clean solder paste is shown in Fig. 7.21. The peak temperature specification of 220°C is driven by the requirement to limit the amount of Pb from the BGA ball into the molten eutectic alloy. The cool-down rate is well known to affect

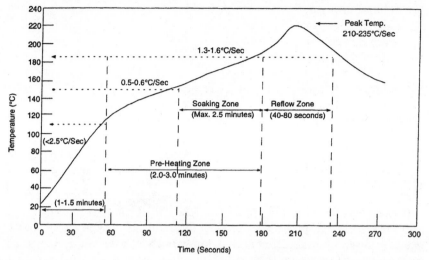

Figure 7.21 A manufacturers' recommended solder profile for Kester 244 no-clean solder paste.

the yield and reliability of the interconnection.[13] Two conditions that are of concern are

1. *Card pickup during oven exit.* Here the solder joints are still molten, and if an operator manually picks up the card to remove it from the conveyor, he or she can distort the resulting solder joint. Cards should not be removed from the conveyor until the solder is solidified.
2. *Uneven cooling of the top and bottom card surfaces.* This case can produce warping of the card, which can potentially cause ball lifting off pads during solidification. This is also known to produce column cracking on CCGA modules.

7.6.3 Ceramic ball-grid array rework

As was mentioned earlier, CBGA modules have a very high assembly yield. However, the rework operation is not as robust as the initial assembly operation. It is important to consider the same critical requirements in initial assembly and apply these to the rework operation. Again, the most important steps are the solder paste volume, card warpage, and control of the reflow profile. The CBGA rework process is outlined as follows:

1. Site thermal profiling
2. Module removal
3. Site dress and touch up

4. Clean, if required
5. Solder application
6. Module placement
7. Module reflow
8. Clean, if required
9. Inspection and electrical test

Specialized rework tools are required which have the following capabilities: hot-gas heating, computer-controlled temperature profiles, calibrated vision system, automatic vacuum pickup and component placement, menu-driven software control, and complete PC data logging. The key process variables are preheat temperature, peak joint temperature, and solder reflow time. To remove the module, a hot-gas reflow tool with a bottom card heater (bias bay) is required. The entire carrier should be preheated to between 75 and 125°C prior to the application of hot gas from the top heater. Preheating is a critical step that minimizes the card warping during removal and limits the thermal shock to the card. The maximum preheat temperature is about 10°C below the glass transition temperature of the card. Typically, the adjacent module is limited to less than 150°C. Moisture-sensitive components require a bake-out of 24 h at 125°C prior to rework to prevent any type of moisture-driven damage to the component during the rework cycle such as popcorning.

Thermal profiling is required for module removal, site flattening, and module reflow. A requirement in thermal profiling is to have a thermocouple read the joint temperature. The thermocouple can be attached with eutectic solder and coated with a thermally conductive epoxy to ensure contact. Another method is to drill a hole through the backside of the component into the solder joint and fill the drilled hole with thermal conductive adhesive. The latter method is highly recommended. Recent experiments have shown that the amount of supercooling can be as great as 20° below the melting temperature of eutectic solder.[15]

A hot-gas tool with vacuum pickup is used to remove the module, and a full carrier preheater is required. In this step the component is removed, while minimizing any type of thermal damage to the card such as pad lifting or warping. After module removal, the site on the card must be dressed. On most pad sites, the 10/90 ball is still on the card pad, held in place with the eutectic solder. All of these Pb-rich balls need to be removed. The remaining eutectic solder on the card pads is nonuniform in coverage. The desired method to dress the card pads is to use a solder vacuum. In this process, pad-by-pad removal of excess solder is done using a vacuum. The resulting surface is similar to a HASL card surface. The manual solder vacuum technique is known to impart very little damage to the card. Cleaning is then required only if water-soluble flux is used during the dressing operation.

Eutectic solder must be applied to either the card pads or to the balls on the module. Solder paste is preferred since it allows for tack to hold the module in place during reflow and to compensate for any z axis variation. As mentioned earlier, the eutectic solder volume needs to be precisely controlled in order to achieve an optimized solder-joint configuration. There are several ways to apply eutectic solder: Screen paste on the CBGA module balls, screen paste on the card pads, use solid preforms, and syringe dispense paste on the card pads. Module screening is a preferred process to apply eutectic solder on the new module that will be reattached to the card. A unique clamshell-type screening fixture is required to accomplish this operation. The weight of the solder paste applied is a critical process control parameter.

When the solder paste has been applied to the module, a split optics prism method is used to place the component precisely on the card pads. Typically, the module is placed with a force between 0.2 and 1.7 kg (0.5 and 4.0 lb), to ensure that the solder paste contacts at a minimum of 50% of the pad surface; however, it must not contact any of the vias. The applied eutectic solder must be reflowed in order to allow for metallurgically wetting. Preheating the entire card to between 70 and 125°C is critical in order to minimize any card warping. The thermal profile needs to use the same requirements as specified for initial attach. Finally, the adjacent modules must be protected from exposure of overheating, resulting in the need for proper thermal profiling prior to reworking the actual product. The same inspection techniques that were implemented for initial attach are also used for rework.

7.7 Cleaning

Since the late 1980s there has been much emphasis on no-clean soldering fluxes and pastes. Over the past few years much advancement has occurred in the field of no-clean materials optimization. Today, many card assemblies are operating without reliability concern in the field using no-clean fluxes. However, when using no-clean fluxes for soldering, the process is very demanding and requires tight process control to ensure that the assemblies meet all specifications. When using no-clean fluxes, components and cards procurements and handling and soldering become very demanding, usually resulting in nonforgiving processes. As mentioned in earlier sections, the assembly of a PBGA flip-chip module requires very low levels of no-clean flux residue during the flip-chip soldering operation. Some no-clean fluxes do not have the activity and require more active water-soluble fluxes. Another benefit of a water-soluble flux is that it can be removed by a water wash, which can enhance the adhesion of the underfill material.

When using water-soluble fluxes and pastes for soldering card assemblies, the soldering process is usually more robust. However, it should also be mentioned that certain no-clean fluxes and pastes can also be cleaned after the soldering operation. There are many important reasons for cleaning at the card assembly level.[17,18] Not only does the soldering process become more robust but also problems related to flux residues are no longer a concern. In a recent article by F. Cala, he cites four benefits to using postassembly cleaning.[19]

1. *Ease and flexibility in the procurement of incoming bare boards and components.* A major requirement when using low residue no-clean soldering processes is that the incoming hardware must be clean and highly solderable. This will usually require ongoing monitoring by the assembly facility to ensure that these requirements are met. If the surfaces of the components and boards are not met, it may be a very costly ordeal to correct the problem at the supplier and could jeopardize critical assembly output.

2. *Absence of postsolder cleaning results in a significantly more demanding and unforgiving process.* One of the goals of postsolder cleaning is that not only is the flux residue removed but also any other types of contamination that may have resulted from either the component or board fabrication, assembly, and handling. Another feature of this operation is that any remaining solder balls on the surface of the assembly are removed.

3. *Cleaning will accommodate current and future device trends and is not at odds with the technology.* Today, a number of devices are extremely sensitive to flux residues such as high frequency devices. For these applications, low residue no-clean fluxes and pastes cannot meet the requirements imposed by these types of devices. Therefore, postassembly cleaning is essential.

4. *Postassembly cleaning significantly decreases the unknowns.* By using a postassembly cleaning operation, the assembly facility eliminates a lot of unknowns in the soldering process.

Even though the use of no-clean materials will continue to grow, certain applications continue to require postassembly cleaning. When offering the cleaning option, more active fluxes can be used that will ensure high solderability, thus resulting in a reliable solder interconnection. Today, many effective and environmentally friendly types of cleaning equipment exist. Many types of CFC alternative cleaning agents are also available, along with closed-looped systems that ensure effective cleaning and are environmentally responsible.

The benefits of cleaning have thus far been well identified. Cleaning is necessary to meet a customer's requirements for quality and reliability, to perform subsequent manufacturing operations, and to give the finished assembly an aesthetically pleasing appearance.

Some of the types of contaminants to be cleaned are

Particulate matter such as lint, debris, metal shavings, etc.

Handling contaminants such as fingerprints, dry human skin, and hair

Processing debris such as oil and greases used in metal cutting operations

Solder paste from screen printing stencils and misprinted boards

Postsoldering residues such as unreacted flux residue, reacted flux residue, and solder debris

Postassembly operations such as router debris

Cleaning is performed during all stages of the manufacturing process that require joining or contact between two surfaces and when soiling occurs. For electronic card assemblies, cleaning is performed at many different stages:

During the manufacture of the printed circuit board

During the manufacture of the electronic component

If a card needs to be rescreened due to poor screen printing initially

To remove solder paste from stencil openings

After soldering components to the printed circuit card

After performing rework

After profiling the assembly to the final form factor

Prior to the attach of heat sinks

Prior to functional testing

Prior to applying adhesives such as encapsulants, potting materials, beading materials, and conformal coatings

There are basically two different methods for cleaning. One is by air blowing and the other uses a liquid contact with enhancements. In air blowing, ionized air is used for electrostatically sensitive assemblies performed in a ventilated capture hood. The nozzle is designed to maximize the amount of debris removal, usually incorporating an air-pressure control device to prevent damage to the assembly. When using liquid contact with enhancements to clean, the efficiency of cleaning increases as more enhancements are added. The most fundamental form of liquid cleaning is by immersion, usually in a static tank. The next level of cleaning incorporates agitation, spray capability, and increased temperature of the cleaning fluid. The most efficient form of cleaning equipment contains in-line tanks using a countercurrent arrangement. In certain applications, a boiling fluid can be employed such as FC70 chlorofluorocarbon chemicals that were predominant over 10 years ago but are now mostly phased out due to environmental concerns. High pressure sprays are added along with high volumes of fluid. Ultrasonics are sometimes also incorporated to enhance the cleaning effectiveness.

In order to have effective cleaning, additions are made to the solvent. Today, water is the most common type of solvent. Specific solvents are designed to remove unique residues. In the design of the solvent, it must be compatible with other assembly materials, must meet safety and water disposal requirements, and must meet reliability and quality criteria. There are two types of residues generated from the soldering operation: nonpolar residues are the by-product from soldering with nonactivated rosin fluxes and polar residues are the by-product from soldering with activated rosin fluxes, which are most often water-soluble fluxes and resin-based fluxes. The nonpolar residues require organic solvents to remove them from the surface of the assembly. Typical organic solvents used are: CFCs, chlorocarbons, HCFCs, hydrocarbons, alcohols, esters, and terpenes. Polar residues are most effectively cleaned with a polar solvent such as water. Sometimes a saponifier is added to the water to enhance its cleaning ability.

There are two basic types of cleaning equipment: batch and in-line, or continuous. Batch processes are used for low volume, highly specialized types of circuit board card assembly. An in-line type of cleaning process is used for a high volume soldering process such as that used to assemble personal computers or for the assembly of engine and transmission controllers which is used in the automobile industry. Since the modern automobile contains so many different types of electronic assemblies, the use of water-soluble fluxes and pastes is declining and there are now more no-clean fluxes and solder pastes used in order to eliminate another costly step in the manufacturing process.

Here are some recommendations to optimize the cleaning operation:

- Minimize the time from soldering to cleaning.
- Use controlled amounts of flux.
- Ensure that soldering process parameters are in control. Excessive heat decomposes flux, creating residues that are difficult to remove.
- The complexity of the assembly determines the cleaning chamber configuration, enhancements, and cleaning parameters. For example, low standoff components, usually less than 10 mils, require:
    Low surface tension cleaning fluid
    High pressure sprays and high volumes of cleaning fluids
    Longer residence times in cleaning sump
    Ultrasonic agitation

There are a number of tests that can be used to measure cleanliness. The simplest is by a visual examination at a specific magnification to look for flux residue and to characterize its appearance. Ionic contamination testing is a more quantitative form of characterizing the amount of active ionic species present on the assembly. This test characterizes only the average ionic contamination over a predetermined surface area of the assembly. More quantitative techniques have been developed to characterize only specific regions of the cards, one of which is surface insulation resistance measurements. Here the effect of surface ionic contamination is assessed to determine if there is any long-term reliability concerns for the assembly. To find the latest specification for cleaning, refer to the IPC web page on the Internet.

References

1. H. W. Manko, *Solders and Soldering,* 2d ed., McGraw Hill, New York, 1979, p. 132.
2. A. F. Schneider, "Understanding the Flux Requirements of J-STD-004 and Its Relationship to the Soldering Requirements of J-STD-001B," Alpha Metals—Internal Technical Report, downloaded from their web site on 1/11/99 (www.alphametals.com).
3. R. Herber, "Rapid Implementation of Advances in Solder Paste," *Inline,* 1999, pp. 12–16.
4. J. S. Hwang, "Solder Paste Technology and Applications," *Solder Joint Reliability: Theory and Applications,* John H. Lau, ed., Van Nostrand Reinhold, New York, 1991, pp. 38–91.
5. C. Bastecki et al., "What Do Time, Temperature, Humidity, and Production Pauses Have in Common?" Alpha Metals and MPM Corporation Technical Report, October 15, 1997, Jersey City, NJ.
6. P. Zarrow, Alpha Metals Documents from ITM an Independent SMT Consulting Firm, January 1999, Jersey City, NJ.
7. "High Performance, Low Cost Interconnections for Flip Chip Attachment with Electrically Conductive Adhesive," Final Report, DARPA TRP No. DE-FC-04094AS98817, January 1998, Arlington, VA.
8. D. L. Santos et al., "A new Electronics Packaging Process Justification via Cost Estimation and Animated Simulation," *Industrial Engineering Research, Conference Proceedings,* Inst. Ind. Eng., Norcross, Georgia, 1977, pp. 626–631.
9. George Westby and Anthony Primavera, Universal BGA/BGA Consortium, Advanced Technologies SMT Laboratory, 1997, Binghampton, NY.
10. L. Gopalakrishnan et al., "Encapsulant Materials for Flip Chip Attach," 48th Electronic Components and Technology Conference, May 25–28, 1998, Seattle, Washington, pp. 1291–1297.
11. M. Ranjan et al., "Die Cracking in Flip Chip Assemblies," 48th Electronic Components and Technology Conference, May 25–28, 1998, Seattle, Washington, pp. 729–733.
12. "Design Standard for Rigid Printed Boards and Rigid Printed Board Assemblies," IPC-D-275, Institute for Interconnecting and Packaging Electronic Circuits, 1999 (from www.ipl.org).
13. C. Milkovich and L. Jimarez, "Ceramic Ball Grid Array Surface Mount Assembly and Rework," Document APD-SBSC-101.0, August 1998, IBM Corporation, Endicott, NY.
14. M. Reis et al., "Attachment of Solder Ball Connect Packages to Circuit Cards," *IBM Journal of Research and Development,* September 1993, vol. 37, no. 5, pp. 597–608.
15. C. G. Woychik and D. Henderson, IBM Corporation, Endicott, New York, private communication, January 1999.

16. Aleksander Zubelewicz et al., "Mechanical Deflection System—An innovative Test Method for SMT Assemblies," EEP-Vol. 10-2, Advances in Electronic Packaging, ASME 1995, pp. 1167–1177.
17. F. Cala, and A. Winston, *A Handbook of Aqueous Cleaning for Electronic Assemblies,* Electrochemical Publications Limited. Asahi House, Port Erin, Isle of Man, 1996.
18. B. N. Ellis, *Cleaning and Contamination of Electronic Components and Assemblies,* Electrochemical Publications, Ayr, Scotland, 1986.
19. F. Cala, "The No-Clean Issue," article from the Alpha Metals Web Site, March 1999 (from www.alphametals.com).

Environmentally Conscious Printed Circuit Board Materials and Processes

Dr. John W. Lott

DuPont iTechnologies
Research Triangle, North Carolina

8.1 Introduction

During the past 8 years, the printed wiring board (PWB) industry has grown from a poor stepchild of the electronics industry, driven almost out of business by overseas competition and tight environmental regulations, to a progressive and aggressive industry continuously evolving with each new technological challenge. It has developed its own recognition of what it needs to do to reduce its environmental impact and accept its responsibility in the environmental as well as technological supply chain. Today it faces many challenges (see Fig. 8.1), both environmental and technical. We will examine, in this chapter, the environmental drivers and the benefits to be derived from making our industry environmentally conscious. We will see that while being environmentally conscious is the right thing to do, it will ultimately be the *only* thing to do with respect to meeting technical and economic challenges. *Environmentally conscious manufacturing* means that we are aware of the impact that our process and materials make and are striving to reduce that impact in a sound technical and economic manner.

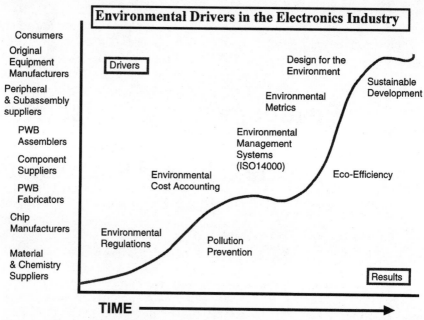

Figure 8.1 Environmental drivers of the industry.

8.1.1 Description of the industry

The printed wiring board industry has been an integral part of the electronics industry since the 1950s, when industrial designers found that they needed to eliminate the cumbersome wiring harnesses used in electronics devices of the time. Most boards were made using primitive liquid resists and gross etchants to form single-sided boards into which tube sockets and, later, leaded components and transistors were soldered. With the advent of dry-film resist, a much more controlled process was possible and the circuit traces became more uniform and reliable. Double-sided and eventually multilayer boards followed. Flexible laminate materials helped eliminate board-to-board or chassis-to-chassis wire harnesses even further and board features continued to shrink. Leaded components eventually disappeared or were minimized so that nearly all components can be surface-mounted, eliminating the need for most through holes. With the advent of microvia technologies, even greater increases in circuit density should occur.

The processes and materials to make either single layer boards or complex flex-rigid multilayer boards have changed, depending only on the characteristics needed in the board. Circuit features, lines, trace, via pads, bonding pads, etc., are all produced either using a subtractive process of etching away unprotected copper or using a semiadditive approach whereby copper is plated on a board selectively leaving thinner areas between the lines which are later etched away. Figure 8.2 shows a generic schematic of a process for making inner-layer circuits and then a multilayer board. Because most of these processes involve the use of chemical reactions to either remove or add copper or even prepare a copper surface for adhesion to another material, there are waste streams associated with each step. Often there are multiple steps leading up to or following a particular fabrication step. In addition, boards must be cleaned before the next step is performed to protect both the board being manufactured and the various baths into which the board must be moved. Each of these types of baths has some waste treatment associated with it. The more aggressive the bath, generally, the more complex the waste treatment required. Inherently, then, the process for making a printed wiring board requires a lot of water, it is usually subtractive at least in part, and the residues from each process step must be captured as solutions.

After a board is prepared, it will be tested and populated with packages and components. These are usually added in an assembly house which must solder on the components in one of several ways. As was noted previously, most components often had leads which were inserted into plated-through holes and the holes filled with solder. Most components now can be surface-mounted, that is, they are attached to pads which may or may not have holes or vias in them. The solder can be applied by a wave, using a hot-air solder leveling machine (to apply a thin uniform solder coating to all exposed copper), solder plating which can be reflowed, or most likely solder pastes which are screened onto the pads and act as a kind of "glue" to hold the component in place while the solder is reflowed using infrared or hot-air heating.

The populated boards are then connected via housing assemblies to create units of a piece of equipment, the CPU for a home computer, for example. Other devices are combined as, for example, cooling fans,

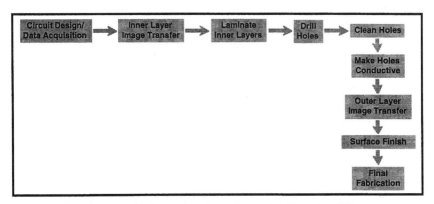

Figure 8.2 Generic schematic for making inner-layer circuits and multilayer.

indicator lights, disk drives, etc., and other servomechanical devices. Thus, the printed wiring board lies between the all powerful chips which perform the millions of calculations needed to drive modern digital devices and the complete equipment. The PWB is, in essence, the ultimate interconnection device for the chips to interact with the world.

8.1.1.1   General nature of industry.   As noted previously, PWB manufacturing is a highly complicated operation involving over 50 process steps which, in turn, require large equipment investments. Even though PWBs are designed for individual, specific applications, the PWB industry is characterized by highly competitive global sourcing with very low profit margins. This has become easier because designs for specific PWBs can be transmitted electronically for fabrication virtually anywhere in the world.

In the United States, the PWB industry has made substantial investments in pollution prevention and control and spends, on average, at least 2.1 to 5% of sales for regulatory compliance and pollution prevention. These investments have provided significant pollution prevention successes. For example, the Institute for Interconnecting and Packaging Electronic Circuits (IPC) and several IPC members won U.S. Environmental Protection Agency (EPA) Stratospheric Ozone Protection Awards for research on eliminating ozone-depleting substances from PWB manufacturing and assembly. Several IPC members have also won EPA 33/50 Pollution Prevention Awards for their aggressive work on preventing pollution, and many IPC members have won state or local awards for their proactive pollution reduction efforts. At least two PWB manufacturers were recognized as "Environmental Champions" under this EPA program.

According to a report published by the EPA,[1] the total world PWB market in 1996 was approximately $21 billion, and U.S. production was greater than $5 billion. As noted previously, the report goes on to say that U.S. domination of this world market eroded from 1980 to 1990, but has come back slightly in recent years. In 1996 there were approximately 700 to 750 independent U.S. PWB manufacturing facilities and approximately 70 captive facilities. Most original equipment manufacturers (OEMs) have shut down their internal PWB operations and now buy their PWBs from independent manufacturers. The vast majority of PWB manufacturers are small- to medium-sized enterprises with annual sales under $10 million (see Fig. 8.3). In the United States, the majority of PWBs (>75%) are produced by independent manufacturers. PWB manufacturing facilities exist in virtually all 50 states and territories.

The report further indicates that since 1980, rigid multilayer PWBs have grown from approximately $700 million in 1980 to almost $3.4

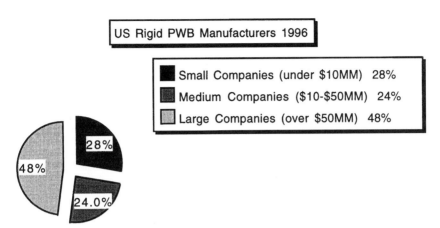

Figure 8.3 Percentage of small, medium, and large PWB manufacturers in the United States.

billion in 1993, accounting for approximately 66% of the domestic market. The rest of the market consists of double-sided rigid boards (25%) and single-sided and flexible circuits (~9%). The PWB industry directly employs approximately 75,000 people, about 68% of whom are in production jobs.

8.1.1.2 Importance of the industry in the United States and other regions. The printed wiring board and its sisters, the chip carrier and multichip module, are the heart of any electronics device. These are the means by which the functioning units (active or passive) are interconnected—to each other and to the outside world—to produce the device's desired function. These are particularly important as integrated circuit chips become faster and have more and more functions in a smaller and smaller footprint. For without the ability to connect the various memory and central processing units to each other and to other electromechanical devices, the chips integration loses its value.

An EPA report[2] on the industry states:

> These electronic systems, in turn, support every other critical technology in the United States. To quote the Council on Competitiveness from their 1991 Gaining New Ground report, "Electronic components are playing an especially important role in driving improvements in information and communication technologies, which in turn are enabling advances in all manufacturing and service industries."
>
> PWBs play a crucial role in these improvements because advances in electronic packaging and interconnections reduce the size and cost of electronic devices while boosting performance. Progress in PWB technology and manufacturing drives U.S. competitiveness in both existing products and new technologies. The U.S. Department of Defense, the U.S.

Department of Commerce, the Japanese Ministry of International Trade and Industry, and the European Community all include electronic systems and components on their critical technology lists.

Electronics drive productivity in almost every industry—one electronics job determines the competitive strength of seven jobs in other industries. Electronics are critical in medical systems, defense technologies, information processing, intelligent manufacturing, propulsion, and materials processing. In addition, a number of new, emerging industries depend on advancing the technical capability of the PWB industry. These include artificial intelligence, biotechnology, digital imaging technology, high-density data storage, high-performance computing, medical devices and diagnostics, opto-electronics, and more. U.S. competitiveness in these new technologies will depend upon advanced PWB technology and manufacturing capability in the United States.

8.1.2 Environmental performance of the industry

Being near the bottom of the supply chain generally means that an industry is closest to basic materials and chemical processes and requires the use of more active chemicals. The PWB industry is no exception. Strong acids and bases are required to etch and plate copper and other metals such as gold, nickel, and tin-lead solder. The many steps that are required to image the precision patterning process that allows etching and plating of fine circuit features also require cleaning steps, development and removal (stripping) of resist images, and various other steps to add or subtract some type of material to the circuit traces.

Traditionally, cleaning was one of the most troublesome steps preceding various steps in the PWB process. Boards were cleaned before lamination of resists, before plating, after removal of resist, and before multilayer lamination. In the past some of this cleaning was accomplished using chlorinated solvents or even chlorofluorocarbons. With the advent of aqueous processable photoresists, organic solvents were no longer used to develop and strip the boards and aqueous cleaning became the method of choice. In addition, as circuit features became smaller, cleaning and also microetching of the base copper became a requirement to ensure a relatively flat and also a clean surface for the resist to adhere to in small spaces. With the implementation of the Clean Air Act banning ozone depleting substances, the remaining chlorinated solvents were eliminated from nearly all shops.

Several publications in the early 1990s indicated that the amount of waste material, both in terms of solutions sent to sewer or wastewater treatment and of solid waste shipped off site, was a significant percentage as compared to the weight of finished product. As part of the Design for Environment Program for the EPA's Office of Pollution Prevention and Toxics, a survey[3] was undertaken to determine what levels of waste were actually being generated and what steps were being taken to minimize or eliminate the waste. The survey found that there was a wide range in the amount of water being used for the same processes, different amounts to wastewater were generated, and the amount of solid waste varied significantly across a very wide range. Pollution prevention drivers, technologies, and methods were also surveyed.

Most PWB manufacturers were still using formaldehyde-based electroless plating for plating on the dielectric walls of plated-through holes. Only a small percentage had switched over to the nonformaldehyde alternatives such as palladium-only systems (14%) or graphite-based systems (very small percentage). This number has probably altered significantly as a result of the IPC/EPA Design for Environment survey "Making Holes Conductive" Project. Other results of the survey showed:

- The range of water use among respondents is very large and there is evidence that some facilities have significantly better water use practices than other facilities. One reason for high water use variability among PWB manufacturers appears to be variable water and sewer use charges paid by the respondents. *A relationship exists between the adjusted production-based flow rates and the cost of water and sewer use.* For facilities that have very high combined water and sewer costs, the adjusted production-based flow rates are very low. Alternatively, facilities with very low combined water and sewer costs have high adjusted production-based flow rates.

- The data indicate that the majority of respondents (63%) must meet local wastewater discharge limitations that are more stringent than the Federal standards. Very few respondents reported any wastewater compliance difficulties.

- The most common regulated pollutants found in PWB wastewater are copper, lead, nickel, silver, and total toxic organics (TTO).

- Two basic wastewater treatment configurations are present at the respondent's facilities: conventional metals precipitation and ion exchange systems. Sixty-one percent (61%) of the respondents reported having conventional metals precipitation systems. Thirty-three percent (33%) of the respondents reported using ion exchange as their basic waste treatment technology and 6.1% installed ion exchange in conjunction with conventional metals precipitation units. (See Fig. 8.4.)

- One-half of the survey respondents have a formal pollution prevention plan. Most facilities have implemented common pollution prevention methods and procedures.

- Low water use rates have been achieved by some survey respondents through the implementation of simple water conservation techniques and/or by using technologies such as ion exchange that recycle water.

Metals Waste Treatment

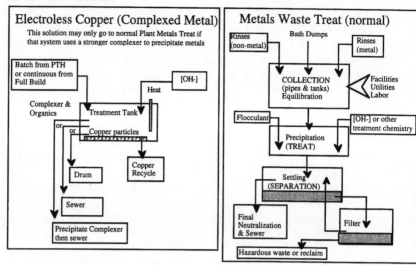

Organic Treatment & Crystallization Examples

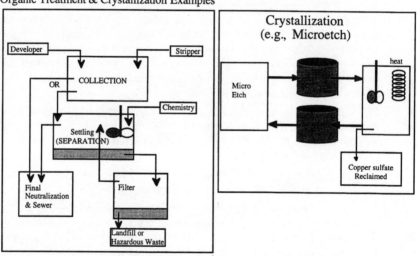

Figure 8.4 Waste treatment generic activities. (Data from TMRC, "Analysis of the Market for Rigid Printed Wiring Boards and Related Materials—for the Year 1996," p. 25, IPC, North Brook, Ill., 1997.)

Regeneration Examples

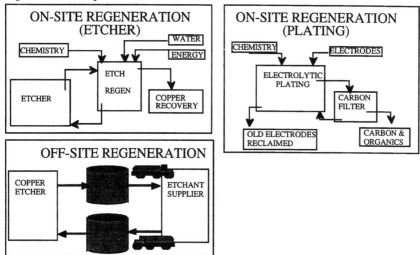

Figure 8.4 Waste treatment generic activities. (*Continued*)

The lowest production-based flow rate among survey respondents was achieved without the use of any sophisticated recycling technology. Rather, they use flow controllers, rinse timers, and reactive or cascade rinsing. The data also indicate that facilities that have implemented the ion exchange technology within their processes have a lower average flow rate than those that have not implemented this technology.

- The data indicate that the use of water conservation methods does not always result in low water use. The four facilities with the highest production-based flow rates do not use ion exchange recycling, but they all indicated that they employ counterflow rinsing, plus some other methods of water conservation. In such cases, it is probable that water is simply being wasted by having unnecessarily high flow rates in their rinse tanks.

- Three-quarters of the survey respondents have implemented recycle, recovery, or bath maintenance technologies that conserve water and/or prevent pollution. The most common of these technologies is the use of porous pots for maintenance of permanganate desmear, ion exchange for water recycle, and electrowinning for metal recovery and reuse. Very few advanced technologies such as diffusion dialysis, membrane electrolysis, or solvent extraction are used.

- Off-site recycling is a commonly used method for PWB manufacturers to manage spent etchant solutions and wastewater treatment sludges.

8.2   Supply Chain and Regulatory Drivers

8.2.1   The electronics industry

The electronics industry is a large and highly visible industry consuming significant energy, water, and resources to produce its products. In addition, its supply chain runs from chips made of highly purified silicon, connected by interconnection devices, components soldered to these devices, subunits created from several devices to finished equipment in cabinets that must not only physically protect and serve as a function of the device but also shield the components from stray electronic and magnetic signals. These interconnection devices are made from ceramics, glasses, organic polymers, precious metals, and toxic heavy metals. They are applied in processes that use toxic chemicals, acidic and caustic solutions, molten metals, corrosive gases, and many other processes that rely on basic chemical reactions to add and subtract materials in precision patterning steps to create the circuits that make up these devices.

Because of its visibility, both as a potential source of pollution and energy consumption, as well as one of the most significant impacts on daily life and industry, the industry has been the focus of much environmental impact examination. But the electronics industry has also been a leader in its environmental proactivity. In this section we will discuss the regulatory impacts on the printed wiring board and assembly industry and some of the proactive efforts that the industry has undertaken to make itself into a model industry with significant advances in sustainable development.

8.2.2   Regulatory compliance issues

All industry is subject to a set of environmental laws promulgated during the 1980s and 1990s by the U.S. government and added to by state and local authorities. We will touch on the more important aspects of these. We would refer you to the original references listed at the end of this chapter for more information. While this chapter presents an overview of these laws, there are subtleties and complexities that require extensive training and knowledge to properly comply with them.

The applicability of many federal regulations is determined, in part, by the chemicals being used at a facility. Individual facilities, however, have their own chemical use patterns. As a result, each facility must identify the universe of rules that apply to it by examining the regulations themselves. Furthermore, implementation of many federal programs is delegated to states with programs at least as stringent as the federal program. Thus, even where federal regulations apply, state laws may impose additional requirements that are not addressed in this document.

8.2.2.1 Clean Water Act. The Clean Water Act provides for development of publicly owned treatment works (POTWs), which are responsible for the virtual elimination of direct discharge of conventional sewage into the nation's waters. It also provides for setting limits on what materials and concentrations of materials may be discharged to these POTWs by various industrial categories.

Under the Clean Water Act (40CFR 403.12), the reporting requirements of the general pretreatment regulations require all industrial users to notify the POT (publicly owned treatment works, that is, the sewer authority) immediately of all discharges that "could cause problems to the POT, including any slug loading," as defined by:

- Pollutants that create a fire or explosion hazard.

- Pollutants that cause corrosive structure damage (in no case can discharges be less than pH = 5.0).

- Solid or viscous pollutants that will obstruct the flow in the POTW or cause interference.

- Any pollutant, including biological oxygen demand (BOD), released at a flow rate or concentration that interferes with the POTW.

- Materials hot enough to cause the POTW to exceed 104°F.

In addition to general restrictions that limit the materials that may enter a POTW, the EPA publishes categorical standards that contain numerical limits for the discharge of pollutants from specified industrial categories. Most PWB manufacturers are specified as "metal finishing" or "electroplating" or both. At this writing, the EPA was in the process of collecting data to rewrite a new category known as the metal products and machining category standard that would encompass both the metal finishing and electroplating categories. It would also change the way limits would be measured and possibly add new materials to the federal limits. Additionally, through sewer use ordinances, local authorities may set tighter limits on and/or add effluents, depending on local conditions, or add materials not on the federal pollutants list (see Table 8.1). Categorical limits are listed by the EPA and apply to the process stream after pretreatment that is part of the regular process stream. Local limits generally apply to the connection point with the community sewer system.

The federal categorical effluent guidelines for metal finishing also contain a list of total toxic organics (TTO), the total concentration of

TABLE 8.1 Other Limits on POTW Effluents throughout the United States

Category	Average limitation [ppm (mg/L)]						
	Northeast	MidAtlantic	Southeast	Midwest	S/Central	Southwest	Northwest
Temperature (°F)	160	147	145	127	132	143	147
Grease, oil, fat	100	84	100	125	133	283	100
pH	5.5–9.5	5.3–10.0	5.5–10.0	5.7–9.5	5.8–9.8	5.5–11.2	5.5–10.5
BOD*	275	317	383	250	275	300	300
COD†	700	450	580	UDL§	UDL§	UDL§	900
Suspended solids	300	333	250	258	288	300	350
Copper	3.1	3.1	1.4	6.0	2.5	6.9	2.5
Lead	1.9	0.9	0.4	4.8	2.2	2.5	1.4
Nickel	2.3	3.3	1.2	5.3	2.8	8.2	2.7
Chromium	3.3	2.9	1.4	7.7	3.7	4.3	3.7
Silver	2.5	1.8	0.2	0.8	2.1	3.6	0.5
Cadmium	0.3	0.6	0.1	0.4	0.4	5.6	1.4
Zinc	2.9	4.2	1.5	6.4	4.0	11.8	3.4
Fluorides	UDL§	UDL§	14.0	UDL§	UDL§	10.0	15.0
Total toxic organs (TTO)‡	5	1.42	21.	3.1	2.1	1.0	1.4

*BOD = biological oxygen demand.
†COD = chemical oxygen demand.
‡TTO = total toxic organs.
§UDL = category not defined by state pretreatment limitations.
Note: The highest in each category is in bold.
SOURCE: James J. Andrus, "Managing PWA Aqueous Cleaning Pretreatment," *Circuits Assembly*, December 1990, pp. 50–61.

which cannot exceed a given specified value. These materials are often also referred to as the priority pollutants list (see Table 8.2). The federal value is 2.13 mg/L. Limits for a number of metal ions often seen in PWB effluent streams are also listed and include copper, nickel, lead, zinc, silver, and cyanide. Again, local sewer authorities will often add to this list and local limits can be lower than the federal standard.

In areas where disposal to a POTW is not practical or prohibited, industrial dischargers are permitted to send treated wastewaters to surface waters under an NPDES (National Pollution Discharge Elimination System) permit. These NPDES permits are site-specific and generally issued by the state or EPA. They will include limits on pH, total suspended solids, BOD5, specific organics, and metals. In many cases aquatic toxicity testing requirements may be required as well.

In a 1990 revision to the Clean Water Act, a revision established these additional requirements:

1. *Notification.* Materials that are considered to be hazardous under RCRA, but can legally be sent to a POTW, must now be monitored and reported [40CFR403.12(p)]. This also required prescribed, written plans for *slug discharges* by "significant users," as well as permits for this situation, reports, and periodic inspections for these same users.

2. *Prohibitions (40CFR403.5.b.1,6-8).* This changed an existing rule concerning streams that represent an explosion or fire hazard to include materials with a flash point of less than 140°F (60°C). It also prohibited discharges to a POTW that might result in toxic gases, vapors, or fumes in quantities that could cause acute worker health and safety problems. The revisions also required the ability to demonstrate that discharges, alone or with other sources in the POTW stream, could not cause a "pass through" or interference. "Pass through" is a situation where the interaction of discharges is such that the result is a "passing through" of materials that should not exit a POTW to surface waters. An interference causes a slowdown or interferes with the POTW treatment process, for example, large oil discharges.

8.2.2.2 Resource Conservation and Recovery Act (RCRA)—hazardous wastes. The Resource Conservation and Recovery Act (RCRA) deals with the federal laws that govern waste generation, disposal, storage, and minimization of "hazardous" waste that is removed by other than the sewer system or as a volatile air pollutant. These wastes are normally shipped outside the plant for recycling, landfilling, or incineration.

Hazardous wastes are controlled by RCRA requirements from the time they are generated, through handling, "transportation to" and "treatment, storage or disposal at" a permitted site, and beyond, that is, paying someone to haul it away and do something with it. The rules

TABLE 8.2 Priority Pollutants

Chemical	CAS#	Chemical	CAS#
4-Nitrophenol	100-02-7	p-Chloro-m-cresol	59-50-7
Ethylbenzene	100-41-4	2,6-Dinitrotoluene	606-20-2
4-Bromophenyl			
phenyl ether	101-55-3	N-Nitrosodimethylamine	62-75-9
2,4-Dimethylphenol	105-67-9	N-Nitrosodi-n-propylamine	621-64-7
1,4-Dichlorobenzene	106-46-7	Chloroform	67-66-3
Acrolein	107-02-8	Hexachloroethane	67-72-1
1,2-dichloroethane	107-06-2	4-Chlorphenyl phenyl ether	7005-72-3
Acrylonitrile	107-13-1	Benzene	71-43-2
Bis(2-chloroisopropyl)			
ether	108-60-1	1,1,1-Trichloroethane	71-55-6
Toluene	108-88-3	Methyl bromide	74-83-9
Chlorobenzene	108-90-7	Methyl chloride	74-87-3
Phenol	108-95-2	Chloroethane	75-00-3
2-chloroethylvinyl ether	110-75-8	Vinyl chloride	75-01-4
Bis(2-chloroethyl) ether	111-44-4	Methylene chloride	75-09-2
Bis(2-chloroethoxy)			
methane	111-91-1	Bromoform	75-25-2
Bis(2-ethylhexyl)			
phthalate	117-81-7	Dichlorobromomethane	75-27-4
Di-n-octyl phthalate	117-84-0	1,1-Dichloroethane	75-34-3
Hexachlorobenzene	118-71-1	1,1-Dichloroethylene	75-35-4
Anthracene	120-12-7	Trichlorofluoromethane	75-69-4
1,2,4-Trichlorobenzene	120-82-1	Dichlorodifluoramethane	75-71-8
2,4-Dichlorophenol	120-83-2	Hexachlorocyclopentadiene	77-47-4
2,4-Dinitrotoluene	121-14-2	isophorone	78-59-1
1,2-Diphenylhydrazine	122-66-7	1,2-Dichloropropane	78-87-5
Chlorodibromomethane	124-48-1	l,l,2-Trichloroethane	79-00-5
Tetrachloroethylene	127-18-4	Trichloroethylene	79-01-6
Pyrene	129-00-0	1,1,2,2-tetrachloroethane	79-34-5
Dimethyl phthalate	131-11-3	Acenaphthene	83-32-9
1,2-Trans-dichloroethylene	156-60-5	Diethyl phthalate	84-66-2
3 Benzo(ghi)perylene	191-24-2	Di-n-butyl phthalate	84-74-2
Indeno(1,2,3-cd)pyrene	193-39-5	Phenanathrene	85-01-8
3,4-Benzofluoranthene	205-99-2	Butylbenzyl phthlate	85-86-7
Fluoranthene	206-44-0	N-nitrosodiphenylamine	86-30-6
3 Benzo(k)fluoranthene	207-08-9	Fluorene	86-73-7
Acenaphthylene	208-96-8	Hexachlorobutadiene	87-68-3
Chrysene	218-01-9	Pentachlorophenol	87-86-5
Benzo(a)pyrene	50-32-8	2,4,6-trichlorophenol	88-06-2
2,4-Dinitrophenol	51-28-5	2-Nitrophenol	88-75-5
Dibenzo(1,h)anthracene	53-70-3	Naphthalene	91-20-3
4,6-Dinitro-o-cresol	534-52-1	2-Chloronaphthalene	91-58-7
1,3-Dichlorobenzene	541-73-1	3,3-Dichlorobenzidine	91-94-1
1,2-Dichloropropylene	542-75-6	Benzidine	92-87-5
Bis(chloromethyl) ether	542-88-1	1,2-Dichlorobenzene	95-50-1
Carbon tetrachloride	56-23-5	2-Chlorophenol	95-57-8
Benzo(a)anthracene	56-55-3	Nitrobenzene	98-95-3

are often complicated, require extensive recordkeeping and documentation, and can lead to significant fines if not followed.

Further, "hazardous" wastes are strictly defined by RCRA as solids, liquids, a semisolid sludge, or containerized gas under pressure that have been abandoned, recycled in certain ways, or considered inherently "wastelike." Wastes are divided into two types: listed and characteristic.

Listed wastes contain wastes that are nonspecific as to the makeup but are generated from any of a number of specific processes. For example, under F-listed wastes we find F006, which are wastes from electroplating sludges. These wastes were originally chosen because they had certain pollutants present in them. These pollutants may or may not be present in your particular waste stream, but they are still considered as listed wastes. The EPA is also considering revisions to RCRA which would define waste strictly based on the pollutants present in that particular stream, not the process. There may also be state and even EPA regional interpretation as to what constitutes a particular listed waste. For example, region III at one time considered resist skins and precipitated resist from stripping solutions as electroplating sludge. This interpretation has now been changed by the EPA and a number of states.

The second type of wastes is characteristic wastes, that is, those having one of several characteristics: ignitability, corrosivity, reactivity, EP toxicity. EP (extraction procedure) toxicity, now known as TCLP (toxicity characteristic leaching procedure), may "catch" some wastes that are not listed (see Table 8.3). These have limits based on a test that determines the leachable amount of the material in the waste.

There are also additional rules for facilities that qualify as "large generators." Most PWB facilities fall under the classification of "large generators" and, as such, must follow certain rules for accumulation, temporary storage (up to 90 days), and disposal of hazardous wastes. Storage and accumulation are beyond the scope of this discussion. All wastes handled by large generators must be manifested using state or federal manifesting forms, which must be sent with the waste to the disposal or treatment facility. The treatment facility must send a copy of the manifest back to the manufacturing site to ensure that the waste has reached its destination and is being treated properly. If notification is not received by the manufacturing site within 45 days of the shipment off-site, then this must be immediately reported to EPA regional administration by means of an exception report (some states require a report after 24 h).

8.2.2.3 Clean Air Act Amendments (CAAA). The Clean Air Act and its amendments are designed to "protect and enhance the nation's air resources so as to promote the public health and welfare and the

TABLE 8.3   TCLP Table

Number	Identity	Concentration (mg/L)	Number	Identity	Concentration (mg/L)
D 1	Ignitability	Less than 140°F	D 23	O-creosol	200
D 2	Corrosivity	pH <2 or >12	D 24	M-cresol	200
D 3	Reactivity		D 25	P-cresol	200
D 4	Arsenic	5	D 26	Cresol*	200
D 5	Barium	100	D 27	1,4-Dichlorobenzene	7.5
D 6	Cadmium	1	D 28	1,2-Dichloroethane	0.5
D 7	Chromium	5	D 29	1,1-Dichloroethylene	0.7
D 8	Lead	5	D 30	2,4-Dinitrotoluene	0.13
D 9	Mercury	0.2	D 31	Heptachlor (and its hydroxide)	0.008
D 10	Selenium	1	D 32	Hexachlorobenzene	0.13
D 11	Silver	5	D 33	Hexachloro-1,3-butadiene	0.5
D 12	Pesticide		D 34	Hexachloroethane	3.0
D 13	Pesticide		D 35	Methyl ethyl ketone	200
D 14	Pesticide		D 36	Nitrobenzene	2.0
D 15	Pesticide		D 37	Pentachlorophenol	100
D 16	Pesticide		D 38	Pyridine	5.3
D 17	Pesticide		D 39	Tetrachloroethylene	0.7
D 18	Benzene	0.5	D 40	Trichloroethylene	0.5
D 19	Carbon tetrachloride	0.5	D 41	2,4,5-Trichlorophenol	400
D 20	Chlordane	0.03	D 42	2,4,6-Trichlorophenol	2.0
D 21	Chlorobenzene	100	D 43	Vinyl chloride	0.2
D 22	Chloroform	6			

*Measure as composite.

productive capacity of the population."[4] Under the CAA, there are six titles which direct the EPA to establish air quality standards and the EPA and the states to enforce and maintain the standards. The new amendments for the CAA became law in November 1990. Many of the provisions deal with hazardous air pollutants (HAP) (see Fig. 8.5).

The first of these amendments deals with materials that fall under title I (ozone nonattainment). Part of this section deals with volatile organic compounds (VOCs). Almost all organic solvents and many chemicals that have been used by PWB manufacturers are included, for example, 1,1,1-trichloroethane, methyl ethyl ketone (MEK), and isopropanol. Many organic stripping solvents, liquid solder masks, and organic degreasers would be affected by control of these solvents. Standards have been established for these types of chemicals by industry categories, much like those for the Clean Water Act (CWA). The EPA has determined these to be Control Technology Guidelines (CTG). Additionally, the air quality in the area in which the industry is located determines the level of emission causing an industry to be qualified as a major source. For example, in areas termed "serious, severe, or

Geographical Area Designation	Federal "Serious"	Federal "Severe"	Federal "Extreme"	State AQMDs California
VOC Emissions to be "Major"*	50 tons/year	25 tons/year	10 tons/year	1 lb/day

AQMD = Air Quality Management District. The California limit shown is for one AQMD, others vary.
* Major Federal sources will require certains types of permitting and technology to control. State limits may be implemented to be lower and impose similar requirements.

Figure 8.5 Hazardous air pollution classifications.

extreme," an industry emitting 50, 25, or 10 tons of VOCs is termed "major" (see Figure 8.5). Major sources have to obtain permits, specifying their emissions and control levels. They are also regulated and required to reduce emissions based on "lowest achievable levels" using "reasonably available technology." In addition, as states and even districts within states develop their own implementation plans (SIP), they may lower the allowed levels of air pollutants. For example, in the state of California one air quality district (San Joaquin Valley) requires a permit for a developing process using more than 2 lb of VOC/day while another (Bay Area) exempts the same process if the weight percent of VOC content of the solution is less than 2.5% or if the VOC *emissions* are less than 10 lb/day or 150 lb/year.[5]

There is also a hazardous air pollutants list which originally contained 189 chemicals subject to regulation. Table 8.4 shows a partial list with chemicals that might be found in the PWB industry. Under federal regulations, these materials are restricted to emissions of less than 10 tons of any one HAP or 25 tons total of any combination of them. Installation of a high degree of control technology will be required depending on the permit conditions and emission rates.

Title VI deals with stratospheric ozone protection, including provisions to limit the use and phase-out of chlorofluorocarbons (CFCs) as well as carbon tetrachloride (by 2000) and methylchloroform (by 2002). There are also labeling and escalating tax requirements to provide negative incentives to speed phase-out.

With titles V and VII, permits and enforcement will again fall principally to the states. The federal ruling requires that state plans include permit fees, reporting requirements, and review of permits. This structure may be much like the current POTW or NPDES permit system. The law also includes strengthening the power of the EPA to enforce regulations with severe penalties.

8.2.2.4 Miscellaneous reporting and recordkeeping regulations. The Comprehensive Environmental Response, Compensation, and Liability Act (CERCLA), also known as the Superfund, authorizes the EPA to respond to releases and threatened releases of hazardous substance

TABLE 8.4   Hazardous Air Pollutants List of Materials Commonly Used in the PWB
Industry

Acetaldhyde	Acrylamide	Acrylic acid
Acrylonitrile	Cresols	Cumene
Dibutyphthalate	Diethanolamine	Ethyl acrylate
Ethylene glycol	Formaldehyde	Methanol
Methyl chloroform	Methyl ethyl ketone	Methyl acrylate
Methylene chloride*	Phenol	Toluene
Lead compounds	Xylenes	
Cyanide compounds		

*Currently the only one in this list that the EPA is considering including in a industry sector regulation.

representing a threat to the environment and public safety. The act also requires companies which release hazardous substances above reportable quantities to report them to the National Response Center.

The Superfund Amendments and Reauthorization Act (SARA) of 1986 created the Emergency Planning and Community Right-to-Know Act (EPCRA), also known as SARA (Title III-313). Its main provisions provide for emergency response plans for fires and potential hazardous releases through reporting and planning with local fire departments and local emergency response commissions (LEPCs). The act also requires manufacturers, including PWB fabricators, with 10 or more employees which use, process, or manufacture specified amounts of certain chemicals to report them each year. Form R covers "releases and transfers" of toxic chemicals to various facilities and environmental media. A great deal of controversy continues around Form R and the toxic release inventory (TRI) which compiles the results of Form R reporting. The definitions of "release" and "transfer" are the subject of continuous discussion and clarification. TRI and EPCRA have recently been the focus of a number of environmental groups who mistakenly believe that TRI represents a source of pollution rather than a use of materials.

8.2.2.5 Pollution Prevention Act of 1990.

The main goal of this act is to separate toxic emission reductions from so-called paper reductions as reported under the toxic release inventory (TRI) under SARA III-313. Additional information from Form R filers began in 1991 including:

Total waste generation. The total amount of material entering any waste stream before treatment or recycling takes place, the amount of chemicals that are treated, and the volume of recycling figures was calculated and reported as well as the percentage change over the previous year. Any toxic release due to a "one-time" event that is not part of the production process began to be reported.

Emission reduction techniques. The source reduction practices used on the chemicals listed previously are reported according to a category:

- Equipment, technology, process, or procedure change.
- Reformulation or redesign of product.
- Substitution of raw material.
- Improvement in general operational phase of the plant. The techniques used to identify source reduction opportunities, for example, audits and employee suggestions.

Anticipated future reductions. An anticipated amount under the calculated waste and amount to be recycled in the next 2 years has to be reported along with a ratio of production in the reporting year to the previous year. This provision has been the most difficult to determine.

8.2.3 Pollution prevention

Pollution prevention was the original response by the industry after complying with these new regulations by treating waste at the end of pipe. *Pollution prevention* is defined as the prevention of the generation of pollutants by minimizing or eliminating the steps or materials that produce them. Early efforts at pollution prevention dealt with minimizing water usage, concentration of wastes for later recycling steps, and utilization of alternative materials to eliminate or minimize wastes. These were chiefly aimed at processes using water. Most of the steps in making a circuit board require large quantities of water and the pollution created in these steps is governed by either the Clean Water Act or, since the sludges produced in cleaning the water before disposal create hazardous waste sludges, RCRA. Later as the Clean Air Amendment Act came into being, processes involving solvents and volatile materials used or created during processing were examined.

Most of the pollution prevention that was done first was considered "low hanging fruit." In other words, the payback for preventing wasted water, energy, or materials was short and usually obvious. As the need for more pollution prevention continued, more sophisticated methods of evaluating various possible options and technologies were needed. This included business decisions based not only on the ability of the option or technology to significantly reduce the pollution and waste, but also on the financial ability of the company to carry the debt during the payback period. Significant resources have been invested in trying to develop means for helping the broad scale of companies that must invest in such technologies.[6] We will discuss various types of pollution prevention later in this chapter

8.2.4 Design for the environment (DFE)

One of the industry drivers for environmentally conscious manufacturing, over and above regulatory concerns and requirements, is design for the environment. Design for the environment is used in two different contexts, one by the EPA (DfE) and one (DFE) by the OEMs or the manufacturers at the top of the electronics supply chain. The EPA definition applies to a methodology for looking at individual process areas or groups of process steps termed by the EPA as "use clusters." Further, it involves determining if there are "cleaner technologies" that can be substituted for this "use cluster." The OEM definition is much broader and actually fits in with the ISO 14000 concept of considering and assessing the environmental impacts of equipment design decisions.

The OEM implementation of DFE means that original equipment manufacturers are beginning to require suppliers of parts, material, equipment, and processes to provide some means of assessing the environmental impact of the supplied element. Some electronics and auto companies have devised "black lists" of materials that must not be in the product or process or should not be used in processes supplied to these manufacturers. Others are attempting to develop various methodologies for weighing these factors for each item or process supplied. One of the more popular means of doing this is using a life-cycle assessment or life-cycle inventory.[7] More recently, consortial efforts[8] have examined selected lists of environmental metrics. These could be either in the form of specific environmental parameters, such as percentage of recyclable material contained, or could relate to the practices of the company providing the material, process, or service (typically referred to EcoEfficient). The concept of using environmental parameters is closer to the more pragmatic enabling of DFE. DFE parameters could be clearly defined and used by manufacturers in making decisions on what products would have the lowest environmental impact or would be most conducive to creating a sustainable industry.

These concepts of assigning an environmental valuation to materials and processes is already being implemented by companies like IBM, Xerox, Motorola, and others.[9] They, in turn, are beginning to ask questions of their suppliers about environmental impacts. Initially these questions will deal with the ability of the supplied materials to be recycled, reused, and refurbished, but eventually they will begin to ask about energy and resource usage in making the materials that are being supplied to them. This will allow them to go even further into truly determining the environmental and resource impacts of the complete systems and products that they make and sell. It will further help them determine which are the most sustainable (that is, those that use up resources versus those that simply "borrow" resources and can be returned to the life cycle of the product at a later date as recycled or reused or refurbished materials).

As noted previously, the EPA definition of DfE is more closely associated with a highly detailed examination of a single process within a manufacturing scheme.[10] After going through an initial scoring scheme for processes within an industry, a "use cluster" or series of process steps associated with one activity is chosen for study. The use cluster is then clearly defined, and all materials used in the processes are inventoried along with any for processes that have been, or are being, developed to have a lower environmental and resource impact. These processes and alternatives are then examined in a CTSA (cleaner technology substitute assessment) to determine cost, health and safety, energy usage, environmental, quality, and performance assessments (see Fig. 8.6). The results are then presented in a nonjudgmental format to allow individual users of the technology to determine which, if any, substitutes would make sense for them. Finally, after the processes have been tested in a demonstration process, implementation of the substitutes are followed up using actual practitioners of the new technologies to determine what was good and bad about the process.[11]

The DFE process is more of a top-down evaluation and will eventually drive the move toward newer and more resource and environmentally "friendly" processes. The DfE process is an evaluation scheme for evaluating, implementing, and disseminating these processes.

8.2.5 Environmental cost accounting

One of the factors which can motivate better practices, processes, materials, and technologies is cost. Even in pollution prevention, many small- to medium-size companies must have a short payback on even simple changes. Early literature on paying for these processes focused on the cost and return on investment payback period. More recently, companies are beginning to look at how these changes will affect or improve yield, reduce hidden environmental and resource costs, and improve their overall cost picture. In order to do this, companies have begun to examine environmental costs more closely.

It has long been recognized that many environmental costs are "buried in the overhead." This is often due not only to the method in which costs are allocated, but also to the unrecognized contribution of various functions to the environmental overhead. Several recent survey projects relating to pollution prevention in both the plating and printing wiring board industries have attempted to quantify this.[12] These references not only give detailed information on equipment and practices for these industries, but also show the costs that can be associated with various processes to deal with manufacturing waste, pollution control, resource usage, and operations. They do not, however, deal with labor costs not specifically related to these processes, such as training to deal with regulations, manifesting and tracking of wastes,

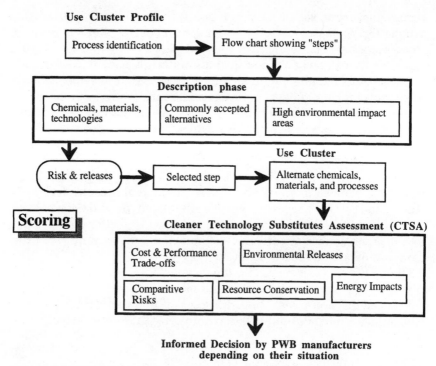

Figure 8.6 Cleaner technology substitute assessment (CTSA).

management of environmental and regulatory requirements, legal and financial liabilities, and other "soft" issues and concerns.

Findings of these surveys, even though they were related to mostly process-related costs, revealed that estimates of costs, as a percentage of sales, were higher than the 1 or 2% that is often quoted. These surveys also were performed before many plants were being required to complete Clean Air Act permits and add technology needed to meet the standards.

Another way of looking at these costs is to allocate them either according to products that accumulate them or to specific processes or materials that have a hidden downstream "cost of use." Allocating costs to products according to what processes and downstream costs they generate is a good way to eliminate those products that are high environmental and resource users. This also increases the overall value of the remaining products by eliminating costs not associated with these products. For example, products that require treatment using a process with a particularly high hazardous waste generation could be either eliminated or the true costs asked for. This would eliminate the practice of having lower impact (or "footprint") products having to "carry" the products with high impact and cost. Another way to accomplish this is to examine the processes and materials that are used for all products. For example, a photoresist that generates a hazardous waste or requires chemical precipitation will cost more to use than one that generates no hazardous waste and can be captured mechanically as particles in the stripper or that is never stripped at all. This approach to pollution prevention and cost-effective manufacturing is only just beginning to be recognized.

8.2.6 ISO 14000—continuous improvement

Much has been written concerning ISO 14000, the environmental management system (EMS) standard. Although most people are familiar with the EMS portion of the standard, they are usually unaware of its other aspects.

In the United States, the U.S. Technical Advisory Group (U.S. TAG) developed the U.S. position on the various ISO 14000 standards. The U.S. TAG is comprised of approximately 500 members representing industry, government, not-for-profit organizations, standards organizations, environmental groups, and other interested stakeholders. The international technical committee (TC207), comprised of members from each country's TAG, developed the ISO 14001 standard which specifies requirements for an environmental management system (the ISO 14001 standard is the only standard in the ISO 14000 series which is designed to be audited, and such a document is called a *specification standard*). There are also several guidance documents being developed by TC 207. Two are guidance documents for the EMS, three are environmental auditing guidelines, and there are other guidance documents for environmental labeling, life-cycle assessment, and environmental performance evaluation.

Basically, ISO 14000 is an ISO 9000 approach to the management of systems used to ensure compliance and to continuously improve the systems that control the environmental impact of a plant. This includes clearly documenting the management systems, controls, tests, and procedures used to maintain these systems. It does not, however, require that the entity or plant seeking certification under 14000 (actually 14001) be in compliance with local regulations. That entity needs only to define these systems and the verification that they are working. The Eco-Management and Audit Scheme or EMAS (the European equivalent of ISO 14000), BS 7750, and several other country-specific EMS standards, require not only compliance but also require, as a first step, a listing of all the regulatory and legal environmental standards that apply to the plant or entity under consideration for certification under the standard.

While a number of companies have already been certified (third-party verification) to the ISO14001 standard, many others have chosen to determine where their existing EMS stands in comparison to the standard in case they decide to certify. Still others have found that the guidance document for implementing 14001 (known as 14000) is useful for helping them to set up their environmental management systems. This can be valuable in itself in that it is often assumed that there are systems in place to maintain compliance when, in effect, compliance is achieved because a single individual in a company knows what needs to be done and does it. The lack of documentation and procedure becomes evident when that individual is no longer at the company. (This is also known as the "What happens if Joe gets hit by a truck?" scenario.)

Additional benefits of having a formalized environmental management system are the ability to audit their effectiveness, to review them and often to improve their efficiency, determine to greater accuracy the resources and costs associated with various environmental and regulatory requirements, and finally to involve all those whose jobs are now recognized as part of environmental management in any improvement processes. An example of a management system scheme is shown in Fig. 8.7.

External benefits of an EMS may be recognition by the community of a environmentally committed company or plant, lower insurance rates due to better controls, and thus less susceptibility to environmental upsets and to fines and legal action. It may also promote more latitude in addressing problems by the regulatory and legal entities such as state and federal EPA as well as the Department of Justice, better access to capital from financial markets, and better employee morale.

There are innumerable references on ISO 14000 in the literature and on the WorldWide Web. The U.S. EPA Design for the Environment program has published two documents specific to the printed wiring board industry on this subject and the National Standards Foundation International (NSF) has also published a report[13] on a demonstration project with a number of case studies accessible from the Internet.[14] There are also numerous texts that deal with implementation of environmental management systems.[15]

8.2.7   Other environmental metrics systems

As noted previously, many regions and countries have various environmental labels, also known as *ecolabels,* that rely to some degree on the perceived environmental impact of the product labeled. A number of other attempts have also been made to define the environmental impact of products for the benefit of whatever customer or user is buying the product. Original equipment manufacturers are particularly interested in the content of subsystems that they use in their

Management System for Hazardous Waste Manifest Tracking

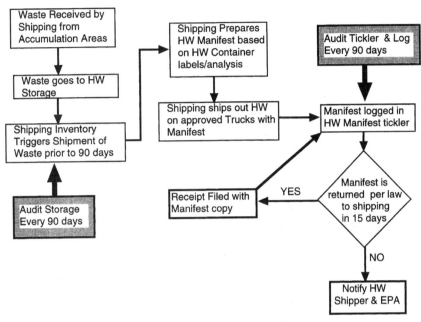

Figure 8.7   Example of an EMS for hazardous waste manifesting.

equipment. They often need information to qualify their products for eco-labels, but also often have to answer questions pertaining to the environmental impact and content of their products. This is becoming more of a need with the initial takeback legislation for electronic equipment already occurring in some European countries.

Several methods for obtaining this information or providing the information have been used and/or proposed. Many companies simply send lists of materials that either aren't allowed in products sent to them or can't be used in processes provided to them. Some companies also have a "prefer not to have" list as well. Volvo, Sony, Mitsubishi, Phillips, and the European Association of Consumer Electronic Manufacturers (EACEM) are examples. Other companies have various rating schemes that they use internally to rate materials and subsystems that they use. For example IBM, Motorola, Hewlett-Packard, and Sun Microsystems all have rating systems.[16] The European Computer Manufacturers Association (ECMA) has proposed a product declaration attribute form for various electronic devices that not only lists materials that are not found in the products, but also lists various attributes such as energy usage and physical emissions (VOCs).[17]

One of the more innovative attempts to develop a set of criteria for environmental attributes was proposed as a project by the Microelectronics Computer Corporation. This project would have developed a set of 10 to 12 criteria that would have been included from material, process, and subsystem suppliers (for example, PWBs)—and OEMs—as part of the normal specifications for that product. One of the problems that exist currently in the sharing of environmental information is the hesitance of suppliers to disclose information that may be proprietary to either their materials or processes. This project had at its core the ability to clearly define the properties needed by the OEMS and methods to calculate and determine these properties. This would have allowed the suppliers to determine the required information without disclosing proprietary information. While many OEMs supported the concept, they felt that such a set of metrics needed to be an international standard.

In addition to looking at the properties of materials in use, several financial institutions are beginning to look at the environmental performance of companies and using this as a way to determine financial risk. This even extends to the perceived ability of companies to survive in a world where resources, energy, and materials will become more limited. The term *eco-efficient* is often used. The World Business Council on Sustainable Development is also a resource examining the sustainability of a company and its value to its shareholders.[18]

8.2.8   Sustainability

The previous topic leads us to the final—and most current—topic on environmental drivers, sustainable development. There are several views of what constitutes "sustainable development," but basically the idea is to *not* mortgage future generations' resources by using them unwisely now. So unlike the bumper sticker, we don't want to "spend our children's inheritance" with respect to the environment and its resources.

There is significant precedence for the most recent publication by the U.S. government of "A Bridge to Sustainability"[19] and an approach more focused on the electronics industry, "White Paper on Sustainable Development and Industrial Ecology."[20] In 1991, the International Chamber of Commerce (ICC) created the "business charter for sustainable development." It comprises 16 principles for environmental management for sustainable development. In 1992, the United Nations Conference on Environment and Development (UNCED) also known as the Earth Summit (or Rio Summit), developed "Agenda 21" and the "Rio Declaration." The "Rio Declaration" is a set of 27 principles for achieving sustainable development supplemented by "Agenda 21," a comprehensive guidance document.

Private sector programs, such as the Chemical Manufacturers Association's Responsible Care™ program, the Global Environmental Management Initiative (GEMI), the Coalition for Environmentally Responsible Economies (CERES) Principles (formerly the Valdez Principles), and the World Business Council on Sustainable Development (WBCSD) developed model codes of conduct that define environmental stewardship in business and manufacturing terms.

With respect to the electronics industry supply chain, the Responsible Care programs in product stewardship and World Business Council on Sustainable development have strong elements of sustainability in their tenets. Responsible Care encourages its members to examine the complete life cycle of a product from idea conception to the product user's final disposition of that product. This type of thinking leads to the idea of closed-loop systems not only in manufacturing processes but also for products produced. In the carpet industry, polymers used in carpets, for example, are now recycled into raw materials to make more carpet, rather than throwing used carpet into landfills. In the same way, the "mining" of metals reclaimed from used or scrap PWBs actually makes the reuse of these metals more economically attractive and uses less energy and fewer resources for "extracting" the metals.

In fact, the whole Moore's law phenomenon in the electronics industry is driving this industry to greater sustainability in two ways. First, as the integrated circuit devices shrink in size and increase in functionality, less resources will eventually be used in producing them (currently the semiconductor manufacturers continue to look at their processes for reductions in resource, energy, and material usage). As these devices shrink, the functionality of interconnect devices, packages, and PWBs will increase as their feature sizes also shrink. The result will eventually be less material and resource use (see the following sections). Second, as the functionality increases and size decreases, the costs tend to also decrease per function, allowing information technology to be used to improve the performance of other systems. Examples of this increasing use of electronic devices to improve other systems can be found in automobile engine controls and sensor devices, smart batteries, and even energy-saving devices on computers. Eventually this technology can spread throughout the entire supply chain to significantly improve the efficiency of each step used in making each "supply" in the supply chain. For example, processes for plating and etching boards are already using more sophisticated control systems to minimize chemical usage in the form of pulse plating and better etchant regeneration.

Interestingly, the Center for Strategic and International Studies issued a report[21] on the findings of the Enterprise for the Environment project. This project, undertaken by a multistakeholder policy dialogue group, attempted to envision an improved "environmental protection system." Among its recommendations are creation of environmental goals and milestones and more flexible and innovative tools to achieve further protection with clear incentives. There were also recommendations concerning environmental management systems, performance-based approaches, as well as economic incentives and drivers. Now that we have examined some of the supply-chain and regulatory drivers, let us examine some of the technical drivers at work in the PWB and assembly industry.

8.3 Technical Drivers and Challenges

The technological progress experienced in the electronics industry has been nothing less than phenomenal. The computing power that once occupied whole buildings can now be held in a laptop computer which can be connected via modem to an almost infinite supply of information and commerce on the Internet. However, a 5-lb laptop may still represent up to 40,000 lb of materials, resources, and waste![22] As the functionality of the various active devices (semiconductor chips) has increased and sizes have shrunk, the supporting structures of the interconnection industry, the chip carriers and PWB assemblies, have followed suit. As these continue to reduce feature and structure sizes, a parallel effort must be made to continue to reduce the waste streams and material and resources used.

8.3.1 Technical drivers move to thinner, finer, and cheaper

As noted previously, the reduction in device size has also meant that the interconnection platforms, chip carriers, multichip modules, and PWBs have had to change to accommodate them. Not only current, but also future, needs have been well documented in a number of technology roadmaps.[23] This has become more important as the number of in/out connections (I/Os) necessary for these active devices to communicate with the outside world and each other has exceeded the ability to connect to them on the edge of the chip. Ball-grid arrays (BGAs) and other array connections which utilized the entire surface of the chip are now needed. This imposes even more requirements on the interconnection platforms (IPs). For example, typical array spacings for the standard BGAs are 1.5, 1.25, (1.27), and 1.0 mm (~40 mils). Newer, even smaller, packages, known as *chip scale packages* (CSP), are addressing the need for even greater component density and closer proximity of the integrated circuit (IC) devices. They often utilize even tighter ball-grid arrays, known as miniature BGAs. Their pitch formats are around 0.8, 0.75, 0.65, and 0.5 mm (~20 mils).

The advent of the microvia (vias less than 4 mils in diameter) and the nondrilling techniques used to produce them will allow this increased density to occur at least in the near future. The techniques used to produce microvias—photodielectrics, laser drilling, and, to some extent, plasma formation of vias—have not only reduced the size of the via, but also improved the accuracy of locating the via with respect to circuit traces and capture pads (where needed). Many of these techniques can be used to form the via in simultaneous processes or to utilize highly sophisticated alignment techniques. This, in turn, can lead to reduction in the size of the capture pad or land, which then significantly increases the density of the circuits that can be manufactured. Thus, the necessity for accommodating higher and higher I/O devices and use of microvias can significantly increase the density of IPs.

The necessity of producing narrower and narrower lines has now also led to the use of thinner and thinner copper traces to permit their production by subtractive etching or semiadditive or additive plating. This allows the use of thinner dielectrics, which again adds to the circuit density increase and now produces thinner, denser (finer) IPs. While the techniques to produce these finer traces require more control of materials and processes and sometimes more sophisticated equipment, the cost per interconnect is expected to decrease.

The overall effect of this, from an environmental perspective, may be that of opposing forces. Overall, less material is left when the final product is complete, but the physical tolerances required and sophisticated controls to achieve them are higher. Thus, initially lower yields and increased waste might be expected. As the processes, equipment, and materials improve, yields will rise and the overall effect should be a lower environmental and resource impact. This will be driven more by cost considerations than environmental benefit at first. Less waste means that more of the starting materials will end up as product and thus a higher profit.

8.3.2 Power requirements and signal speed

Power requirements for equipment have risen as functionality has increased until some of the major material usage on some assemblies has come to be the cooling devices. Reducing feature size and distances, as well as resistance and the overall electronic environment of the interconnections, should alleviate some of the power needs. In addition, several companies have begun to develop ICs with at least some copper traces, thus speeding up the devices and reducing the amount of power needed.[24] In addition, there is a significant need and effort being made to develop passive components integrated into the structure of the IPs, particularly in PWBs. These buried capacitors and resistors not only address the need for lower power and higher clock speed, since they can be placed within the board as an integral part of the circuit traces, but also will decrease the thickness and weight of the overall construction.

8.3.3 Increased portability

Again the same needs for more functionality as computers become more mobile and have increased abilities to communicate via cellular or other radio linkages to computer networks, phones, faxes, the Internet, or even production equipment require thinner, lighter, and cheaper materials. All of these are addressed as indicated here. In addition, battery technology starts to play a more important role since more and more of the weight of a portable device is the battery power. Battery technology continues to evolve as manufacturers move to lighter materials with a higher storage density. In addition, batteries are becoming increasingly smart. They have on-board packages to regulate power, indicate recharge needs, and even regulate the recharging process.[25]

8.3.4 Technology turnover rate

Technology turnover will increase as more informational and computing power is crammed into smaller and smaller packages and moves into broader and broader applications. This will continue the same need for higher density and functionality per volume. In addition, different operating environments will be encountered as applications move into more device controls (for example, smart appliances, cars, motors, agriculture, etc.). The turnover rate is already affecting how OEMs will be designing future equipment. Equipment may be designed on a modular basis so that a central chassis can handle multiple upgrades. In addition, cascading technology, that is, the use of an obsolete technology in a higher function utilized in current lower technology, may proliferate, for example, computer parts utilized in household appliances or game machines. These types of drivers may require modular designs of IPs or even redundant "programmable" circuits that would enable the IP assemblies to be used for multiple product usages.

8.3.5 Recycling, reuse, and refurbishment

To continue this theme on the back end of equipment life cycle, the electronics industry is focusing more and more attention on end-of-life issues for its equipment. Many active components can be recycled as noted previously and reused in lower functionality applications or simply refurbished for further use in new equipment. The IP structures, however, are currently only being "recycled" by removing the metals through either thermal degradation of the polymers[26] or incineration to capture the energy value of the organics. Use of organic structures that can either be transformed back into reusable organic materials in the same or similar forms is more highly desired. Some plastics can be converted directly back to raw materials that were used to make them and thus recaptured[27] The opportunity exists for makers of polymers to develop closed-loop systems with their customers that have a high degree of sustainability, particularly with those customers who have already initiated some level of take-back programs.[28]

This requirement for the ability to reuse, recycle, or refurbish the materials and components in electronics equipment means that IPs will have to consider more carefully the materials that are used. This means not only how they are used initially, but also how they will be recycled. Brominated polymers, for example, which have allegedly been implicated at sources of dioxin in incinerators, have found themselves being considered for elimination as materials of use in PWBs in Europe. Materials used for adding physical support to PWB substrates must be considered not only for their support, but also for their ability to be processed in microvia applications and finally to be recaptured in a closed-loop or at least a "cascade" recycling scheme.

8.4 Pollution Prevention Options for Environmentally Conscious Manufacturing

With the preceding sections on regulatory, supply chain, and technology drivers, it is easy to see that there will be increasing responsibility for manufacturers. Those who are going to survive as viable producers of PWBs or other IPs must consider what their environmental "footprint" will be, not only from their own direct input, but also from the choices of materials and processes that are specified.

The following sections will discuss both the technical and management options that are available for developing an environmentally conscious manufacturing framework. They are aimed principally at existing processes and materials. Emerging materials and technologies will be discussed in the following sections. We will also attempt to provide some places to look for additional help. In general, pollution prevention can be accomplished by source reduction, recycling (internally or externally), or elimination/substitution.

8.4.1 Management of environmental and pollution systems

In order to improve any system or process, you must first understand it. There are many ways to do this. A simple diagram of the process steps, their layout, and how they are plumbed would be a first step that often can be done by simply checking with the facilities manager. Next, an analysis or inventory of the process steps with respect to what goes into each and what comes out of each is also useful. This would include not only product, but waste streams, recycle streams, and even lab analysis samples. Finally, the management systems used to control these various aspects through procedure documents, waste forms, analysis practices, audits, etc., should be evaluated.

By doing these first steps, several things are accomplished:

1. You may discover some very obvious things to correct.
2. You will determine the source of most of your waste.
3. You will see a number of candidates for pollution prevention.
4. You will find some flaws in your environmental management system or you may find that you do not have one at all.
5. You may discover that you are paying a lot more for pollution and/or waste than you should be.
6. You may also discover that some of the material or process choices that you have made may have significant unrecognized downstream costs associated with them.

Let's examine some of these possibilities:

1. *Some obvious things.* Water overuse is a typical problem in many PWB facilities. The need for good rinsing between processes is almost always essential, but simply having a high rate of water rinsing is not necessarily the answer. Quick fixes, such as timers and cascade rinse tanks, are recognized as good practice.[29]

In the case of segregation of waste streams, many waste streams are incompatible with each other or at worst cause you to treat with the wrong and/or more expensive technology than is required.

In the case of material waste, inadequate maintenance procedures may be requiring you to change process chemistries more often than is necessary.

2. *Source of wastes.* By inventorying your waste streams and, better yet, assigning correct costs to them, you can create a hierarchy for determining where to best spend time and resources improving the situation. If this is linked to item 6, that is, if you can tie a particular waste source to a particular material or process, you might be able to consider a process or material with less impact on your waste and costs. For example, many of the new direct plating chemistries designed to replace electroless plating for plated-through holes don't require the use of high complexer concentration. The complexer makes the copper in the electroless plating bath rinses harder to remove in waste treatment. Another example is resist stripping solution. If you

can use a resist that breaks into filterable particles in your stripper and then doesn't easily redissolve, then you can treat most of the resist solution by simply filtering off the resist, rather than having to precipitate it later. In addition, your stripper solution lasts longer!

3. *Sources of pollution prevention projects.* The diagramming of your existing process steps and their waste streams should be your first step when considering pollution prevention projects. In addition, the inventory of materials used and processed will give you a clearer overall picture of the possibilities. Also, don't overlook your employees as sources of suggestions for pollution prevention. They will be more willing to participate in pollution prevention if they are the source of some of the ideas. In addition, they are the most knowledgeable people about the process and the actual situation around each process.

4. *Environmental management system.* This is very important. As noted previously, an EMS offers you many advantages. It ensures that no matter who is running any piece of equipment or process, the system takes care of any regulatory and environmental concerns. You may not even realize that you have a system in place, because it currently depends on people who know their jobs well enough to do the right thing. For the system to always be effective, the procedures, records, tests, etc., that make the system work should be documented. Those involved with each system should be trained to stick to the system. For example, suppose that Charlie, who runs your stripper in photoimage, knows to let Sharon in waste treat know when he's using a stripper with MEA rather than the usual plain caustic stripper. Sharon will then be aware that her system is going to receive a batch of water with a high concentration of copper and complexer in it and can treat it accordingly. If Charlie moves to another area and his replacement doesn't know to do this, then Sharon could have a waste treat upset. By having a written procedure for equipment bath dumps and training people who come into the area, you have an ongoing system to handle potential problems. You can also check and see if they really work by auditing them from time to time and developing improvements which are *written into the procedure.*

5. *Cost of waste and pollution.* This, of course, is very important whether you are the operator or the owner. Significant savings can be captured by monitoring the costs of environmental and regulatory systems. Regulatory failure can cost you significant fines or even criminal prosecution; environmental failures can cost you money and impact the environment. Once you know your costs, you can assign priorities to what needs to be fixed. There are a number of government documents on how to assess not only the cost of environmental and regulatory systems, but also how to determine payback periods, costs of capital, and some examples of what others are paying for such systems.

8.4.2 Implementation of pollution prevention, process of selection, and cost estimates

There are many forms of pollution prevention that can be implemented by any shop. The first step is to inventory the water usage in a plant as well as the waste streams associated with them. Next, several options should be developed based on these usages and waste streams to either reduce the waste and flow or reuse them. Let's look at a few examples of pollution prevention and see what might go into the process of selection.

Example 1. Low Hanging Fruit Most companies have at least thought of looking at simple changes that can reduce costs of water, for example, using cascade rinses and flow limiters on spray rinses to minimize water usage. It is often overlooked that, by using multiple cascade rinses, a much lower flow rate can be used to effect the same level of cleanliness.[30] This not only reduces water costs, but also reduces the amount that has to be waste treated. A simple cost estimate would be to determine the current water usage and the cost for treating this water and then measuring the same flow after implementation of reduction methods. The cost reduction would be the reduced water costs, as well as a waste treatment cost, based on the amount of solution treated. There are several other examples of these simple low cost changes in work practices.[31]

Example 2. Multiple Usages of Process Waters There are several examples of this. One company[32] utilizes its microetchant from an electroless copper line preclean step. When the copper concentration reaches a threshold limit then fresh etchant is added using a simple feed-and-bleed system. This not only prevents frequent bath dumps, but also maintains the copper concentration in a range where the microetch rate is more consistent.

Next this overflow goes via gravity feed to a microetch used to prepare the imaged copper for pattern plating. The bath can utilize an etchant with higher copper concentration, since it must only remove a small amount of copper (4 to 6 μin). This bath, in turn, flows into a bath used for electroless copper rack-stripping. Once this tank is full, it is pumped into a holding tank where it is used on weekends to clean extraneous copper from the electroless tank walls.

Finally, the copper concentration is now so high that the solution can be pumped into an electrolytic plate-out cell and the copper recovered and sold at the current price of copper metal. The remainder of the solution can then be waste treated.

The cost analysis included the following elements:

Annualized savings:	
Reduced materials handling/labor time associated with bath maintenance*	$11,000
Potassium persulfate (microetchant) purchases	$29,463
Sulfuric acid purchases	$1,500
Estimated waste treatment chemicals	$2,800
Total	$44,763
Capital costs:	
Photocell (for monitoring copper threshold)	$0[†]
Plumbing and pumps	$1,200
Estimated installation (labor) costs	$250
Total	$1,450

*Labor savings may be somewhat misleading in that unless the labor savings result in less personnel or more production, this number may not really have a monetary impact on the company.
†The photocell was provided by a vendor, its actual cost is approximately $1,200 which would add another 10 days to the payback time.
Note: Payback time was estimated at 12 days (determined by a simply dividing the savings by 365 to determine the savings per day and then dividing that into the costs).

Example 3. Water Cleaned Up for Reuse in the Same or Similar Process   Water reuse has become of significant interest to manufacturers in areas of high water cost or low availability. Among a number of methods to clean up water for reuse, membrane filtration has been used for many years even for drinking water applications. One recent paper dealt with a combination of technologies to clean up process water for reuse and capture some of the polluting materials.[33] This particular case compared reverse osmosis (RO) membranes alone, a system with the RO combined with ion exchange columns, and a system using simple conditioning and filtering to clean water used in critical rinsing for use in noncritical rinsing. The effective cleaning rates were 75 to 85%, 70 to 80%, and essentially 100%, respectively. Savings in water costs and improved yield due to better rinsing water quality depended on local water costs and production/water volumes.

To summarize:

1. For simple changes, simple calculations can be used to determine costs and payback times. For more complex changes, more accurate calculations on cost factors based on internal rates of return, etc., should be used.
2. Labor costs should be used with care, since the labor savings may not actually result in less labor costs overall.
3. Agreements with material suppliers may be based on boards produced and not on amounts of chemicals actually used, so some considerations should be made in this case as to cost savings.

4. All costs need to be included in making an assessment of the net impact of the pollution prevention project. These should include
   - Costs:
     Capital cost of the equipment
     Installation costs of equipment, including plumbing changes for segregation of wastes
     Startup and operation costs, including consumables and energy costs
     Facilities costs if new construction is required
     Assignable labor costs requiring *new* labor
   - Financial benefits:
     Reduced water and sewage costs (sometimes these are lumped together by municipalities)
     Reduced utility fees including connection fees and/or time-of-day discounts
     Rebates from sewer authorities for water use reductions
     Reduced process chemical usage and waste treatment costs
     Increased capacity due to increased usage of same facilities
   - Other considerations:
     Determine impact on sewer use, wastewater, and air permits
     Ensure that changes do not negatively affect quality of products
     Improve relationship with local water and regulatory authorities

8.4.3 More complex pollution prevention methods

As was noted in the previous sections, once simpler source reduction, reuse, and recycling projects have been completed, pollution prevention must be more completely evaluated since the cost will most likely be higher and the payback period longer. These processes can be classified as closed-loop systems for reuse, chemical recovery, chemical maintenance for reduced usage, and substitute technologies. A very comprehensive evaluation of these technologies, as used in the plating industry, can be found in Ref. 34. A document focused specifically on the printed wiring board industry can be found in Ref. 35 and Ref. 35a.

8.4.3.1 Closed-loop systems. Closed-loop systems are generally of the type where a chemical bath is plumbed into a regeneration system which regenerates the chemistry and sends it back to the process bath. Examples are etchant regeneration systems[36] in which either ammoniacal or cupric chloride, or a combination of these etchants, is regenerated using combinations of either solvent extraction plus electrowinning or electrodialysis and electrowinning (or electrolytic plating out of copper). Another example is regeneration of the solution used to strip solder plating from tab and connector fingers.[37] Even process solutions used for developing and stripping dry-film photoresists can be reused with a minimum of added chemicals using closed loop systems.[38]

8.4.3.2 Recovery systems. Chemical recovery is very similar to closed-loop system technology except that rather than reusing the materials in the same process bath, the chemicals are recovered for recycling or reuse in another industry. This is usually true because the recovery process cannot separate the impurities or by-products of the process sufficiently to allow reuse in the original process. Most typically this is accomplished by one or more of the following technologies: ion exchange, electrowinning, reverse osmosis or ultrafiltration membranes, or electrodialysis.

Ion exchange units make use of polymers with active sites that can exchange one negative ion for another (anions) or more commonly positive ions (cations) for each other. This is often used for removing copper and nickel from printed wiring board solutions. The polymer beads are housed in a "column" through which the solution flows and exchanges copper or nickel ions for hydrogen ions. Once the capacity of the column to remove metal ions is reached, the metal is removed by treating the polymer beads with acid, thus capturing a very concentrated solution of the metal ions. These solutions are often further treated by plating out the metal on an electrode of the same metal or on stainless steel or other surface from which the pure metal (copper or nickel) can be physically removed and sold as pure metal. This process is also known as *electrowinning.*

Membranes of various types can be used to concentrate process solutions by allowing molecules of a particular size to pass through the membranes while restricting others, thus concentrating them in the treatment stream. The same is true for diffusion dialysis or electrodialysis systems. They utilize an ion exchange membrane that allows the anions (that are to be recycled) and hydrogen ions to pass through either to a water stream flowing in the opposite direction or to a charged direct current applied to opposite sides of the membrane, respectively. These anions may then be recycled in the process solution, while the metals left in the treatment stream can be concentrated or treated as indicated previously for recovery.

8.4.3.3 Chemical maintenance for reduced usage. As indicated previously, improved chemical maintenance can not only reduce the frequency of bath replacements (see Sec. 8.4.2, Example 2), but also improve the quality of the bath performance. Chemical maintenance is taken to mean maintaining a processing bath without application of additional chemicals. An example would be a feed-and-bleed process where used solutions are replaced with new in an overflow type of system. Only if the overflow is "cleaned up" by one of the preceding regeneration or recovery processes would this be considered "maintenance." Sources of contamination are the chief reason for baths to require maintenance. These sources can include drag-in from other baths, breakdown, or by-products of the chemicals or chemical reactions occurring in the bath, or impurities present on the parts (oil and grease, for example). In addition the bath may simply become depleted of some of its active or inactive (surface agents) ingredients.

Proper analysis of the various baths is obviously a necessity for controlling the chemistry and thus the efficiency of the bath. Some analyses are also designed to measure the contamination levels. Many baths can be monitored on a continuous, periodic basis by a number of techniques. This type of analysis, along with a continuous cleaning process, tends to keep the concentrations of all the ingredients close to the optimum level and removes contaminants in a continuous manner. Again, these operations would tend to provide a more consistent level of performance and reduce product scrapped due to low quality.

Many baths can be "cleaned" of contaminants using simple filtration devices or carbon filtering. For plating baths, this means that some of the minor organic components used to control the quality of the plating will be removed and must be added back.

8.4.3.4 Substitute technologies. As the feature size and demands on printed wiring boards have increased, the technologies to produce them have changed and are continuing to change. This, along with some regulatory drivers, has caused manufacturers to reexamine some of the processes used for making boards. Examples of substitute technologies are dry-film photoresists to replace solvent-coated liquid resists, aqueous cleaners to replace solvent cleaners, aqueous processable photoresist to replace solvent and semiaqueous processable resists and soldermasks, and most recently direct plating for through-hole and blind vias to replace formaldehyde-based electroless plating baths. Currently, the industry is also evaluating various replacements for the solder coating used on copper lines as an etch resist and used as a copper surface oxide protectant during storage.

Considerable effort has gone into the latter two replacement technology assessments (see Sec. 8.2.4), requiring not only technology performance assessments but also environmental, health, and safety risk assessments and cost evaluations.

As the size of features decreases, other technologies will be employed to meet this demand. Currently, a number of photodielectric technologies are being tested (see Secs. 8.5 and 8.6). Direct imaging of photoresists (thus eliminating the need for a phototool and its accompanying chemistries) and even elimination of photoprinting of resists[39] are also being considered.

8.4.4 Other opportunities

While the preceding sections have given a brief overview of some options for reducing or eliminating pollution in plant process operations, there are additional considerations for reduction of waste. These include selling waste materials to recyclers for reuse or recycling rather than sending the waste to an incinerator or landfill, recycling in-plant as was noted for some process solutions such as etchants and plating baths, and finally working with materials suppliers on cooperative efforts to reduce materials usage.

Recycling materials outside of the plant can include not only chemicals or materials recovered in recovery or closed-loop processes but also scrap materials. Even waste treatment sludge can often be recycled for the metals contained within it. Those plants that cannot regenerate their own process solutions can often work with suppliers who can regenerate them at their own facility and then ship them back to the manufacturer's plant. Materials can also be recycled, most printed wiring board shops send scrap solder and even solder dross back to suppliers for recovery and reuse of the solder. Even scrap circuit boards, while often recycled only for their metal content, have found a market as decorative items for clipboards, book covers, and other novelty items.

Even process solutions that contain materials that can't economically be recycled may be useful in a process that utilizes the characteristics of that solutions, such as acidity. This often happens where a printed wiring board process solution is too contaminated for use in a board process, but could be utilized by another industry that isn't affected by the contaminants.

As mentioned previously, many plants are now working with their suppliers in cooperative agreements to supply not chemistry, but completed parts. In other words, rather than selling manufacturers chemistry or materials, suppliers are charging based on the number of good parts processed. This allows both the supplier and the manufacturer to concentrate on using the minimum amount of material and splitting the savings generated. The result is usually a reduction in the amount of material utilized and requiring disposal or treatment.

8.4.5 Sources of information for pollution prevention

There are several texts with information on pollution prevention and process optimization:

George Cushnie, *Pollution Prevention and Control Technologies for Plating Operations,* National Center for Manufacturing Sciences, 1994.

Guidelines for Implementing Hazardous and Non-Hazardous Waste Minimization, IPC-WM-72, The Institute for Interconnecting and Packaging Electronic Circuits, Northbrook, IL, August 1992.

Plus a number of publications by the EPA:

"Printed Wiring Board Pollution Prevention and Control: Analysis of Survey Results," EPA 744-R-95-006, Environmental Protection Agency, September 1995.

"Guidelines to Pollution Prevention—The Printed Circuit Board Manufacturing Industry," EPA 625-7-90-007, Environmental Protection Agency, June 1990.

"International Waste Minimization Approaches and Policies to Metal Plating," EPA530-R-96-008, Environmental Protection Agency, (NTIS/PB96-196 753), Rockville, MD, 1996.

Waste Minimization Manual, Government Institutes, Inc., 1990.

The EPA Manual for Waste Minimization Opportunity Assessments, Government Institutes, Inc., Rockville, MD, 1988 (EPA document EPA 625-7-88-003, July 1988).

Environmental Regulations and Technology—The Electroplating Industry, EPA 625-10-85-001, Environmental Protection Agency, September 1985.

On the Internet the following web sites have a large number of manuals, booklets, and case studies applicable to pollution prevention in the printed wiring board industry:

The PWB Resource Center, *www.pwbrc.org,* maintained by the EPA, IPC, and National Center for Manufacturing Sciences (NCMS).

The National Metal Finishers Resource Center, *www.nmfrc.org,* similar to PWB Resource Center, but focused on metal finishing and plating and maintained by AESF and NCMS.

Enviroene, *es.epa.gov/,* includes waste reduction guides for the PWB industry.

U.S. EPA Design for the Environment, *www.epa.gov/opptintr/dfe/index.html,* includes most of the Design for the Environment Publications generated for the PWB industry.

Department of Energy DFE, *terrassa.pnl.gov:2080/DFE/.*

Office of Solid Waste, *www.epa.gov/epaoswer/hazwaste/minimize/intl-wm/.*

President's Council on Sustainable Development, *www.whitehouse.gov / PCSD / Publications.*

MCC's page on PWB Resources, *www.mcc.com / projects / env / pwb_resources.html.*

8.5 Evolutionary Improvements in PWB Processes

As was noted previously in "Substitute Technologies," Sec. 8.43, some of the most significant decreases in pollution, waste, and energy use reduction may occur as technology drives the design and fabrication of printed wiring boards to finer features and greater functionality. In this section we will examine some of the current technology changes that are occurring or have just occurred. By "evolutionary," we mean that a process is only somewhat modified from that used in the past and the infrastructure and technology is already in place. We will examine water-borne liquid resists, permanent resists, tin-lead coating replacements, direct plating, and no-lead solders.

8.5.1 Aqueous-borne resists

The drive for low VOC liquid applied materials probably saw its first manifestation in low VOC fluxes. These fluxes were traditionally applied using a low boiling solvent such as an alcohol. These are now almost universally water-borne. Liquid resists and solder masks have always had a large environmental and cost burden attached to their use. Like their brothers, solvent-based inks and paints, they are typically 40 to 60% solvents. These solvents have more recently been regulated by VOC limits under the Clean Air Act Amendments of 1990 as well as local regulations requiring extremely low limits on their emissions. This, in turn, required users to implement typically expensive ongoing pollution abatement solutions, such as thermal oxidizers (incinerators), on the plant site with the associated capital, installation, maintenance, control, and permit costs. This has driven the manufacturers of liquid resists and solder masks to attempt to develop aqueous-borne coatings containing little or no solvents. These materials still require significant drying capabilities due to their water content, so that there is still a significant environmental impact in terms of the energy required. One or two manufacturers currently offer an all-aqueous solder mask, but currently there are no publications benchmarking their performance versus their solvent-borne or dry-film competitors.

One class of aqueous-borne resists that require considerably less drying are the electrodeposited resists. In this process, the resist is held in solution in the form of micelles of the resist composition using a very small amount of high boiling organic solvent in the water. The micelles are charged and thus attracted to an oppositely charged copper on clad material. The micelles are thus attracted to copper and coat it, forming a thin layer of protective resist. The process is self-healing, since pinholes are sources of high current density, and thus will be "plated" with the resist micelles. The thickness of the resist is controlled by the ability of the resist to be attracted to the copper through the resist already in place, so that the resist thickness is rather thin and cannot be "plated" to thicknesses appropriate for electroplating applications. The process is used exclusively for print-and-etch applications. After the resist is applied by electrodeposition, it is dried, exposed, and then developed. This imaging is then followed by etching. It has found limited application in the United States, but is more popular in Asia.

8.5.2 Permanent resists

The advantages of a permanent resist can be found in the elimination of steps that normally follow the metal pattern defining steps. For resists that would be used for print and etch applications, the permanent resists would not have to be stripped and would function as part of the adhesive bond between the interlayer dielectric and the copper circuit. The stripping step would be eliminated as would the oxide surface preparation steps needed to make the copper stick better to the innerlayer dielectric during press lamination. The waste treatment for all the process streams for these two steps would also be eliminated.

Early examples of this approach could be found in additive plating applications. One U.S. company utilized a resist that was designed for use in an electroless plating bath. The resist was applied over a catalyzed surface and then imaged with circuit lines. These were developed out to leave channels for the electroless plating to form the circuit lines. The electroless copper was then allowed to plate up to the top of the resist, thus forming a nearly planar surface. This process was only used on the outer layers of boards. It did improve the ability to restrict solder using a thin solder mask since the outer surface was planar.[40]

A permanent resist for conventional print and etch was investigated by both the IPC October Project and a DARPA Environmentally Conscious Electronics Manufacturing Initiative Project.[41] There were a number of technical challenges besides the need to have the resist function as a permanent part of the dielectric. It also had to be thinner than most dry-film resists and have all of the physical and mechanical characteristics of a dielectric. In the final analysis, there were insufficient cost drivers to pull this type of resist into the market.

Interest in permanent photodielectrics used to make microvias supplanted the need for this type of resist.

8.5.3 Direct plating alternative

Direct plating and similar processes were developed to replace the electroless copper plating used in making the insides of inner-layer connection holes conductive. To connect two sides of a board electrically or to connect two adjacent layers together, a hole or via must be made in the polymeric or inorganic dielectric between the two circuit layers. Since the dielectric is usually nonconductive, electroless copper plating was originally developed to plate on the nonconductive surface. The process actually derives from spray-plating and silver-plating applications used for decorative purposes.[42]

The electroless plating process for making vias, through-holes, and interconnections utilizes a number of preparation steps to clean both organic and inorganic residues from the hole walls. There are also steps to microroughen the surface of the dielectric in the hole or via. A catalyst, usually a platinum or palladium colloid, is deposited on the surface. This is reduced to the metal, often with a tin-reducing agent. This catalyst then is used to catalyze electroless copper from a highly caustic solution containing copper salt complexed by an organic complexing agent such as EDTA. The complexer keeps the copper in solution in the highly caustic solution where it would normally not be very soluble. A reducing agent, usually formaldehyde, participates in the reaction that converts the complexed copper ions into copper metal at the surface of the catalyst or on copper already present. A thin layer of electroless copper is all that is needed, since this step is usually followed by an electroplating step to complete circuitry on the outer surfaces of the circuit board.

Some of the environmental concerns with electroless plating are the use of formaldehyde, a carcinogen, the use of a complexers which can interfere with the operation of the metals waste treatment area if not segregated and the waste treatment required. A number of plating chemistry suppliers have developed alternatives to the traditional formaldehyde electroless plating. These fall into roughly four categories: electroless using nonformaldehyde reducing agents, and electroplating on surfaces prepared by carbon or graphite seeding, palladium seeding, and conductive polymer surface preparation. These technologies were evaluated both from an environmental, health, and safety perspective as well as by performance and economics.[43]

These evaluations were also followed up with interviews with commercial board manufacturers using the various technologies.[44] Among the findings of this report were the comments that facilities implementing these technologies needed to take a "whole process" or systems approach. This was taken to mean that process changes had both upstream and downstream implications and might require process changes in other existing processes. This was shown by the need for management commitment and to the operations personnel to overcome problems that inevitably occur during installation of new processes.

The results of these studies showed that all of the technologies, with the exception of conductive polymers (which weren't included in the performance demonstrations), had equivalent electrical performance and stability over time. While the alternatives to formaldehyde electroless had lower environmental, health, and safety concerns, they still contained materials whose toxicity require the care normally associated with working with industrial chemical processes. Several of the carbon/graphite and palladium processes, however, could be used in horizontal, enclosed processors. These were totally enclosed and prevented exposure of operators and offered better control of emissions. They also allowed the process to be installed in a very small footprint. There were a number of processes that also used much less water than some of the other processes. The waste treatment of these alternatives was simpler in many cases, since copper complexers were not used.

In summary, a number of commercially available alternatives for formaldehyde-based, copper electroless plating have been tested and are in production lines in many PWB shops. They give at least equivalent electrical and quality performance. They often have environmental, health, and safety as well as economic benefits. It should be noted that not all processes work well in all shops and local variations mean that all aspects of a given process should be reviewed carefully for any complications.

8.5.4 Surface protection and metal etch resists

Tin-lead solder has been used in this country for many years in applications other than electrically and mechanically connecting components to circuit boards. One of these applications is for surface protection of the copper on completed boards being shipped to assemblers or stored prior to assembly. The copper on these boards is prone to oxidation and contamination from various organics, and assemblers often required a "solderable surface" of PWB manufacturers. Recently, many organic and inorganic surface protection materials have been made available and can be applied by a simple dip step to protect the copper from oxidation and contamination. The U.S. EPA is currently working on a Design for the Environment project similar to that for direct plating.[45] The chief reason for eliminating the solder coating is to minimize the use of solder, especially since it is often in excess of that needed for bonding components to the board.

Another application that uses a consumable coating of solder is that of metal etch resist. In the pattern plating processes, solder has been plated over the copper of circuit lines. After the photopolymer resist between the lines is removed, the thinner copper electrically connecting the plated lines must be etched away, resulting in the need for a metal etch resist during this etch. Some shops had experimented with differentially etching the thinner background copper, that is, not using a metal resist but etching both the plated copper and the thin copper between lines. This method only removes a small amount from the plated lines, since the copper in between lines is very thin (see Fig. 8.8). As circuit lines become thinner and narrower, the control of the etching sprays has not been sufficient to control this well. Instead, a number of alternative metals are being used. Most often, this is tin that is either electrolessly or electrolytically applied. This requires the use of etchants that will not attack the tin, tin with the proper undercoat or tin with the proper surface structure. The tin can then also act as the surface protection and is readily solderable. The chief advantage of these applications, from an environmental perspective, is the elimination of lead and lead solutions.

8.5.5 No-lead solders

There now appears to be a growing drive to eliminate/minimize lead from solders. Several Scandinavian countries have indicated that they will ban imports of lead in the next 5 to 10 years. More recently, the Japanese have also indicated that they will follow this line. While extensive work on various solders has been done, only recently have these materials found their way into commercial production quantity products.[46] Cost and reliability have been the chief concerns of most potential users of alternative solders. However, as feature sizes on circuits continue to shrink, there is also evidence that many solders, including tin-lead combinations, may reach their limit in providing the mechanical and physical properties needed to do the job.[47] So while it could be argued that lead solder is not an environmental concern due to the extensive recycling infrastructure for lead and solder, technical reasons may cause its displacement.

There are a number of reviews of the various no-lead solders in the literature, so we will not attempt to cover this material in depth. Suffice it to say that all aspects of solder replacement need to be evaluated, just as they were in the DfE projects for direct plating and surface protection noted earlier.

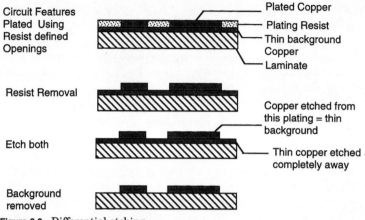

Figure 8.8 Differential etching.

8.6   Technologies for Revolutionary Improvements in Processes

As stated earlier, the drive to thinner, cheaper, and lighter can act as a catalyst for technologies that are also environmentally beneficial. For example, thinner and lighter actually mean that the overall circuit density, as well as the functionality of PWBs, will continue to increase. This, in turn, means that less material is used and technologies to make denser features, with both smaller lines and vias, will be developed and used. This allows the development of not only new materials but also new processes to accomplish this. Materials, processes, and equipment that might not have seemed economically practical for lower densities may become not only economically attractive, but also technically necessary. This can mean that materials that were once consumable will take on an additional functionality or be eliminated. It may also spur development, where practical, of dry processes whose unused materials can be recaptured All of these possibilities can have enormous environmental benefits. Rather than having to depend on increasing the eco-efficiency of existing processes, new processes can be developed which provide the technology needs while also eliminating environmental burdens. Listed in the following sections are materials and processes that are, at this writing, either becoming part of the infrastructure or are under development.

8.6.1   Organic printed circuit boards

8.6.1.1 Microvias. While this subject is covered in depth in another chapter of this book, we will mention it with respect to environmental impacts. Microvias are becoming the next significant enabling technology for PWBs. The ability to not only significantly shrink the size of both the via hole and the pad, but also place blind and buried vias on almost any layer, has been shown to significantly increase the potential density of a PWB. Assuming a reasonable process capability, this should immediately decrease the amount of material and waste generated in making boards. The improved accuracy of the three techniques used to make these vias should also improve yields and reduce waste boards. Depending on the technology used, there could also be a reduction in the amount of energy used to make the board. The three technologies have the advantages/disadvantages listed in Table 8.5 and Fig. 8.9.

In addition, a combination of laser with a photodielectric material could be used to make channels for circuit lines with vias produced by lasers through the same or similar material. The vias and channels for circuit lines could then be metallized in a single step using any of a number of techniques such as electroless plating or conductive pastes (see Fig. 8.10). Photopolymers have also been developed so that two wavelengths of light can expose different layers. This exposure was followed by a development step to produce the channel and via in a single step already in register due to the artwork used[48] (see Fig. 8.11). In the case of these latter materials and processes, the consumable resist used to pattern circuit lines in register with the vias has been eliminated by a photodielectric which acts as the circuit line imaging element and also the dielectric.

8.6.2 Polymer thick-film materials

Polymer thick-films (PTF) materials are organic-based pastes or film-like materials containing either conductive or resistive particles.

TABLE 8.5   Environmental Comparison of Photo, Laser, and Plasma Microvias

Technology	Materials	Environmental advantages	Environmental disadvantages
Laser	Laser, dielectric	No plasma gas No developer solution	Air emissions from vaporized dielectric
Photo-polymer	Photopolmer dielectric, artwork phototool for vias	No air emissions Solvent is recyclable	Small amount of dissolved photopolymer, used phototool
Plasma	Resist to pattern holes in copper on dielectric, artwork photo-tool for vias		Resist process streams, plasma gas, reacted dielectric gas used phototool

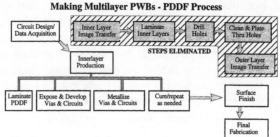

Making Multilayer PWBs - PDDF Process

• **Reduce material, energy & water usage**
  – Additive rather than subtractive
  – Multiple functionality of dielectric
  – Reduced process steps & equipment
  – Dry film "vs." liquid materials

Figure 8.9 Environmental advantages of microvias.

These materials are placed on PWBs in either circuit lines, vias, ground plane, shielding layers, or resistive patterns and cured on the dielectrics being used (see Fig. 8.12). The process of application can be transfer printing, screen printing, or metal stencil printing or squeegeeing directly into surface features. The PTF may or may not have a volatile vehicle or solvent present. The environmental advantages are that the processes can be nearly completely dry and additive in that the PTF is applied only where it is needed. In addition, unused material can be recovered and reused. Material that eventually becomes unusable can also be recycled.

As noted previously, vias and channels developed out of photodielectric materials can be filled with a PTF or via fill material and then cured to make conductive lines and vias. Alternatively, vias or even through holes can be filled with via fill materials using either squeegeeing or a metal mask or stencil to direct the paste into the holes. Electroless and/or electrolytic copper can be plated over many of these cured inks to add circuit lines.

8.6.3 Via plug examples

Since the process is, strictly speaking, additive, it can also be used to put down metal for many purposes only where it is desired. Examples, as noted earlier, include ground planes or shielding for boards or even individual layers (see Fig. 8.13).

8.6.4 Buried passive components

Buried passive materials have been available in ceramic materials for some years in the form of screenable pastes which are fired and then

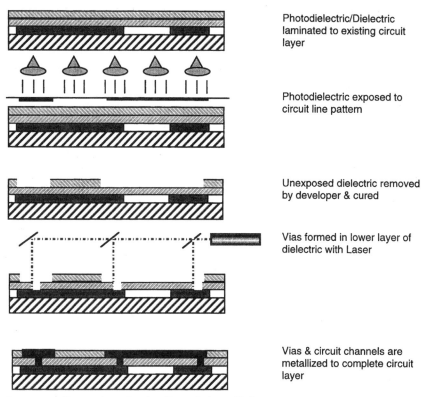

Photodielectric/Dielectric laminated to existing circuit layer

Photodielectric exposed to circuit line pattern

Unexposed dielectric removed by developer & cured

Vias formed in lower layer of dielectric with Laser

Vias & circuit channels are metallized to complete circuit layer

Figure 8.10 Process to make circuits and vias with laser.

laser trimmed to the dimensions, providing the desired electrical characteristics (see Fig. 8.14). PTF materials other than conductors can also be used as well for buried passive components such as resistors or capacitors.[49] The volume of these materials can be determined either by the rheology and stencil or screen pattern used. Alternatively, when used in conjunction with photodielectrics, it would be possible to image the desired volume and then develop out a "space" from the photodielectric as desired. This would be especially useful if the photodielectric were a positive working photopolymer. Alternatively, the volume could also be constructed using laser ablation or vaporization.

Preformed buried components can also be made to exact tolerances and then are embedded in the appropriate dielectric or photodielectric materials. Technologies to form resistors either from buried layers of low conductivity metals[50] or using a metal formed at the annular ring of a via[51] have also been developed and used in some applications.

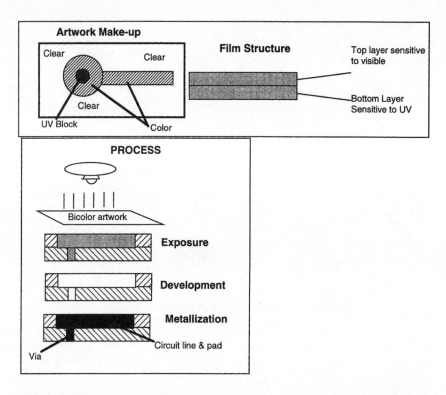

Figure 8.11 Process using bi-color artwork to simultaneously produce circuit channels and vias.

8.6.5 Flex and flex—rigid packaging

The use of flexible dielectric materials has reduced the weight and volume of a number of circuit applications. Originally, polyimide flexible materials were developed for high heat/cold cycle performance circuits. Because they could be manufactured in much thinner dielectric thicknesses and could be combined with sections of rigid circuitry and folded over on themselves, they were also used in applications requiring a very small volume such as automotive, aerospace, and camera applications. In addition, the mechanical strength of these materials eliminated the need for a fill material such as glass fiber. Thus, this dielectric could be etched chemically or laser-drilled more easily than rigid materials such as glass epoxy.

8.6.6 Photodielectric elements

Although only in its infancy, the ability to use light as the medium for signal transfer versus electrically through conductors has been recog-

Example 1 – Shielding Layer

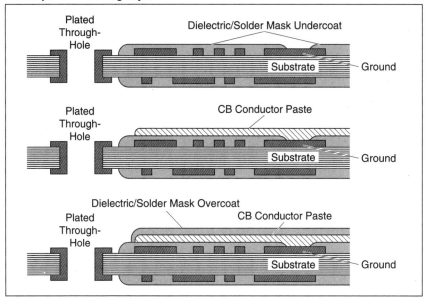

Example 2 Ground plane layers

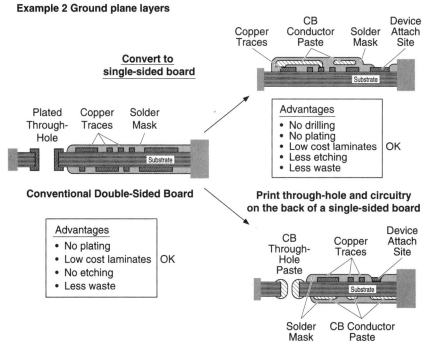

Figure 8.12 PTF shielding and ground plane examples.

Example 1 - Buried Vias

ViaPlug for Innerlayers with Buried Vias

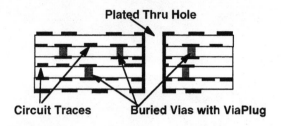

Plated Thru Hole

Circuit Traces   Buried Vias with ViaPlug

Example 2 Buried Plated Through Hole

Via Plug for Increased Density

Typical Multilayer with Additional Redistribution Layers & Thru Holes

Same Board with Via Plug to Eliminate Space on Surface

- Dry Process for Additional conductivity
- Elimination of PTH thru redistribution layer
- Increased surface density

Figure 8.13   Via plug examples.

nized as the ultimate solution for handling high data throughput. Both glass fiber and polymers have been developed for use as wave guides, with glass initially being the solution of choice for long cables. For on-board and interassembly connections, however, polymers are now showing better cost/performance characteristics.

One process with an especially low environmental impact involves the use of holographic photopolymers. The process utilizes photopolymeric materials manufactured to be extremely clean to prevent internal scattering due to extraneous particles. This dry film is coated to high thickness accuracy for highly defined light guides that can be formed in large format or multiple repeat patterns of smaller units.

▭	**Passive Element**
▬	**Conductor (PTF)**
◼	**Via Plug(PTF)**
▨	**Photodielectric**

Figure 8.14   Examples of buried passive components.

Photopolymer Light Guides

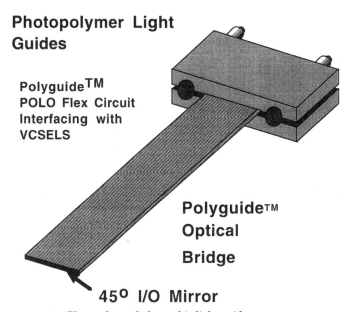

Polyguide™ POLO Flex Circuit Interfacing with VCSELS

Polyguide™ Optical Bridge

45º I/O Mirror

Figure 8.15   Photopolymer holographic light guides.

The process relies on internal diffusion of specific monomers to modulate the optical properties of the areas of the photopolymer to be used as light guides (see Fig. 8.15). The process is completely dry with no etching or molding. Guide formation is completely light induced at 15 to 45°C, followed by photo and thermal fixing. This produces light guides with low scatter waveguide sidewalls. In addition, these same materials, as well as other families of photopolymers designed for holographics, can be imaged holographically to produce mirrors, gratings, and other optical elements used in modifying the light from the guides.[52]

Since these materials are used with, or as part of, typical printed wiring boards, they must also survive the processing conditions used in making a board. The most critical properties of these materials and constructions, outside of their ability to manipulate light, are resistance to degradation due to high temperatures such as those experienced during solder operations and moisture sensitivity. These properties, as well as the specifications for light manipulation listed later here, have been demonstrated in at least two DARPA programs.[53,54]

Photopolymeric holographic devices are also being utilized in display applications such as mirrors, gratings, and even displays themselves. Using holographic exposure techniques, these devices can also be produced by completely dry processes. It may be expected that eventually the display capabilities of holographic photopolymeric materials will be coupled with the light guides carrying the signals to generate the display images.

8.6.7 Organic chip packages

Ceramic chip packages have been the chief infrastructure for this component of manufacture, but due to cost and weight drivers, manufacturers are now beginning to look to PWB technologies to produce small chip packages with high levels of functionality in a lighter, less expensive organic medium. Microvias are one of the chief enablers of this approach, and many of the technologies that we have mentioned here are contributing, and will continue to contribute, to the advancements in this area. Again, the chief environmental driver will be the use of less material for the same functionality, and less weight will thus lower power consumption for transport and in some cases for final operating use.

8.6.8 Chip-mounting options—solder placement

As chip functionalities increase, the ability to have sufficient I/O connections on the chip and to the PWB becomes one of the major hurdles. Many chips now utilize area array or flip-chip bonding, often directly to the circuit board. There are a number of processes utilized to build up sufficient solder in a very precise and fine pattern for later bonding to the chip and board. Many of these involve using sacrificial resists and extensive plating. One of the more unique methods utilizes a photopatterned thin, temporary adhesive to precisely attach uniformly sized solder balls (see Fig. 8.16). Again, this is a complete dry additive process. The solder balls are applied to the photoadhesive by a mechanical process and solder balls not attached to an imaged spot are collected and reused. In a continuous roll process, the support material holding the adhesive can be recycled.

In a program sponsored by NIST, a unique ink jet technology was developed to very precisely place molten solder balls.[55] It may be possible to use this technology in this application as well.

PhotoPatterning of Solder Bumps

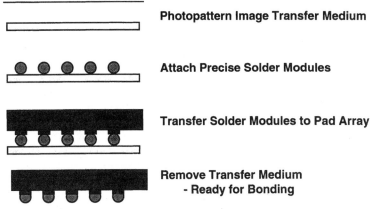

Photopattern Image Transfer Medium

Attach Precise Solder Modules

Transfer Solder Modules to Pad Array

Remove Transfer Medium
- Ready for Bonding

Figure 8.16 Tacky dot process.

8.6.9 Laser direct imaging of resists

Following the pattern established by the printing industry, PWB manufacturers are looking to resist suppliers to develop resists that can
be imaged by direct laser imaging. There are at least two possible
techniques for using lasers to image resists. One utilizes a scanning
process, switching the laser off and on with a computer to "write" the
pattern by polymerizing the resist selectively. The second possibility
would use a positive resist to simply draw out the features to be developed, This technology is now possible because of:

1. The drop in laser price as laser technology has advanced along with
   the computers that control them.
2. The improvements in the quality of lasers.
3. The wavelengths available with lasers.

Suppliers of photoresists have already developed a number of resists
with special imaging chemistry that takes advantages of these lasers.[56]

The environmental benefit of this technology is the elimination of the
phototool, its preparation steps and chemistries, and the associated
waste streams which contain silver. In addition, assuming that the
resist itself has no dirt present, the laser should not have the problem
of either phototool-related opens or shorts due to dirt on the phototool
or scratches in the artwork. This should improve the yield at the photoimaging step. In addition, many lasers have a self-alignment feature
that will also decrease yield losses due to misalignment, especially of
lines to vias and layer to layer.

8.6.10   Other possibilities

As noted earlier, researchers are working on processes to imprint depressions into copper foil to define circuit lines.[57] Basically, this process utilizes technology similar to that used for microembossing grooves and features in compact disks to make circuit lines and padless microvias. The process uses a special dielectric material with a filler which will not interfere with the microembossing process. The lines are defined by ridges and the vias by "points" in the ridges which are made in a metal master that has been electroformed to define these features. The master is then pressed into the foil-clad laminate material. The "points" form vias through the laminate and the ridges emboss U-shaped channels where circuit lines will be formed (see Fig. 8.17). A nonphotosensitive resist is applied to the embossed surface where it fills the vias and U-shaped channels. The excess is removed by squeegee or a similar process and the unwanted copper foil is etched away, leaving the circuit lines and vias. The resist is then washed away and, if necessary, additional metallization is added to the vias. These vias are in perfect registration since they are part of the embossing pattern which makes the circuit lines at the same time and can thus be padless. The process could be used to make very high density circuits and in theory, at least, the resist could be recovered and reused.

Others have developed methods for using sophisticated sand blasting to form microvias.[58] These techniques could also be employed to form the circuit line channels, with the environmental advantage being that, combined with a PTF metallization, the process could be completely dry and the developing media—sand—would be completely recycled.

Added to this is the work going on to develop low cost, somewhat unsophisticated organic integrated circuit chips for low level functions.[59] It would then be possible to use some of the techniques for making PWBs to incorporate the chips right into the board itself.

In fact, one could envision a process to make highly functional PWBs that was completely dry and additive.

8.6.11   Ceramic technologies

While the use of ceramics in some applications that require low cost and weight are being taken over by organic packaging, some applications will still require extensive ceramic technology. Many of the desires that are only now being developed for organic PWBs have been an integral part of ceramic packages for a number of years. Such processes as buried resistors and capacitors, additive metallization (using screen-printed ceramic-based inks), and microvias have been in place in ceramics for a number of years. As the density, functionality,

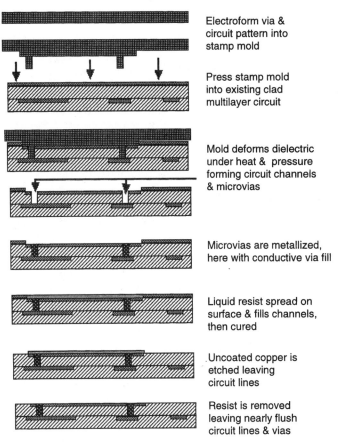

Electroform via & circuit pattern into stamp mold

Press stamp mold into existing clad multilayer circuit

Mold deforms dielectric under heat & pressure forming circuit channels & microvias

Microvias are metallized, here with conductive via fill

Liquid resist spread on surface & fills channels, then cured

Uncoated copper is etched leaving circuit lines

Resist is removed leaving nearly flush circuit lines & vias

Figure 8.17 U-shaped lines and microvia process.

weight, thickness, and electrical requirements become tighter, however, ceramic technology will have to keep pace with new developments. Some of this work has been accomplished using photoceramics which can be exposed and developed to make fine vias and circuit channels. One particularly sophisticated technique for making conductive vias is known as *diffusion patterning* (see Fig. 8.18).[60] It essentially uses only a small amount of water and diffusion within the ceramic to create very fine conductive vias.

8.7 Evaluation of Technologies

As newer technologies become available, processes to evaluate their impact from a technical, economic, quality, and environmental (or

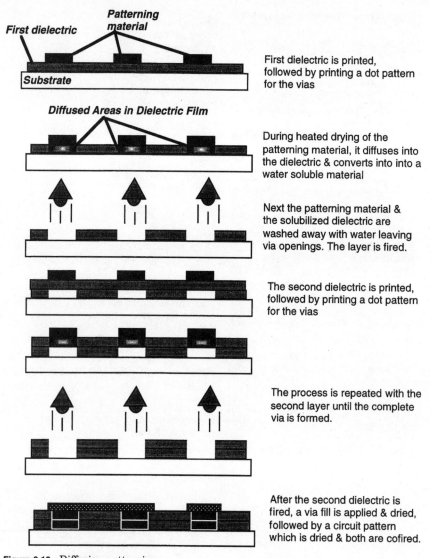

First dielectric

Patterning material

Substrate

First dielectric is printed, followed by printing a dot pattern for the vias

Diffused Areas in Dielectric Film

During heated drying of the patterning material, it diffuses into the dielectric & converts into into a water soluble material

Next the patterning material & the solubilized dielectric are washed away with water leaving via openings. The layer is fired.

The second dielectric is printed, followed by printing a dot pattern for the vias

The process is repeated with the second layer until the complete via is formed.

After the second dielectric is fired, a via fill is applied & dried, followed by a circuit pattern which is dried & both are cofired.

Figure 8.18 Diffusion patterning.

societal) impact will be needed. As we have noted, a number of efforts are under way to evaluate the real costs of environmental aspects of materials and processes. There are also efforts to evaluate materials by some method of design for environment metrics. As resources become more scarce, the need for closed-loop systems and the concept of sustainability will be a real factor in the marketplace. Currently, there are a number of methods to evaluate various technologies and industries using life-cycle assessment and eco-efficiency. The earliest efforts are being generated by large original equipment manufacturers, some European financial institutions, as well as several organizations devoted to sustainability. Those organizations promoting and studying sustainability have recognized that sustainability, or the ability to make the products we need while depleting our natural resources the least, will become a competitive issue in an ever-increasing service- and information-based economy.

8.7.1 Technology assessments

Nearly every link in the supply chain in the electronics industry has developed, or is developing, a technology roadmap. One of the first was the Semiconductor Industry Association roadmap. This plan for future needs also included sections on the environmental implications and thus the design needs with respect to environmental impacts. The IPC also developed an early roadmap in 1994, which, while it mentioned environmental concerns, covered them in a separate shorter IPC Environmental Roadmap."[61] MCC also published its landmark Life Cycle Assessment of a Computer Workstation that examined nearly every aspect of the electronic supply chain including the PWB industry. This was followed by two Electronic Industry Environmental Roadmaps. The first combined many of the aspects of the other industries' supply-chain roadmaps, while the second focused more on the needs of the ultimate products at the top of the supply chain and, in particular, their final disposition. The National Electronics Manufacturing Initiative initially sponsored by ARPA under the President's National Scientific and Technology Council as part of the Office of Science and Technology Policy took a higher level look at the complete supply chain from both a technical and manufacturing/infrastructure approach. All of these roadmaps are updated on a periodic basis and have become the focal point for a number of government funding efforts to support critical needs throughout the supply chain. Other regions have now begun to generate their own roadmaps including Japan, Hong Kong, etc.

The value of these roadmaps is not only to plan technological changes but also to see where the gaps are. In addition, they offer us the opportunity to help guide some of the work needed into areas that have the least environmental impact and the greatest opportunity for sustainability.

8.7.2 Modeling evaluations

There have been efforts to evaluate current and emerging materials and processes from a environmental cost and impact standpoint. One particular model utilizes a PWB design feature to call up particular process step choices and then not only determine an estimate of materials costs, but also an environmental cost and resources impact.[62]

As was mentioned previously, the EPA Design for the Environment program has its own set of "screening" tools to provide a high level assessment of a "use cluster" environmental, health, and safety impact. More sophisticated tools have been developed by a number of universities and some private companies.[63]

There are also tools for assessment of pollution prevention options. There are number of EPA documents to walk a team through the process and, as noted earlier, Tellus Institute has developed a computer program for actually determining the costs of the various options.

8.7.3 Environmental benefits

While many countries and regions have generated various eco-label programs to stimulate environmental awareness by both consumers and manufacturers, many of these label programs are significantly flawed. A few even have the characteristics of a nontariff trade barrier. Many have requirements that measure effects on one media such as surface waters, while ignoring impacts on other media such as air.

ISO 14000 can be used to stimulate some environmental benefits, even if this is only setting up systems within manufacturing sites to minimize out-of-control processes. The more proscriptive EMAS system does guarantee compliance with local environmental rules, something that many third-world countries apparently haven't managed by enforcing existing environmental regulations. ISO 14000 does have as part of its suite of guidelines and standards elements to improve on eco-labeling criteria, label criteria testing procedures, and auditing of environmental management systems. One of the guidelines also seeks to develop useful protocols for implementing life-cycle assessment.

Eco-efficiency is a term that has more recently been seen as one standard to gauge environmental benefits from pollution prevention and efficient use of resources. Eco-efficiency is a cross between good economic practices (low waste and low energy consumption) and sustainability. The World Business Council on Sustainable Development and other similar organizations, mostly based in Europe, are developing this concept. The purpose of measuring eco-efficiency is to drive industries toward low waste and low resource and energy consumption.

There are also several efforts to reduce the use and impact of "toxic materials." Unfortunately, there is no clear definition of a "toxic material." While there appears to be a very distinct definition of "highly toxic materials", "toxicity" with respect to materials that need to be minimized is not so easily defined. The newest concern for many is the defining of endocrine disrupters. These are chemicals or materials which mimic or disrupt the natural information or message exchange between cells in the body via the hormones. There are also lists of chemicals that contain materials that have a high degree of persistence toxicity and bioaccumulation, or PTBs, that are often linked to endocrine disrupters. Again, unfortunately, there is either a shortage of knowledge around many chemicals[64] or the schemes for attempting to rank chemicals (see, for example, the EPA's Priority Pollution List[65]) have too narrow and unbalanced an appraisal of the materials involved. Models developed by several universities that have more sophisticated modeling of these characteristics are being tested. Currently, lists of high production volume (HPV) chemicals are being developed so that a hierarchy of more intensive testing can begin. There are also European and Asian lists with materials on them that have a very low toxicity or environmental impact. Individual companies also have lists that are derived from many sources but these are usually not researched to determine their true hazard. One of the main problems with materials whose toxicity is questionable is that the efforts to reduce their use does not consider that these materials may be used in ways that restrict their exposure to operators, the public, and the environment or that the materials may be in closed-loop systems and used in the same system over and over again.

Toxicity reduction efforts need to be more clearly defined so that clear metrics are used to assign priorities. Materials used in closed-loop systems that are fully recovered and recycled should not be penalized. For example, while there is concern over lead in paint, lead-acid batteries are recycled by the millions every year without posing a health or environmental threat. Materials that are used in tightly controlled industrial systems should also be permitted under these conditions.

While many of these tools can measure the environmental benefit of existing technologies, it may be more difficult to assess new technologies that are being developed. Many of these technologies, such as the ones we described in the section on revolutionary processes, may play a major role in reducing the environmental impact. They will do this by replacing the existing technology completely and may eliminate the need for certain types of pollution prevention equipment. In fact, they may eventually eliminate the regulatory drivers (and burdens) currently placed on the PWB and other electronic industry sectors.

8.7.4 Public benefits

Industry as a whole, and the electronics and printed wiring board industry in particular, have significantly improved their environmental, health, and safety performance. The voluntary 33/50 program of 1991 significantly reduced 17 of the most highly used chemicals on this

TABLE 8.6   33/50 Program Chemicals*

Benzene	Cyanide and cyanide	Methyl isobutyl ketone
Cadmium and cadmium	compounds	Nickel and nickel compounds
compounds	Lead and lead	Tetrachloroethylene
		compounds
Carbon tetrachloride	Mercury and mercury	Toluene
Chloroform	compounds	1,1,1-Trichloroethane
Chromium and chromium	Methylene chloride	Trichloroethylene
compounds	Methyl ethyl keton	Xylenes

*EPA Voluntary Program to Reduce National Pollution releases and off-site transfers of 17 chemicals—33% by 1992 and 50% by 1995.

list (see Table 8.6). With very few exceptions, no PWB manufacturers use ozone-depleting substances (ODS) in their products or their processes. Chlorinated solvents used for resist development and stripping ended before the ODS provisions of the Clean Air Act came into being. Most PWB shops at least recycle their etchant solutions either off-site or on-site and most also recycle their wastewater treatment sludge for metals recovery. Many have metals reclamation systems or rinses for plating and/or etchant lines and all have some sort of pollution prevention programs in place.

There are several efforts within the industry at this writing to continue to improve the performance of the industry. The IPC trade association is working to develop environmental cost-account benchmarking to drive improvements through cost recognition of pollution. They are also working with the EPA to model an effort like the "Goals 2000" program used by the metal finishers. This program asks companies to agree to specific waste, energy, and pollution reductions by a specific date and is strictly voluntary. There continues to be pressure from the customers of PWB manufacturers to improve environmental management and to be aware of certain metrics associated with the products sold to OEMs. While the metrics used are not yet in place, there continues to be interest in establishing a short list of metrics related to the life cycle of components of products, for example, how much of the materials can be recycled, reused, refurbished, etc.

Many of these metrics will drive sustainability. It is recognized that we may eventually have to build our products using an ever-decreasing supply of virgin raw materials. It is, therefore, very important to start now to build an infrastructure that will use the minimum of resources in the first place, use materials and processes that can be recycled, and capture as much of the materials back into the supply chain with the least use of new resources.

8.8 Summary

While pollution prevention has, and continues to be, an important element in environmentally conscious manufacturing in the printed wiring board industry, there will be an ever-increasing effect from newer technology developments. These developments will include not only substitute technologies for older processes and materials, but also new ones that are not-in-kind. For example, direct plating may replace electroless plating, but may eventually be replaced by a completely different kind of metallization such as polymer thick film. Much of the decision process will be driven by individual trade-offs and timing requirements of each manufacturer.

There must also be an increased awareness of the life cycle of the products from the electronics supply chain. This applies not only to the PWB manufacturers, but also the OEMs. While OEMs are often aware of the life cycle of the finished product, they are not as aware of the life cycle of the individual components nor of the effect that their design decisions may impose on component manufacturers. Thus, the life-cycle awareness has to be a whole industry or complete supply-chain effort, not just the OEMs alone. Significant improvement is also needed in the environmentally conscious manufacturing "tools" available at all levels of the supply chain.

Finally, it is evident that pollution prevention efforts, acceptance of newer technologies, and recognition of "real" environmental costs have significantly strengthened the PWB industry. As third-world companies who have not had to deal with these problems now have to deal with them, this strength will become an advantage. By continuing to work with the entire supply chain, partnering with the regulatory community, academic institutions, and consortia, the printed wiring board industry can continue this need to become both technologically and environmentally strong.

References

1. Printed Wiring Board Industry and Use Cluster Profile," Design for the Environment Printed Wiring Board Project, EPA 744-R-95-005, U.S. Environmental Protection Agency, September 1995.
2. Ibid., p. 1-1.
3. "Printed Wiring Board Industry Pollution Prevention and Control: Analysis of Survey Results," Design for the Environment Printed Wiring Board Project, EPA 744-R-95-006, U.S. Environmental Protection Agency, September 1995.
4. "Profile of the Electronics and Computer Industry," EPA Office of Compliance Sector Notebook Project, EPA/310-R-95-002, U.S. Environmental Protection Agency, 1995, p. 91.
5. "Printed Circuit Board Manufacturing," Compliance Assistance Program, California Environmental Protection Agency (Sacramento), Air Resources Board, Compliance Division, July 1998, pp. 600–612, 614.
6. See, for example, "Total Cost Assessment: Accelerating Industrial Pollution Prevention Through Innovative Project Financial Analysis," EPA -741-R-92-002, Environmental

Protection Agency, May 1992, and David W. Vogel, "Using True Environmental Cost Information to Integrate Environmental Strategies into Core Business Procedures," presented at AESF Week 1998, January 26–28, 1998, Buena Vista, Florida.

7. A life-cycle inventory is actually the first step in producing a life-cycle assessment. This requires that the product be followed from its inception to its eventual disposal or recycling. The mass balance of what goes into the product as well as the energy and other resources is determined as well as the waste and product streams resulting. This is the inventory. The effect or impact of each of these inputs and outputs on the environment is the actual assessment.

8. Microelectronic Computer Corporation of Austin, Texas, has initiated a program to develop environmental metrics. It was unfunded at this writing. The World Business Council on Sustainable Development has examined a number of efforts to determine EcoEfficiency and Supply Chain Management.

9. A. D. Veroutis and J. A. Fava, "Framework for the Development of Metrics for Design for the Environment Assessment of Products," *Proceedings of the 1996 IEEE International Symposium on Electronics and the Environment (ISEE-1996),* 1996, pp. 13–18. F. Herman, L. Scheidt, and H. Stadlbauer, "Inclusion of Environmental Aspects and Parameters in Product Specification & Product Assessment: A Report on ECMA TC 38," *Proceedings of the 1996 IEEE International Symposium on Electronics and the Environment (ISEE-1996),* 1996, pp. 305–306.

10. This process has been done for not only the printed wiring board industry but also for the printing and dry cleaning industries. For an overview of the implementation of this process, see the EPA document: "Design for the Environment—Building Partnerships for Environmental Improvement," EPA-600-K-93-002, Environmental Protection Agency, September 1995.

11. For the making holes conductive (direct metallization process), see EPA-744-R-97-001, Environmental Protection Agency, February 1997, "Implementing Cleaner Technologies in the Printed Wiring Board Industry: Making Holes Conductive."

12. See, for example, "Printed Wiring Board Pollution Prevention and Control: Analysis of Survey Results," EPA 744-R-95-006, September 1995, "Printed Wiring Board Pollution Prevention and Control: Analysis of Updated Survey Results," EPA 744-R-98-003 August 1998, and G. C. Cushnie, Jr., *Pollution Prevention and Control Technology for Plating Operations,* National Center for Manufacturing Sciences, Ann Arbor, Michigan, 1994.

13. "Environmental Management System—Demonstration Project," Final Report, NSF International, Ann Arbor, Michigan, December 1996 (*www.nsf.org*).

14. For example, "Building an Environmental Management System: H-R Industries' Experience," EPA 744-F 97-010, Design for the Environment, Printed Wiring Board Case Study 8 and Case Study 9, available at *http://www.epa.gov/dfe* or from the Pollution Prevention Clearinghouse, U.S. EPA, 401 M Street, SW (7409), Washington, DC 20460.

15. Cascio, Joseph, ed., *The ISO 14000 Handbook,* CEEM Information Services with ASQC Quality Press, Reston, VA, 1996. "Environmental Management Systems: An Implementation Guide for Small and Medium-Sized Organizations," NSF International, November 1996. Tom Tibor and Ira Feldman, *ISO 14000: A Guide to the New Environmental Management Standards,* Irwin Professional Publishing, New York, 1996.

16. See, for example, J. Andersen and H. Choong, "The Development of an Industry Standard Supply-Base Environmental Practices Questionnaire," Conference Record of the 1997 IEEE International Symposium on Electronics and the Environment (ISEE), San Francisco, California, 1997, pp. 276–281. E. Craig, "Integrating Environmental Considerations into Supplier Management Processes at Sun Microsystems, Inc.," Conference Record of the 1997 IEEE International Symposium on Electronics and the Environment (ISEE), San Francisco, California, 1997, pp. 282–289.

17. "Product-related Environmental Attributes—Technical Report TR-70," ECMA Standardizing Information and Communication Systems—CDRom Version 4.0, February 1998, 114 Rue du Rhone, CH1204 Geneva, Switzerland or download at *www.ecma.ch.*

18. J. Blumberg, A. Korsvold, and G. Blum, "Environmental Performance and Shareholder Value," Publication of the World Business Council on Sustainable Development, available from E&Y Direct, PO Box 934, Bournemouth, Dorset BH8 8YY, UK or download in PDF format from *www.wbcsd.ch.*

19. "A Bridge to a Sustainable Future—National Environmental Technology Strategy," National Scientific and Technology Council (Office of Science and Technology Policy), 1600 Pennsylvania Ave., Washington, D.C. available at *www.nttc.edu/env/envstrat.html.*

20. Available via download from the Institute of Electrical and Electronics Engineers, Inc., EHS Technical Activities Board website at *www.ieee.org/tab/ehswp.html* or from IEEE at 445 Hoes Lane, Piscataway, NJ 08855-1331.

21. "The Environmental Protection System in Transition—Toward a More Desirable Future," The Center for Strategic and International Studies, 1997. (Email *books@csis.org* or download from */www.csis.org/.)*

22. Plenary Address by Ray Anderson of Interface, Inc. at the 1999 International Symposium on Electronics and the Environment.

23. "The National Technology Roadmap for Electronic Interconnection," Institute for Interconnecting and Packaging of Electronic Circuits, Northbrook, Illinois, 1997; NEMI Roadmap, SemaTech Roadmap. Electronic Industry Roadmap, MCC, Austin, Texas.

24. IBM and Motorola press releases.

25. "1996 NEMI Roadmap," National Electronics Manufacturing Initiatives, Inc., Herndon, Virginia.

26. R. Allred, "Tertiary Recycling Process for Electronic Materials," ARPA Environmentally Conscious Electronics Manufacturing Workshop, Raleigh, North Carolina, September 19–21, 1995.

27. "Recycling: A *PET* Project," *DuPont Magazine Online,* November 23, 1997; see also *www.DuPont.com/polyester/products/petrtec/index.html.*

28. For example, Hewlett Packard already has a "take-back" program to recycle the plastics in its printer toner cartridges: J. Heusinkveld, "Product Stewardship: Providing Customer Focused Solutions for Environmental Responsibility," presentation at the 1998 International Symposium on Electronics and the Environment, Chicago, Illinois, May 4–6, 1998.

29. See, for example, Chaps. 3 and 6 in J. B. Kushner and A. S. Kushner, *Water and Waste Control for the Plating Shop,* 2d ed., Gardner Publications, Inc., Cincinnati, OH, 1981.

30. See, for example, Jeff Erb, "Water Reuse Project," *IPC Expo 1998 Proceedings,* p. S07-2-3.

31. See, for example, "Pollution Prevention Work Practices," EPA document EPA 774-F-95-004, Environmental Protection Agency, July 1995 (Design for the Environment—Printed Wiring Board Case Study 1).

32. "A Continuous-Flow System for Reusing Microetchant," EPA document EPA 774-F-96-024, Environmental Protection Agency, December 1996 (Design for the Environment—Printed Wiring Board Case Study 5).

33. J. M. Hosea, "Water Reuse for Printed Circuit Boards—When Does It Make Sense?," *Proceedings of the Technical Conference—IPC EXPO '98, April 1998,* pp. S07-1-1-4.

34. George Cushnie, *Pollution Prevention and Control Technologies for Plating Operations,* National Center for Manufacturing Sciences, Ann Arbor, Michigan, 1994.

35. "Printed Wiring Board Pollution Prevention and Control: Analysis of Survey Results," EPA 744-R-95-006, Environmental Protection Agency, September 1995.

35a. "Printed Wiring Board Pollution Prevention and Control: Industry Survey and Literature Review," EPA 744R-98-003, August 1998.

36. "On-Site Etchant Regeneration," EPA 744-F-95-005, (Printed Wiring Board Case Study 2), Environmental Protection Agency, July 1995.

37. "Opportunities for Acid Recovery and Management," EPA 744-F-95-009, Environmental Protection Agency. (Printed Wiring Board Case Study 3), September 1996.

38. "Waste Minimization Through Solution Recycling," *Circuitree,* December 1992, p. 10.

39. George D. Gregoire, "Innovative Uses For IMPRINTED, U-Shaped PWB Traces & Microvias," *IPC Expo 1998 Proceedings,* April 26–28, 1998, pp. S01-1-5.

40. J. E. Gervay, "Current Status and Future Prospects for Dry Film Photoresists in Additive Processes," *Proceedings of the IPC Fall Meeting,* Denver, Colorado, Paper IPC TP 746, September 1983, pp. 1–7.

41. D. Vaughan, C. Weathers, and J. Lott, "Advances in Environmentally Conscious Manufacturing Through the Use of Permanent Photoimageable Materials in PWB Fabrication," IPC Printed Circuits Expo 1996, San Jose, California, March 1996, pp. S10-3.

42. W. Goldie, *Metallic Coating of Plastics,* vol. 1, Electrochemical Publications Ltd, Middlesex, England, 1968, p. 3.

43. "Printed Wiring Board Cleaner Technologies Substitutes Assessments—Making Holes Conductive," EPA-744-R-97-002 a & b, Environmental Protection Agency, 1997.

44. "Implementing Cleaner Technologies in the Printed Wiring Board Industry: Making Holes Conductive," Design for the Environment Printed Wiring Board Project, EPA-744-R-97-001, 1997. See also "Direct Metallization Report," the City of San Diego, San Diego, California.

45. See *www.epa.gov / opptintr / dfe'pub / surface.htm.* The Swedish Environmental Consortium IVF is also working on a similar project, including a life-cycle assessment of surface protections materials.

46. B. Trumble, "Printed Circuit Assembly with No Lead Solder Assembly Process," Conference Record of the 1997 IEEE International Symposium on Electronics and the Environment (ISEE), San Francisco, California, 1997, pp. 25–27.

47. A. W. Gibson et al., "Environmental Concerns and Materials Issues in Manufactured Solder Joints," Conference Record of the 1997 IEEE International Symposium on Electronics and the Environment (ISEE), San Francisco, California, 1997, pp. 246–251 (and references therein).

48. "Dichromic Exposure to Produce Photovias & Circuit Channels for Multilayer Circuits" (1986), U.S. Patent 4,463,111.

49. Dr. John J. Felten and Dr. Yueh-Ling Lee (private communication, 1999).

50. B. Mahler, "Planar Resistor Technology in High Speed Computer and Telecom Applications," IPC Printed Circuits Expo 1998, April 1998, p. S05-6.

51. "Annular Circuit Components Coupled with Printed Circuit Board Through-Hole," U.S. Patents 5,708,569 (1998), and 5,603,847 (1997).

52. B. L. Booth et al., "Polyguide Polymer Technology for Optical Interconnect Circuits and Components," SPIE Photonics West, San Jose, California, February 1997.

53. K. H. Hahn et al., "Gigabyte/s Data Communication with the POLO Parallel Optical Link," *1996 Proceedings of 46th Electronic Components & Technology Conference,* May 1996, pp. 301–307.

54. K. S. Giboney, "Parallel-Optical Interconnect Development at HP Laboratories," SPIE Optoelectronic Interconnects and Packaging IV, paper 3005-26, February 13, 1997.

55. "Solder—Getting Laser Bumps," *Electronic Engineering Times,* November 18, 1996.

56. M. Ehlin, "High Speed Photoresists for Laser Direct Imaging—Meeting the Needs of the New Century?," *Proceedings of the Technical Conference—IPC EXPO '98,* April 1998, p. S12-2.

57. See, for example, G. D. Gregoire, "Innovative Uses for Imprinted, U-Shaped PWB Traces & Microvias," *IPC Expo 98 Proceedings,* March 1998, pp. S01-1-3, and D. Klapprott and C. Cedarleaf, "Imprint Patterning—Emerging Materials and Process Technology," *Printed Circuit Fabrication,* vol. 21, no. 6, June 1998, pp. 46–50.

58. T. Aoyama et al., "Sand Blasting—A Novel Approach to Microvia Processing," *Printed Circuit Fabrication,* vol. 21, no. 1, January 1998, pp. 18–22.

59. R. DeJule, "All Polymer Integrated Circuits," *Semiconductor International,* April 1998, p. 42.

60. C. R. S. Needes et al., "Diffusion Patterning—Materials and Processing," *ISHM Proceedings,* 1993, pp. 463–468.

61. Electronic Interconnection Environmental & Safety Technology Roadmap," The Institute for Inteconnecting and Packaging Electronic Circuits, Northbrook, Ilinois, March 1994.

62. P. A. Sanborn, J. W. Lott, and C. F. Murphy, "Material-centric Process Flow Modeling of PWB Fabrication and Waste Disposal," *Proceedings of IPC Printed Circuits Expo,* San Jose, California, 1997, pp. S-10-4-4–S-10-4-12.

63. See, for example, the survey in C. Mizuki, P. A. Sanborn, and G. Pitts, "Design for Environment—A Survey of Current Practices and Tools," *Proceedings of 1996 IEEE International Symposium on Electronics and the Environment,* May 6–8. 1996, Dallas, Texas, pp. 1–6.

64. While actual toxicity data are not always readily available for all chemicals, there are a number of models that use structure activity relationships (SAR) to predict the toxicity behavior of these chemicals. Some models are better than others.

65. See "Priority Pollutants" in the document at *www.epa.gov/epaoswer/hazwaste/minimize/tool/drftpc/pdf.*

9

Reliability and Performance of Advanced PWB Assemblies

Thomas J. Stadterman

U.S. Army Material Systems Analysis Activity
Aberdeen Proving Ground, Maryland

Michael D. Osterman

University of Maryland
College Park, Maryland

9.1 Reliability and Performance of Advanced PWB Assemblies

This chapter discusses reliability issues related to advanced printed wiring board (PWB) assemblies, otherwise known as circuit card assemblies (CCAs). Up-front reliability assessments (that is, physics-of-failure analysis) of CCA designs can increase reliability, which can dramatically reduce the total cost of ownership and risks. With tighter design cycles, industry can no longer afford the traditional approach of testing-in reliability. To achieve simulation-based CCA qualification, software can be used to model and assess structures that occur in CCAs. In this chapter, the design process for CCAs is discussed in terms of physics-of-failure (PoF) analysis activity. This chapter outlines information resources necessary to support this analysis activity, including environmental, materials, manufacturing defects, and failure mechanisms.

Section 9.2 introduces the PoF design process and Sec. 9.3 outlines the basic theory of reliability, which can be used to probabilistically model failures. Section 9.4 describes potential environments for CCAs and the problems they could cause. Knowledge of the environmental conditions and usage of CCAs is essential in designing a reliable circuit card. Small variations in the environment can have great consequences in the life, or reliability, of the CCAs. Section 9.5 describes the stress analyses that can be performed on a CCA to determine the stresses at various failure sites. Stress analyses discussed in this chapter include thermal and vibration. Once the stresses at the failure sites are determined, failure models are used to estimate the time-to-failure for each failure mechanism and site. Section 9.6 describes common failure mechanisms associated with component interconnects, board metallization, and external connections. In addition to the failure mechanism description, common failure models are discussed. Section 9.7 discusses manufacturing defects, which can greatly affect the reliability of CCAs. Sections 9.8, 9.9, and 9.10 provide examples of reliability analyses of electronic CCAs. Finally, section 9.11 summarizes this chapter and presents the concluding remarks.

9.2 Physics-of-Failure Design Process

Physics of failure (PoF) is an approach used to develop reliable products which involve knowledge of root-cause failure processes to prevent product failures by incorporating robust design and manufacturing practices. The physics-of-failure approach incorporates reliability into the design process by establishing a scientific basis for evaluating new materials, structures, and electronics technologies. Information to plan tests and screens, and to determine electrical and thermomechanical stress margins are identified by this approach. Physics of failure encourages innovative, cost-effective design through the use of realistic reliability assessment. This physics-of-failure approach involves:

- Identifying potential failure mechanisms (for example, chemical, electrical, physical, mechanical, structural, or thermal processes leading to failure), failure sites (for example, component interconnects, board metallization, and external connections), and failure modes (for example, electrical shorts, opens, or deviations which result from the activation of failure mechanisms).

- Identifying the appropriate failure models for specific failure mechanisms and sites, including inputs associated with material characteristics, damage properties, relevant geometry at failure sites, manufacturing flaws and defects, and environmental and operating loads.

- Determining the variability for each design parameter when possible.

- Computing the effective reliability function.

- Accepting the design if the estimated time to failure meets or exceeds the requirement.

Figure 9.1 shows a flowchart of the PoF design process. The figure describes the stakeholders in the performance of the process, inputs of the process, types of analyses performed, and the outputs of the process.

9.2.1 Incorporating physics of failure into the electronics design process

When examining the electronic design process, it is clear that a central element is the ability to examine and reduce risks. With the advent of simulation software, electronic product development is moving from the design-build-test mentality to a spiral development model.[1] The spiral model consists of four key phases through which the development process cycles. These phases include

1. Development and verification phase
2. Plan phase

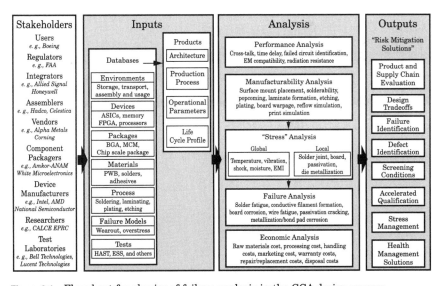

Figure 9.1   Flowchart for physics-of-failure analysis in the CCA design process.

3. Objectives, alternatives, and constraints phase

4. Risk evaluation phase

The spiral model consists of cycling through these phases, continuously improving and refining the product's concept and design.

Key components to the risk evaluation phase include technology risk, market risk, and supply-chain risk. The questions addressed by those risks are provided as follows:

- *Technology risk.* Can the product be built?

- *Market risk.* Will the demand for the product exceed the cost to develop it?

- *Supply-chain risk.* Are there qualified industries capable of supporting the manufacture and production of the product?

Cost is obviously the deciding factor in the risk-assessment process, but failure analysis capabilities are critical in evaluating supply-chain and technology risk.

The life expectancy of a circuit card assembly (CCA) is determined by examining known failure sites and evaluating known and available failure models for prominent failure mechanisms (the PoF approach). The failure models can be classified as either overstress (catastrophic failure) or wearout (damage accumulation failure). In general, simulation (thermal, vibration, and moisture) of the CCA storage, transport, and use conditions are needed to provide appropriate inputs for evaluating the failure models. In order to simulate the CCA, a computer representation of the CCA must be constructed.

To best impact the CCA's reliability, the PoF analysis should be performed in the early design (concept phase) and carried into the detailed design phase. Boeing Commercial Airplane Group has estimated that up to 40% reduction in cost and weight could be achieved if reliability analysis could be integrated into the design phase.[2] Benefits of a PoF analysis include the following:

- Identification of design flaws
- Identification of weak or problem parts
- Identification of destruction limits
- Identification of wearout limits
- Estimation of failure-free operating periods

PoF analysis is an important element in the development of CCAs. A flowchart of how reliability assessment fits into the electronic design process is presented in Fig. 9.2. This figure shows the PoF analysis being performed after the circuit schematic and partitioning are completed, but before the routing is started.

9.2.2 Outputs of a physics-of-failure analysis

The first and primary output of the PoF analysis is the determination of failure locations, types, and severity. Failures in CCAs have been documented and investigated by industry. A scientific understanding of how materials react to loads allows for the development of algorithms and analytic models that can be used to assess failure. In many cases, analytic models exist for specific geometry, material, and load combinations, which frequently occur in CCAs. Defect identification is another potential output of a PoF analysis. In some cases, failures occur due to defects that are introduced in the manufacturing process. A PoF analysis can be used to determine the type, location, and severity of defects based on the CCA's architecture and the manufacturing processes.

PoF is also critical in performing accelerated life tests, which are tests conducted at stress levels higher than expected operating stress levels to reduce the length of a test. For accelerated life testing to be

Reliability Assessment Tools

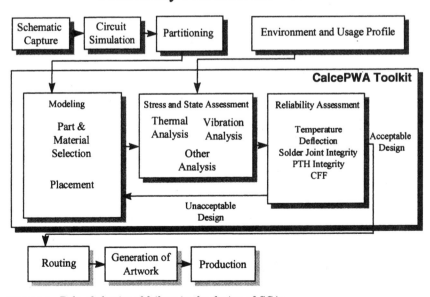

Figure 9.2 Role of physics of failure in the design of CCAs.

effective, a PoF analysis is needed to determine how the CCA will react under these high-stress conditions and to map the outcome at accelerated conditions to actual use conditions. This analysis technique can be used to help determine the test conditions and how to fixture the CCA to excite only relevant failures. In addition, PoF can be used to identify failures that occur only as a result of the accelerated test conditions, not actual-use conditions. Accelerated-life testing can help decrease development costs by reducing the time to conduct qualification tests.

Along with product qualification, PoF is critical to CCA screening. Screens are used to remove CCAs that fail earlier than expected due to manufacturing defects. Screening involves examining the CCAs during or after manufacture to find the types, locations, and severity of defects or flaws introduced from the manufacturing process. Screens can be carried out by CCA inspection and/or stress-based screening methods. To apply a stress screen, it is necessary to know what failures result from defects in the CCA and how the application of external loads can precipitate failure of flawed CCAs. PoF is essential for screening by providing time-to-failure estimates based on material, geometry, and load conditions. A manufacturing flaw may be represented by an inappropriate geometry and/or a changed material property. A PoF analysis is necessary to determine if a stress-based screen is viable.

Information about the type, location, and severity of failures and defects facilitates the product and supply chain evaluation. For example, the use of alternative parts from different suppliers may be evaluated to determine whether the change allows the CCA to meet its requirement. The use of surface-mounted as opposed to insertion-mounted packages is an example of such a process.

PoF analysis also provides a basis for performing environmental stress management. Stress management solutions can be evaluated in terms of their effectiveness in reducing harmful loads and increasing useful life of the CCA. For example, the response of a CCA to various levels of vibration could be used to determine the amount of damping required to adequately protect the CCA.

Finally, PoF analysis allows for the development of guidelines for CCA monitoring and devices to provide early warnings for impending equipment failure. By understanding the type, location, and severity of failures in the electronic product, stresses may be monitored and the remaining life estimated in real time. In an alternative approach, devices such as fuse may be developed based on expected failures. Under field-loading conditions, these devices would be calibrated to fail before the expected CCA failure, thus becoming an integral part of an early warning system. Since the material, architecture, and use conditions of CCAs are generally unique, the design of the fuse or fuses requires a physics-of-failure assessment of the CCA.

9.3 Reliability Theory

Basic reliability theory can aid in the PoF analysis by providing a method to probabilistically model failure mechanisms. Most engineers have a general understanding of the concept of reliability. Products that always work are reliable, while products that fail unexpectedly are unreliable. With any product there is an actual reliability and a perceived reliability. From a business standpoint, the perceived reliability can be as critical, if not more critical, than the actual reliability. The weight given to reliability assessment activities should be based on a risk analysis. For example, if failure results in potential loss of life, reliability activities may dominate the design process. However, if the failure results in minor inconvenience, a low level of reliability activities may be justified. In this section, the concept of reliability is introduced from an engineering standpoint and examples of its use are provided.

9.3.1 Reliability mathematics

By definition, reliability $R(t)$ is the probability that a system, component, or device will operate without failure, under stated conditions, for a specified period of time. Since failure and survival are mutually exclusive events, they can be expressed by

$$R(t) + Q(t) = 1 \qquad (9.1)$$

where $R(t)$ is the reliability at time t and $Q(t)$ is the probability of failure at time t. By considering a product sample size, N_o, and the number of failures at a given time, $N_f(t)$, the number of surviving items at time t can be expressed as

$$N_s(t) = N_o - N_f(t) \qquad (9.2)$$

From this information, the reliability of the product can be expressed as

$$R(t) = \frac{N_s(t)}{N_o} = \frac{N_o - N_f(t)}{N_o} \qquad (9.3)$$

The plot of the ratio of number of failures, $N_f(t)$, to the total number of products as a function of operating life, is called the *probability density function* (PDF) for failure. From Eqs. (9.1) and (9.3), the PDF for failure is defined as

$$f(t) = \frac{1}{N_o} \frac{d\,[N_f(t)]}{dt} = \frac{d\,[Q(t)]}{dt} = \frac{d\,[1-R(t)]}{dt]} = -\frac{d\,[R(t)]}{dt} \tag{9.4}$$

By integrating this equation from $\tau = 0$ to $\tau = t$ and recognizing that $N_f\,(t<0) = 0$, it can be written

$$\frac{N_f(t)}{N_o} = \int_0^t (\tau)\,dt \tag{9.5}$$

By assuming that the product can fail at some future time, it can be written

$$\int_0^\infty (\tau)\,d\tau = 1 \tag{9.6}$$

From this information, it can be seen that the probability of failure becomes a cumulative distribution function (CDF), expressed as

$$Q(t) = \int_0^t f(\tau)\,d\tau \tag{9.7}$$

and the reliability is expressed as

$$R(t) = \int_0^\infty f(\tau)\,d\tau \tag{9.8}$$

Another useful value is the instantaneous failure rate often referred to as the *hazard rate*. The hazard rate can be thought of as the probability of a failure in the next instant of time, given the item is currently operating. Hazard rate can be mathematically defined as

$$h(t) = \frac{d\,[N_f(t)]}{dt} \frac{1}{N_s(t)} \tag{9.9}$$

or it may be expressed as

$$h(t) = f(t)\,/R(t) \tag{9.10}$$

Since $f(t)$ can be defined in terms of reliability, hazard rate can be expressed by

$$h(t) = \frac{f(t)}{R(t)} = -\frac{dR(t)}{dt} \frac{1}{R(t)} \tag{9.11}$$

From the preceding definition, it can be seen that the reliability can be expressed in terms of the hazard rate. By noting that $R(0) = 1$ and manipulating this equation, reliability can be expressed by

$$R(t) = \exp\left[-\int_0^t h(\tau)\,d\tau \right] \tag{9.12}$$

9.3.2 Hazard rate curve

When populations of products are examined, failures can be broken down into early life, useful life, and late stages of life. Failures in the early life stage are generally related to defects caused by the manufacturing process. The number of failures related to manufacturing problems generally decreases as the defective parts fail, leaving a subpopulation of reduced-defect products. During the early stage, the hazard rate tends to decrease with age. During the useful life, failures may occur due to random overstress conditions, accidents, and mishandling. In a properly designed product, the hazard rate over the useful life is assumed to be very low and constant because of the random occurrence of these overstress events. An increasing hazard rate during the useful life period indicates premature wearout of the product caused by design weaknesses, sustained high-stress levels, or latent defects that escaped the early life stage. As the product approaches the late life stage, it degrades due to repetitive or sustained stress conditions that cause damage accumulation in the product. The hazard rate during the late life stage increases dramatically as more and more products fail due to wearout. When plotting the hazard rate over time as depicted in Fig. 9.3, these stages form the "bathtub" curve if constant stress dominates the useful life region (low level stress curve) or the "roller coaster" curve if wearout starts to occur in the useful life region (high level stress curve).

9.3.3 Common failure distributions in reliability

The time to failure due to a failure mechanism at a particular site can be modeled by using common distributions, including the exponential, Weibull, and lognormal. If time to failure for a particular failure mech-

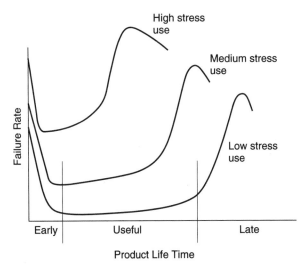

Figure 9.3 Product hazard rate versus time for different stress levels.

anism is known to follow a distribution, testing can be used to estimate the parameters of the distribution. In addition to modeling time to failure, statistical distributions can be used to probabilistically model the failure mechanism. Instead of deterministically solving a failure mechanism model for one time-to-failure value, probabilistic modeling provides a time-to-failure distribution. This distribution can be determined computationally by representing input parameters with distribution functions and performing a Monte Carlo simulation. In other cases, a statistical distribution, based on test data for a particular failure mechanism and site, may simply be associated with a failure model. Common failure distributions are presented in the following sections.

9.3.3.1 Weibull distribution. One of the most useful distributions is called the Weibull function, developed in the early 1950s by Waloddi Weibull.[3] A Weibull distribution can be used in a wide range of applications, because the distribution can take the form of many shapes. By changing the shape parameter, the Weibull distribution can be used to model the decreasing hazard rate in early life, constant hazard rate, and increasing hazard rate. Weibull functions can be expressed by two or three parameters. The two-parameter Weibull function used to define the PDF is expressed as

$$f(t) = \frac{\beta}{\eta} \left(\frac{t}{\eta} \right)^{(\beta-1)} e^{-(t/\eta)^{\beta}}$$

(9.13)

where β and η are nonzero positive model constants. When used as the PDF, β is referred to as the shape parameter, while η is referred to as the characteristic life or scale parameter. From Eq. (9.7) and using Eq. (9.13), it can be written

$$Q(t) = 1 - e^{-(t/\eta)^{\beta}}$$

(9.14)

When $t = \eta$, it is observed that 63.2% of a population have failed.

The importance of the Weibull function can be seen when the characteristic life is fixed and the shape parameter β is varied. From the plot in Fig. 9.4, it can be seen that the failure rate decreases for $\beta < 1$, is constant for $\beta = 1$, and increases for $\beta > 1$. As the shape parameter becomes increasingly large, failures occur closer to the characteristic life and the failure rate increases dramatically as time approaches the characteristic life. If the failure PDF is expressed by the two-parameter Weibull function, the hazard rate is defined as

$$h(t) = \frac{\beta}{\eta} \left(\frac{t}{\eta} \right)^{(1-\beta)}$$

(9.15)

Hazard Rate

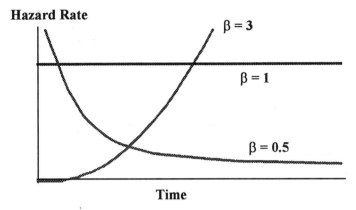

Figure 9.4   Two-parameter Weibull hazard rate curves.

and the reliability is defined as

$$R(t) = e^{-(t/\eta)^\beta} \tag{9.16}$$

In addition, the mean time to failure (MTTF) is defined as

$$\text{MTTF} = \int_0^\infty e^{-(t/\eta)^\beta} = \eta\Gamma\left(\frac{1}{\beta} + 1\right) \tag{9.17}$$

where $\Gamma(x)$ is the gamma function. If $\beta = 1$, the Weibull distribution function coincides with the exponential distribution.

The two-parameter Weibull distribution, as discussed here, can be extended to a three-parameter distribution with the third parameter called a *guarantee or threshold parameter*.[4] This distribution can be used to model a failure mechanism at a particular site that has a failure-free period. The probability density function is given as

$$f(t) = \frac{\beta}{\eta}\left(\frac{t-u}{\eta}\right)^{(\beta-1)} e^{-(t-u/\eta)^\beta} \tag{9.18}$$

where the parameter u is the failure-free period and t must be greater than u. The probability of failure before u is zero. The three-parameter Weibull hazard rate is expressed as

$$h(t) = \frac{\beta}{\eta}\left(\frac{t-u}{\eta}\right)^{(1-\beta)} \tag{9.19}$$

The reliability can be expressed as

$$R(t) = e^{-(t-u)/\eta)^\beta} \tag{9.20}$$

and the mean time to failure, MTTF, is defined as

$$\text{MTTF} = u + \eta\Gamma\left(\frac{1}{\beta} + 1\right) \tag{9.21}$$

where $\Gamma(x)$ is the gamma function. Figure 9.5 shows an increasing hazard rate for a three-parameter Weibull distribution.

9.3.3.2 Exponential distribution. The exponential distribution, with its constant hazard rate, is commonly used to model random overstress failures. One interesting phenomenon of the exponential distribution is that the failure process represented has no memory.[5] In other words, the probability of a failure between time t and time $t + Dt$ is independent of time t. An item modeled by an exponential distribution that survived until time t has no memory of its past. The exponential function used to define the PDF is expressed as

$$f(t) = \lambda e^{-\lambda t} \tag{9.22}$$

where λ represents the rate at which failure will occur and λ must be greater than zero. If the failure PDF can be expressed by the exponential function, the hazard rate is constant at the value λ, expressed as

$$h(t) = \lambda \tag{9.23}$$

and the reliability is expressed as

$$R(t) = e^{-\lambda t} \tag{9.24}$$

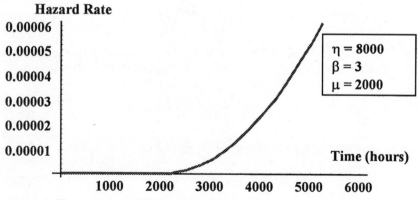

Figure 9.5 Three-parameter Weibull hazard rate curve.

The mean time to failure, MTTF, is defined as

$$\text{MTTF} = \frac{1}{\lambda} \qquad (9.25)$$

The exponential distribution can also be extended to a two-parameter distribution with the second parameter again called a guarantee or threshold parameter.[4] This distribution can be used to model a failure mechanism with a failure-free period, followed by a constant failure rate. The two-parameter exponential can also be used to model the tail of an increasing hazard rate in which little information is known, or can be estimated, beyond the failure-free period and the very beginning of the increasing hazard rate. In this case, the PDF can be expressed by

$$f(t) = \lambda e^{-\lambda(t-u)} \qquad (9.26)$$

where the parameter u is the failure-free period and t must be greater than u. The probability of failure before u is zero. The hazard rate for the two-parameter exponential is

$$h(t) = \lambda \qquad \text{at } t > u \qquad (9.27)$$

and the reliability can be expressed as

$$R(t) = e^{-\lambda(t-u)} \qquad (9.28)$$

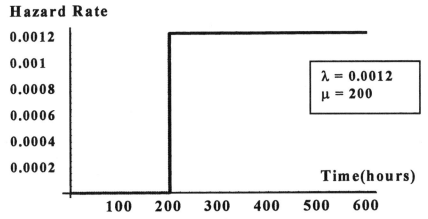

Figure 9.6   Two-parameter exponential hazard rate curve.

The mean time to failure, MTTF, is defined as

$$\text{MTTF} = u + \frac{1}{\lambda} \tag{9.29}$$

Figure 9.6 shows the hazard rate of a two-parameter exponential distribution.

9.3.3.3 Lognormal distribution.

Another important distribution for modeling time to failure is the lognormal distribution. A time to failure is lognormally distributed if its logarithm is normally distributed. Lognormal is useful for modeling failure processes that are the result of many small multiplicative errors such as time to failure due to fatigue cracks.[4] The distribution is also useful for modeling time to failures that are the product of many independent random variables.[5] The hazard rate increases over time, then decreases. The lognormal is useful for modeling data where less than 50% of the items have failed. The PDF is expressed as

$$f(t) = \frac{1}{\sigma t \sqrt{2\pi}} \; e \left[-\frac{1}{2\sigma^2} (\ln t - \mu)^2 \right] \tag{9.30}$$

where σ and μ are model constants. The mean time to failure, MTTF, is defined as

$$\text{MTTF} = e \left(\mu + \frac{1}{2}\sigma \right) \tag{9.31}$$

Figure 9.7 shows a hazard rate for a lognormal distribution.

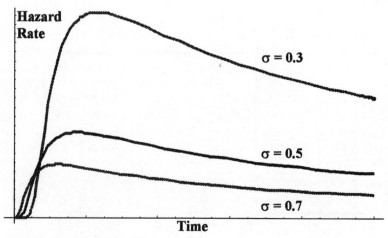

Figure 9.7 Lognormal hazard rate curves.

9.3.4 Sample use of time-to-failure distributions

Suppose that CCAs are failing due to one type of failure mechanism and the data in Table 9.1 is available. At what time will the reliability of the product drop below 90%? The first step in the process is to determine the distribution. If the failure mechanism's time to failure is known to follow a particular distribution, that distribution can be used. If the distribution is not known, a "goodness of fit" analysis can be used to check how well a distribution fits the data.[6] Assuming that this particular failure mechanism can be described by a two-parameter Weibull distribution, the first step is to determine the Weibull parameters, β and η.

By observing that the probability-of-failure equation defined in Eq. (9.14) for a two-parameter Weibull distribution can be linearized, the following equation can be written

$$\ln \{ \ln [1 - Q(t)]^{-1} \} = \beta [\ln (t) - \ln (\eta)] \qquad (9.32)$$

This equation can be solved using special Weibull graph paper or by applying the method of least squares. In either case, you should arrive at $\beta = 2.5$ and $\eta = 3000$. It should be noted that since β is greater than unity, a wearout failure is responsible for the failure. Using these Weibull parameters, the calculated reliability at 1200 days is $R(t = 1200 \text{ days}) = 0.9$.

Failure distributions can also be used to analyze test data. Suppose a test is conducted for 2000 h with 10 parts, and the failure information is recorded in Table 9.2. At this point, failure times exist, but the probability of failure is unknown. One method to estimate the probability of failure is median rank. The median rank method is determined by solving the binomial distribution equal to 0.5 and solving for the unreliability. The reader should consult a reliability text for more information on median rank and maximum likelihood functions.[4,5,6] Maximum likelihood can provide better estimates of the Weibull parameters, but is more complicated than median rank. Table 9.3 shows the results of using the median rank method to determine the probability of failure

TABLE 9.1   Examples of Circuit Card Assembly Failures During Operation

Percent failure	Time (days)
0.0621	1000
0.162	1500
0.289	2000
0.479	2500

TABLE 9.2 Examples of Circuit Card Assembly Failures during Test

Item	Failure time (h)
1	1231
2	1500
3	1670
4	1850
5	1964
6–10	Did not fail during the test

TABLE 9.3 Calculated Median Ranks

Device number	Time to failure (h)	Median rank
1	1231	0.0670
2	1500	0.1632
3	1670	0.2594
4	1850	0.3557
5	1964	0.4519
6		
7		
8		
9		
10		

for each of the failure times. Using a least-squares fit algorithm, we can estimate the following Weibull parameters:

β: 4.584

η: 2193.53

γ: 0.000000

The correlation coefficient (that is, the goodness of fit) equals 0.999191, which indicates that the Weibull distribution is acceptable for modeling the times to failure.

From the discussions in this section, the application of reliability theory and distributions have been presented. To realize the PoF approach, the failure models must be related to physical properties of the CCA and the environment. The environment of the CCA is discussed in the next section.

9.4 Effect of Environment on CCA Reliability

It has long been recognized that useful life of circuit card assemblies (CCAs) is governed by the environment and use conditions. Each year, hundreds of millions of dollars are spent designing and qualifying

CCAs for use in field environments. These activities have spawned a service industry devoted to testing electronic equipment and an equipment manufacturing industry for developing test apparatus. In order to assess the reliability of CCAs, engineers must consider the product and its entire product life cycle. The product life cycle includes manufacturing and assembly conditions, storage conditions, handling and transport conditions, and use or operational conditions. Key environments for electronic equipment can be found in Table 9.4. Environmental conditions that influence the reliability of CCAs include temperature, vibration, mechanical shock, moisture, ionic or chemical contamination, electrical bias, electromagnetic radiation, voltage spikes, and radiation. Table 9.5 lists some key environmental parameters for CCAs.

Uncontrolled storage conditions may expose CCAs to temperature extremes, moisture, and ionic contamination. Handling can present

TABLE 9.4   Key Environments for Circuit Card Assemblies

Office
Consumer
Telecommunications
Commercial avionics
Military avionics
Automotive under hood
Military ground and ship
Industrial motors
Space

TABLE 9.5   Key Environmental Parameters

Environment	Parameter
Temperature	Maximum
	Minimum
	Average
	Number of temperature cycles per year
Relative humidity	Maximum
	Minimum
	Average
	Number of humidity cycles per year
Vibration	Mode
	Wave form
	PSD curve
Shock	Acceleration PSD
	Pulse shape
	Maximum g force
	Time of pulse
Contamination	Ionic concentration
	Industrial gas concentration

high mechanical loads, as well as expose CCAs to electrical shock, moisture, and ionic contamination. Transport may expose the CCAs to severe vibration loads. In some instances, transport environments may be more severe than the use environment. The external environment affects use conditions as well as the internal environment produced by operation of the CCAs. Careful consideration is required to prevent damage incurred by these environments. In the remainder of this section, the most common environmental conditions are individually discussed.

9.4.1 Temperature

There is no question that temperature plays an important role in the reliability and performance of CCAs. Temperature influences reaction rates, changes material behavior, and may degrade operational characteristics of circuits. A review of the influence of temperature on microelectronic devices has been presented by Lall et al.[7] CCAs experience high temperatures during manufacturing and assembly. Temperature extremes are also common during transportation, uncontrolled storage, and use. General concerns at high temperatures include the following:

- Dimensional changes
- Increased chemical reactions
- Reduction of material strength
- Changes in electrical behavior of electronic components (resistors, capacitors, etc.)

Specific failure mechanisms that are accelerated by high temperature include the following:

- Delamination of fiber bundles in the printed wiring board
- Delamination of the board metallization
- Depolymerization of the epoxy
- Metal migration of the board metallization
- Corrosion of the solder joints, plated-through holes, vias, board metallization, and connectors

High temperature effects on electronic products have been extensively examined by McCluskey et al.[8]

In addition to high temperature, low temperature also affects the reliability of circuit boards. CCAs can be exposed to low temperature extremes during transportation, uncontrolled storage, and use. General concerns at low temperature include the following:

- Dimensional changes
- Changes in electrical behavior of electronic components
- Stiffening of vibration dampers
- Freezing of trapped moisture
- Embrittlement of materials

Specific failure mechanisms accelerated by low temperature include the following:

- Solder-joint fatigue from stiffening of vibration dampers
- Board fiber bundle fatigue from the stiffening of vibration dampers

Probably the most significant aspect of temperature is not high or low temperature extremes, but the change of temperature (that is, temperature cycling).[7] A CCA is composed of many different materials that have wide ranges of coefficients of thermal expansion (CTE). These wide ranges of CTE along with extreme changes in temperature result in significant thermomechanical strains.[9] Such changes in temperature may occur during manufacturing (for example, vapor depositions and soldering operations), external environmental changes, and normal power cycling. Specific failure mechanisms accelerated by temperature cycling include the following:

- Solder-joint fatigue
- Plated-through hole or via barrel fatigue
- Plated-through hole or via barrel buckling
- Delaminations between the layers of the composite
- Delaminations between the composite and the metallization

There are several approaches for dealing with the temperature in CCAs. A list of mitigation activities includes the following:

- Keep temperatures well below formation temperatures.
- Reduce the temperature cycle range.
- Avoid rapid changes in temperature.

The selection of the heat-transfer technology and equipment can be used to constrain CCA temperatures below the formation temperatures and to reduce the CCA temperature cycle range. Thermal analysis can then be performed to determine the effectiveness of this technology and equipment (thermal analysis will be discussed in Sec. 9.5.1). The goals of heat transfer include removing the heat dissipated by the components, constraining component temperatures below the rated operational values, and minimizing thermal gradients between components.[9] To control component temperatures, the heat must be transferred from the component to a heat sink. Typical heat transfer ranges from 0.2 to 20 W/cm². Other factors that influence the heat-transfer technology and equipment include cost, weight, location, accessibility, maintainability, fan noise-level limits, and power consumption of thermal control mechanisms. These factors should be considered when selecting the heat-transfer technology and equipment.

9.4.2 Vibration

A simple definition of *vibration* is an oscillating motion where a structure moves back and forth. If the motion is repetitive, it is called *periodic motion*; if the motion never appears to repeat itself, it is called *random.* As CCAs have become more ubiquitous, the effect of vibration has become an increasing concern. Electronics can now be found in vehicles, commercial aircraft, helicopters, missiles, and all types of hand-held devices (for example, phones, beepers, radios, etc.). With CCAs used in these applications, vibration-induced failures are becoming more likely. CCAs can experience vibrational stresses during manufacture, screen tests, transportation, and usage. Depending on the application of the CCA, one of these phases will produce the most severe vibrational stress.[9] During the use phase, where and how the CCA is mounted on a vehicle, aircraft, etc., will determine the level of vibrational stress.

As mentioned previously, vibration may be periodic or random. The most straightforward form of periodic vibration is the simple harmonic motion, which can be described by a sine wave plot of displacement versus time.[9] The equation for displacement is as follows:

$$x(t) = A \sin (\omega t) \tag{9.33}$$

where A is the amplitude of displacement and ω is the circular frequency in radians per second. The period, τ, which is the time required for the motion to repeat, is usually expressed in seconds and is defined as $\tau = 2\pi/\omega$. The frequency, f, is the reciprocal of the period and is expressed in cycles per second, or hertz, Hz, and is defined as $f = \omega/2\pi$.

Since random vibration is nonrepetitive, it is called a *nondeterministic phenomenon.*[9] Random vibration can have either narrow or broadband characteristics, and these refer to the range of frequency in which the disturbance is transmitted. In addition to never repeating itself, random vibration can theoretically have amplitude spikes that approach infinity. The intensity of the spikes cannot be determined or predicted. Because of this uncertainty in the intensity and frequency of the vibration, random vibration can be best modeled probabilistically.

To gain an understanding of the intensity and frequency of the vibration, the standard deviation and spectral density must be measured. The distribution of peaks in random vibration can be modeled with a normal or Gaussian distribution. The mean value of this distribution is as follows:

$$\bar{y} = \frac{1}{T} \int_0^T y(t)dt \qquad (9.34)$$

and the mean square value is as follows:

$$\overline{y^2} = \frac{1}{T} \int_0^T y^2(t)dt \qquad (9.35)$$

Both the mean and the mean square can be used to represent the average value of peaks in a random distribution. The variance of the distribution, which is a measure of the scatter from this average value, can be represented by the following equation:

$$\sigma_y^2 = \frac{1}{T} \int_0^T [y(t) - \bar{y}]^2 dt \qquad (9.36)$$

and when this equation is expanded and integrated, it produces the following:

$$\sigma_y^2 = \overline{y^2} - (\bar{y})^2 \qquad (9.37)$$

The variance can be calculated from the mean square and mean of the random vibration distribution. It is common for random vibration to have a mean value equal to zero, which implies that the variance is equal to the mean square value. For $\bar{y} = 0$

$$\sigma_y^2 = \overline{y^2} \qquad (9.38)$$

and the standard deviation from the mean of the distribution is

$$\sigma_y \sqrt{\overline{y^2}} = y_{\text{rms}} \qquad (9.39)$$

When $\bar{y} = 0$, the standard deviation is equal to the root mean square (rms) of the distribution. This rms value is sufficient to characterize the signal. The normal distribution is used to characterize the magnitude of the peaks in this random signal. Table 9.6 shows the probability of

TABLE 9.6   Percentage of Time a Vibration Magnitude Is within Various Ranges

Range of magnitude	Percentage of time
$\pm\sigma$	68.27
$\pm2\sigma$	95.54
$\pm3\sigma$	99.74
$\pm4\sigma$	99.96

peak random vibrations for various magnitudes (in multiples of rms values). For CCA vibration design, the 3σ values should be considered.[9] To obtain the 3σ values, multiply the rms acceleration, displacement, or stress values by three. A CCA designed for a rms random vibration level of 5g should be able to withstand vibration of 15g. Vibration levels in excess of 15g will only occur 0.3% of the time. Random vibration environments for CCAs are normally defined by power spectral density (PSD) curves, which represent vibrational input and response in terms of acceleration squared as a function of frequency. PSD is expressed as

$$P_0 = \lim_{\Delta f_0 \to 0} \frac{G_0^2}{\Delta f_0} \tag{9.39}$$

The units for PSD are g^2/Hz.

Because either harmonic or random vibration causes rapid reversals of stress in CCAs, this can lead to the following problems[10]:

- Loosening of fasteners
- Wire chafing
- Touching and shorting of electronic parts
- Optical misalignment
- Material cracking or rupture
- Electrical noise
- Intermittent disconnecting of separable contacts
- Fretting wear of separable contacts
- Interconnect fatigue failures
- Shorting or fracture of parts due to dislodged particles or parts

Many vibration problems can be minimized by avoiding coincident resonance frequencies which magnify acceleration forces rapidly. This can be accomplished by knowing and adjusting the natural frequencies of the electronic box and CCA. Determining natural frequencies can be accomplished by performing vibration analysis or testing. In addition to avoiding coincident resonance frequencies, the goal of vibration design should be to increase the natural frequencies of the CCAs. In general, higher natural frequencies of CCAs result in low displacements and a greater resistance to vibration loads.

Increasing the natural frequencies can be accomplished by stiffening the CCA, decreasing the weight of the CCA, or by changing the way the CCA is supported in the electronic box. The way a CCA is supported in an electronic box is very influential in the CCA's response to vibration. A CCA that is not firmly attached to a box allows easy connector engagement, but it is not desirable for vibration. Supports desirable for vibration firmly attach the CCA to the electronic box. These supports reduce deflections, translations, and edge rotations. Firm supports increase the CCA natural frequencies. In addition to CCA supports, rib stiffeners can be used to increase the natural frequencies. The ribs stiffen the CCA but do not add significant weight. Ribs have a relatively high modulus of elasticity because they are usually made of steel, copper, or brass. Ribs may be soldered, bolted, welded, or cemented to the PWB, and may be undercut to allow circuitry to pass beneath them. Ribs must be used properly for them to be effective. Effective rib placement carries the load directly to a support, while poor rib placement carries the load to a free end. Figure 9.8 shows effective and ineffective placement of ribs on a CCA. Vibration analysis to determine the natural frequencies will be discussed in Sec. 9.5.

9.4.3 Mechanical shock

Shock is a transient condition where the equilibrium of a system is disrupted by a sudden applied force or by sudden change in the direction or magnitude of velocity.[11] Shock can occur during assembly, handling, or transportation of the CCA. During usage, shock can occur if the CCA is mounted in a vehicle, aircraft, or hand-held device. Shock can also occur if two CCAs collide during vibration, explosions, or wind

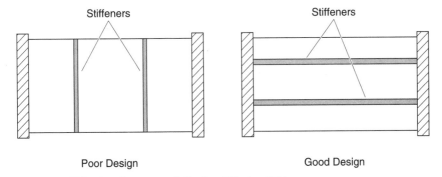

Poor Design   Good Design

Figure 9.8 Effective placement of ribs for stiffening CCAs.

gusts.[9] Shock loading produces significant forces, but over a short period of time. After shock loading, the motion of the CCA consists of a damped harmonic vibration. The intensity of the response of a CCA to a shock load is determined by how close the CCA's natural frequencies are to the applied shock load.

There are many different methods to represent shock motion and its effects.[11] One of the most popular is the pulse shock, which represents acceleration and displacements in the form of simple shapes such as half sine wave, square pulse, and triangular pulse. Although these shapes do not represent real-life shock, they can be used to model shock loads and assess the susceptibility of the CCA to shock load. Another common way to address shock is through the velocity shock. This approach is concerned with CCAs that experience a sudden change in velocity such as from an assembly falling. In this case, the assembly and CCA's velocity abruptly goes to zero.

When a CCA experiences a shock load, it initially deforms in the direction of the shock pulse. As the pulse diminishes, the CCA will vibrate at its natural frequencies with the first natural frequency being the most predominant. There are four basic types of shock failures including the following[11]:

1. Fractures or permanent deformations due to high stresses
2. Loosening of bolts and supports from the high acceleration levels
3. Impact between adjacent CCAs due to high displacement
4. Electrical malfunctions of the CCA components due to the deformation of component's internal structures (may only malfunction during the shock load)

Specific fracture failures of CCAs include cracked solder joints and fractured leads on large, heavy components. Fatigue is usually not a problem with shock loading, because several thousand or more stress cycles are most often required for fatigue to occur. Shock failures are usually less than a few hundred stress cycles, so stress concentration factors can be ignored.

The most common technique used to protect CCAs from shock loading is the use of shock isolators. Shock isolators reduce the magnitude of the shock transmitted, which protects the CCA from excessive displacements and failure. The shock isolators must be designed for anticipated shocks, and if the isolators are designed incorrectly, then shock amplification can occur. Another factor which influences shock isolator design is that the isolators must work for vibration, as well as shock. Unfortunately, isolators that are good for shock are generally not effective for vibration, and vice versa. Isolators must be designed for both shock and vibration environments. Another problem is that as the shock isolator acts as a vibration isolator, it produces heat. If this heat is not dissipated rapidly, it may destroy the isolator itself. Before shock isolators are designed, the natural frequencies of the CCA or the electronic assembly are required. The analyses to determine the natural frequencies are addressed in Sec. 9.5. The ratio of the first natural frequency of the CCA and the natural frequency of the shock pulse must be known to determine the potential shock amplification.[11]

In addition to shock isolators, the magnitudes of the shock transferred to the CCA can be reduced by separating the natural frequencies of the electronic assembly and the CCA. If these natural frequencies are separated, even if the CCA is less than the electronics assembly, the magnitude of the shock load will be reduced. When the CCA has 2 times the natural frequency of the assembly, or vice versa, then severe dynamic coupling effects are reduced. This is called the *octave rule*, because the natural frequencies are separated by one octave.[11]

9.4.4 Moisture

The presence of moisture on the surface of CCAs can produce loss mechanisms and electrical shorting. In addition, moisture accelerates metal corrosion. The absorption of moisture by dielectric materials can lead to changes in capacitance and lower insulation resistance. Further, moisture absorption can produce geometry changes which, in turn, can produce localized stresses. Plastic components, underfills, and board laminate materials are susceptible to moisture absorption. The combination of temperature and humidity influences the rate at which materials and components may absorb moisture, as well as the build up of surface moisture. Concerns with moisture include

- Material swelling
- Change in capacitance
- Promotion of biological growth
- Promotion of corrosion

Moisture is a concern in the manufacture and assembly of CCAs. The presence of moisture in plastic packages has been identified as a factor in popcorning.[12] Popcorning occurs during a reflow soldering process when moisture trapped in a package vaporizes, causing the package to rapidly expand. Generally, this results in a degradation of the package case due to fractures of the case and around the die and/or lead frame. Special handling and additional processes can be applied to reduce or eliminate the risk of popcorning.

Freezing conditions present another concern related to moisture. Moisture trapped in nonmetallic materials may lead to delamination, solder-joint fracture, and metallization fracture in CCAs during a freeze/thaw cycle.

Moisture promotes corrosion of metal surfaces leading to increased contact resistance in electrical contacts. In addition, moisture also promotes galvanic mechanisms at joined materials. Anodic corrosion sites in materials due to moisture and stress can produce stress cracks and further degrade the performance of electronics. The penetration of moisture into plastic packages can also lead to corrosion at exposed metal regions on the surfaces of the packaged chip. In particular, die bondpads are susceptible to failure due to corrosion.[13]

The risk posed by moisture may be mitigated through the application of a postcost (conformal coating). *Conformal coating* is a thin polymer layer that provides a moisture barrier for printed wiring assemblies. Considerations on conformal coating include cost, effectiveness, application process, rework process, and mechanical effectiveness. Several polymeric materials have been used as conformal coating. Paralyene is the most expensive material, but provides the best material properties.

9.4.5 Ionic contamination

Handling and the outdoor environment provide ready sources of ionic contamination in CCAs. Salt, which is present in coastal and marine environments, has a highly deteriorating effect on interfaces of dissimilar metal by promoting galvanic activity and producing corrosion products. Other sources of ionic contamination include flame-retardant materials, body oils from handling, and chemical residues left over from the manufacturing process. Figure 9.9 presents sources and types of contamination to which a CCA may be subjected.

Some of the effects of ionic contamination include generation of corrosion products, loss of material, pitting, and migration. Corrosion products on metal surfaces can increase contact resistance. Material losses as a result of galvanic activity can produce open circuits and mechanically weaken parts. Pitting can lead to loss of mechanic strength and may produce localized variations of contact resistance, while metal migration can produce short circuits. The PoF approach requires that environmental and use conditions be considered in the assessment of CCA integrity and reliability. Numerical simulations or testing is required to quantify the severity and magnitude of the environment on the CCA. Characterization of the stress resulting from the environment will be discussed in the next section.

9.5 CCA Stress Analysis

The underlying philosophy of physics of failure (PoF) is based on the fact that certain physical situations (environmental, conditions, oper-

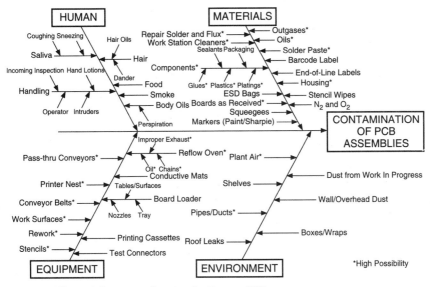

Figure 9.9 Potential sources of contamination on CCAs.

ating conditions, and product architecture) can result in CCA failure. The failure of the CCA can be allocated to an identifiable site and mechanism that is the basis of root-cause analysis. Stress analysis is necessary to determine the severity of a physical situation within a larger system. For the case of a CCA, stress analysis provides information about the individual parts that comprise the CCA.

Since the most predominant failures in CCAs involve a mechanical failure at the interface of joined materials (that is, solder joints, plated-though holes), stress evaluation of an electronic product includes temperature and vibration analysis. The corrosion and metal migration in electronic products is associated with moisture. Thus, it is necessary to estimate the level of moisture in the electronic products. Other methods may be necessary to determine the amount of caustic chemicals due to manufacturing and environment and to estimate the level of harmful radiation. The results of the stress analyses are critical to the failure analyses activity. In this section, key stress analyses for CCAs are discussed.

9.5.1 Thermal analysis

The objective of the thermal analysis is to determine the temperature of the circuit board and the components based on the heat generated by the components, method of cooling, and the environmental conditions. Outputs of a thermal analysis include: component case, component junction, and board layer temperatures. Because temperature extremes, temperature gradients, and cyclic temperatures influence the useful life of CCAs, several thermal management strategies exist. Thermal management strategies are based on three fundamental modes of heat transfer which include conduction, convection, and radiation. Conduction plays a significant role in any thermal analysis strategy.

Heat transfer by conduction is governed by the intrinsic property of material to allow heat to move from an area of high concentration to an area of low concentration. The rate at which heat moves through a media is governed by its thermal conductivity. A list of conductivity for typical board materials is provided in Table 9.7. Thermal analysis involves solving the heat-transfer equation:

$$\frac{\partial}{\partial x}\left(K_x \frac{\partial T}{\partial x}\right) + \frac{\partial}{\partial y}\left(K_y \frac{\partial T}{\partial y}\right) + \frac{\partial}{\partial z}\left(K_z \frac{\partial T}{\partial z}\right) + \dot{q} = \rho c \frac{\partial T}{\partial \tau} \qquad (9.41)$$

where K_x, K_y, K_z = media conductivity in x, y, and z directions
$\dot{q}$ = heat source per unit volume
c = specific heat of the media
ρ = density of the media
T = temperature distribution in the media

In most cases, steady-state temperature results are sufficient to assess failure conditions. An example would be that a temperature cycle can be adequately defined by only determining the steady-state conditions for the high and low ends. An evaluation of the heat capacity of the structures on the CCA is necessary to determine if transition will play a significant role. While there may be time lags in heating and cooling between pieces of CCA, the differences are not generally assumed to be large.

If only the steady-state temperature condition is considered, the general conduction heat-transfer equation can be simplified to

TABLE 9.7   Conductivity for Typical Board Materials

Material	K_x, K_y (W/m°C)	K_z (W/m°C)
Epoxy fiberglass	0.39	0.2
Copper	386	386
Aluminum	204	204
Polymide glass	0.35	0.35
Kovar	16.3	16.3
Solder	50.6	50.6
Molybdenum	138	138
BeCu C17400	130	130

$$\frac{\partial}{\partial x}\left(K_x \frac{\partial T}{\partial x}\right) + \frac{\partial}{\partial y}\left(K_y \frac{\partial T}{\partial y}\right) + \frac{\partial}{\partial z}\left(K_z \frac{\partial T}{\partial z}\right) + \dot{q} = 0 \qquad (9.42)$$

where K_x, K_y, K_z = media conductivity in x, y, and z directions
$\dot{q}$ = heat source per unit volume
T = temperature distribution in the media

For CCAs the conduction heat-transfer Eq. (9.41) can often be simplified. Since components are mounted on the top or bottom surfaces of printed wiring boards, it is often convenient to assume that $q(x,y,z) = 0$. For a wide range of physical situations, the in-plane boundary conditions on the circuit card can be defined as either adiabatic (no heat loss) or isothermal (constant temperature):

adiabatic boundary conditions:

$$\left.\frac{\partial T}{\partial x}\right|_{(x^0, y^0)} = 0 \text{ and } \left.\frac{\partial T}{\partial y}\right|_{(x^0, y^0)} = 0 \qquad (9.43)$$

isothermal boundary condition:

$$T(x_0, y_0) = T_0 \qquad (9.44)$$

For the top and bottom planar surfaces of the CCA, one can assume either no heat loss, or define heat loss as a function of an ambient temperature. In the latter case, the energy balance at a surface can be expressed as

$$-K_z \frac{\partial T}{\partial z} + q_s = h (T - T_{\text{ref}}) \qquad (9.45)$$

where q_s = surface energy generated per unit area
h = convection heat-transfer coefficient
T_{ref} = reference temperature

The fundamental problem involved in convective heat transfer is the determination of the appropriate convection heat-transfer coefficient. The convection heat-transfer coefficient depends on the physical configuration of the PWB and the type of convective cooling process, that is, forced or natural. Depending on the physical situation, the determination of the heat-transfer coefficient may be accomplished by using either theoretical formulations or empirical equations. It can be assumed to be zero at the top and bottom board surfaces if the conduction heat-transfer mode dominates.

Two common convection heat-transfer configurations are depicted in Figs. 9.10 and 9.11. For the configuration depicted in Fig. 9.10, the convection heat-transfer coefficient would be based on buoyancy forces produced by changes in the density of air that is local to the circuit

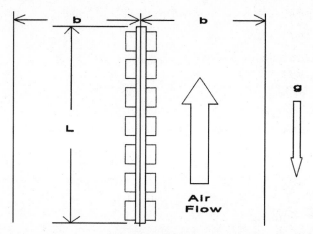

Figure 9.10   Vertical orientations for natural convection and improved cooling.

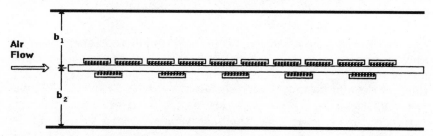

Figure 9.11   Convection cooling by forcing airflow across the surface of the CCA.

card. The buoyancy forces produce a natural flow of air that helps cool the board. Blockages or restrictions will reduce the flow and reduce the convection cooling path. Natural convection is appropriate for low power applications. As the amount of heat to be removed increases, a more active method of thermal management may be necessary. At the extreme, this could involve direct impingement cooling or the application of actively cooled heat sinks. For moderate conditions, a forced airflow over the CCA may be sufficient. This technique is depicted in Fig. 9.11. As with natural convection, blockages and restrictions to airflow can result in reduction of heat-removal capability. While convection and conduction paths are generally the primary heat-transfer modes, in closely confined enclosures radiation heat transfer may provide a strong influence to temperature. Since radiation is a surface phenomenon, it is possible to express the radiation heat transfer in a manner similar to the convection heat-transfer model. In these models, the heat-transfer rate of radiation is a function of the surface temperatures, the surface finishes, and the amount of visible surface area.

In terms of the component temperatures, thermal resistance networks are often used to evaluate case and junction temperatures. The component case and the junction temperatures can be evaluated by calculating the substrate temperature directly below the component. Once the substrate temperature is known, the component case and junction temperatures may be calculated in the following manner. Using a resistance network approach, the case temperature of a component can be defined as

$$T_{case} = T_{substrate} + q\,R_{cs} \tag{9.46}$$

where $T_{substrate}$ = temperature of the board surface directly below the component (°C)

q = heat dissipated by the component (W)

R_{cs} = thermal resistance between the case and the substrate, which is defined next

The case to substrate resistance is a function of the electrical interconnects and the gap between the bottom surface of the component and the top surface of the board. One method of approximating the case to substrate resistance is to create a simple resistance network. Such a network model is depicted Fig. 9.12 , where the interconnect and gap resistances are represented as parallel paths. In this model, the case to substrate resistance can be expressed as

$$R_{cs} = \cfrac{1}{\left(\cfrac{1}{R_{bcs}}\right) + \left(\cfrac{1}{R_{lcs}}\right)} \tag{9.47}$$

where R_{bcs} = thermal resistance of the gap material

R_{lcs} = thermal resistance of electrical interconnects

For the gap resistance, R_{bcs} can be defined as

$$R_{bcs} = \frac{t_{cs}}{K_{cs}\,A_c} \tag{9.48}$$

where t_{cs} = specified case to substrate distance

K_{cs} = conductivity of the specified case to substrate material

A_c = component area

The effective interconnect resistance, R_{bcs}, can be defined as

$$R_{lcs} = \frac{l_{pin}}{n_{pin}\,K_{pin}\,A_{pin}} \tag{9.49}$$

where l_{pin} = effective length of a single interconnect

n_{pin} = number of pins

K_{pin} = conductivity of a single interconnect

A_{pin} = effective area of a single interconnect

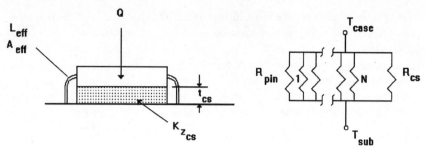

Figure 9.12 Thermal resistance model for conduction heat transfer between the component and the PWB.

The junction temperature of a component is evaluated by the following expression:

$$T_{\text{junction}} = T_{\text{case}} + q\,R_{\text{jc}} \tag{9.50}$$

where R_{jc} = junction-to-case thermal resistance is specified as a component attribute

Many manufacturers provide tables defining the junction to case resistance. One should be careful in using these numbers, since these value are often related to the method used to make the evaluation.

Once a thermal analysis problem can be expressed, there are usually several ways of solving it. Solution approaches are shown in Table 9.8. In addition to simple conduction, convection, and radiation, other important heat-transfer technologies may need to be analyzed. These heat-transfer methods include forced air over the CCA, cold rails, and cold plates.

9.5.2 Vibration analysis

The vibration analysis is used to determine the response of the CCA due to the random oscillating motion of the structure that contains the CCA. When calculating the vibrational life of solder joints, which is the most prevalent failure site associated with vibration on CCA, the most critical step is determining the fundamental mode shapes and frequencies. The dynamic board curvature, and the components' solder joint and lead stresses, are inversely proportional to the square of the CCA's natural frequency. By only slightly increasing the natural frequency of the CCA, the maximum deflection amplitude is rapidly decreased, thereby reducing the stresses and strains on the solder joints. Any error in the natural frequency calculation dramatically alters the calculated solder-joint fatigue life.[14]

TABLE 9.8   Solution Approaches for Thermal Analysis

Approach	Modes	Transient	Comments
Finite difference	Conduction, convection, radiation	Yes.	Detailed analysis most frequently used.
Finite element	Conduction, convection, radiation	Yes.	Detailed analysis. Offers advantages of multiple elements shapes.
Resistive network	Conduction, convection, radiation	Yes, but usually used for steady-state analysis.	Coarse analysis. Requires considerable approximations.
Bulk analysis	Convection, radiation	Yes.	Crude analysis. Requires considerable approximations.

While calculating the natural frequency of a CCA, the boundary conditions are critical. Classical boundary conditions include free, simple, or clamped. *Simple supports* restrict the out-of-plane motion of the CCA, but allow rotational motion. *Clamped supports* restrict both out-of-plane and rotational motion, while *free ends* allow unrestricted motion. Clamped supports provide high natural frequencies, but are not physically possible or practical for CCAs that must be plugged into the electronic assembly. For rack-mounted CCAs, wedge lock supports are usually used. While wedge locks attempt to restrict the rotational motion, their actual support characteristics are between a simple and clamped support. Some edge conditions may change as the result of the magnitude of the vibrational load applied. During high magnitude vibration loads, simple supports may act more like free edges and clamped supports like simple supports.

The natural frequency of a CCA can be experimentally or numerically determined. Experimental determination requires the placement of a strain gauge or accelerometer on the CCA, attaching the CCA on a dynamic shaker, and measuring the response of the CCA to a known input. Numerically, the natural frequency of the PWB can be determined using first-order approximations or finite-element modeling.

9.5.2.1 First-order approximations for natural frequency. One method of approximating the natural frequency of a CCA is to model its response using a trigonometric or a polynomial series. The Rayleigh[11,15] method, a popular method of obtaining an approximation of the natural frequency, is based on using a trigonometric expression to model the behavior of the CCA under vibration loading. The Rayleigh method assumes the deflection curve which satisfies the boundary condition for a particular vibration problem. Based on this approximation, the strain energy and the kinetic energy in the plate can be estimated. Recognizing that the strain energy must equal the kinetic energy allows for the approximation of the natural frequency. This procedure is relatively straightforward for rectangular plates. The Rayleigh method usually produces a natural frequency that is slightly higher than the true natural frequency.

Using the Rayleigh method, the natural frequency for various boundary conditions can be calculated. Results have been reported by other authors and are represented here.[9,11,15] For all boundary conditions on rectangular CCAs, the natural frequency is defined by

$$f_n = C \sqrt{\frac{D}{\rho}} \tag{9.51}$$

where C = dependent on the defined boundary condition
$\quad\quad D$ = the board rigidity as defined by

$$D = \frac{Et^3}{12\,(1-\nu^2)} \tag{9.52}$$

where E = the effective modulus of elasticity of the board
$\quad\quad t$ = the thickness of the board
$\quad\quad \nu$ = the Poisson's ratio of the board

Table 9.9 shows values of C for various methods of supporting the CCA (adapted from Ref. 15). Figure 9.13 provides the references for A and B used in Table 9.9.

The Rayleigh method can also model CCAs with ribs for stiffeners using equations similar to those in Table 9.9. For an example of a CCA with simple supports on the edges and a rib stiffener, the natural frequency is calculated by the following[11]:

$$f_n = \frac{\pi}{2} \sqrt{\frac{1}{\rho} \left(\frac{D_x}{a^4} + \frac{4D_{xy}}{a^2b^2} + \frac{D_y}{b^4} \right)} \tag{9.53}$$

where D_x and D_y are the bending stiffness along the x and y axis, respectively, and D_{xy} is the torsional stiffness.

The application of the Rayleigh method provides a quick assessment of natural frequency and is particularly useful in the early design stages. Although helpful, the Rayleigh method should only be used if its limitations are understood. In the Rayleigh method, the following assumptions are made:

- Board is a homogeneous isotropic plate, not multilayered.
- Board shape is perfectly rectangular with no edge or corner cut-outs.
- Boundary conditions are classical (that is, simple, clamped, or free).

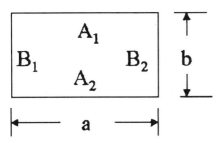

Figure 9.13 Reference figure for the Rayleigh method.

TABLE 9.9 Calculated Values for C Based on Boundary Conditions (see Fig. 9.10)

Fixture[*] $A_1 A_2 B_1 B_2$	C
FFFF	$0.5\pi[2.08a^{-2}b^{-2}]^{-0.5}$
FFFC	$0.56\,a^{-2}$
FFCC	$3.55\,a^{-2}$
FCCF	$\pi(a^{-4} + 3.2\,a^{-2}b^{-2} + b^{-4})^{0.5}/5.42$
CCCF	$\pi(0.75\,a^{-4} + 2\,a^{-2}b^{-2} + 12\,b^{-4})^{0.5}/3$
CCCC	$\pi(3\,a^{-4} + 2\,a^{-2}b^{-2} + 3\,b^{-4})^{0.5}/1.5$
FFSS	$0.5\pi a^{-2}$
SSSF	$0.5\pi(0.25\,a^{-2} + b^{-2})$
SSSS	$0.5\pi(a^{-2} + b^{-2})$
FFCS	$0.78\pi\,a^{-2}$
CCCS	$0.5\pi(2.45\,a^{-4} + 2.9\,a^{-2}b^{-2} + 5.13\,b^{-4})^{0.5}$
SSCS	$0.5\pi(2.45\,a^{-4} + 2.32\,a^{-2}b^{-2} + b^{-4})^{0.5}$
SCCS	$0.5\pi(2.45\,a^{-4} + 2.68\,a^{-2}b^{-2} + 2.45\,b^{-4})^{0.5}$
SSCC	$\pi(16\,a^{-4} + 7.7\,a^{-2}b^{-2} + 3\,b^{-4})^{0.5}/3.53$
FCSS	$0.5\pi(a^{-4} + 0.608\,a^{-2}b^{-2} + 0.126\,b^{-4})^{0.5}$
FSCC	$0.5\pi(2.25\,a^{-2} + 0.25\,b^{-2})$
FSCS	$0.5\pi(2.56\,a^{-4} + 0.57\,a^{-2}b^{-2})^{0.5}$
FSSF	$\pi(a^{-2} + b^{-2})/11$
FSCF	$0.5\pi(0.138\,a^{-2} + 0.251\,a^{-2}b^{-2})$

[*]S = simple, F = free, and C = clamped.

- Boundary conditions are applied over the entire length of an edge.
- Mass of the components is smeared over the CCA.
- Component stiffness is neglected.

Numerical simulations are generally required for complex geometries, CCAs with heavy components, or nonclassic boundary conditions.

9.5.2.2 Finite element analysis for calculating natural frequencies. Finite element analysis (FEA) is a very powerful tool for vibration analysis.

Complex shapes or materials can be modeled, which makes this technique superior to the Rayleigh method. FEA also has the ability to model elastic boundary conditions and can include the mass and stiffness of the components on the CCA. Although FEA has many benefits and is a very powerful tool, the user has to have a theoretical background in FEA and experience using FEA code. Common FEA tools include NASTRAN, ANSYS, ABAQUS, ADINA, and Pro/Mechanica. Small errors in the modeling phase can lead to unrealistic results during an FEA.[9].

Although the generalized FEA tools listed here require extensive knowledge and experience, FEA tools designed specifically for analyzing only CCAs are much easier to use. The University of Maryland CalcePWA software provides a customized FEA package for the vibrational analysis of CCAs.[16] The analysis is based on flat-plate theory with anisotropic elements with equivalent elastic modulus, density, and Poisson's ratios. The boundary conditions are simple, clamped, wedge-locks, or rotational and can be applied as an edge or a point support. The stiffness matrix and the consistent mass matrix of each element are determined using material property information and variational methods. The elemental matrices are then assembled into the respective global matrices for the whole structure by using boundary conditions and element connectivity information. Once the global stiffness and mass matrices are calculated, the natural frequencies and the mode shapes can be determined using eigenvalues and eigenvector extraction techniques.[14]

General or custom FEA packages can be used to assess the effect of components on a CCA. Components stiffen the CCA because they add support to the CCA, which increases their natural frequency, but they also add weight which decreases the natural frequency. Depending on the size and location of a component, its stiffness benefits may be overcome from the added weight. A good rule of thumb is to consider the stiffness of a component if its length is one-quarter of the shortest board length.

9.6   CCA Failure Mechanisms and Models

Failure is defined as the inability of a system to perform its intended function. Failures in CCAs result from a complex set of interactions between the stresses that act on a CCA and the materials of a CCA. *Failure mechanisms* are the processes by which a material or system degrades and eventually fails. Three basic categories of failure include overstress (that is, stress-strength), wearout (that is, damage accumulation), and performance tolerance (that is, excessive propagation delays). Overstress and wearout failure categories are commonly related to irreversible material damage; however, some overstress failures can be related to reversible material damage (that is, elastic deformation). Overstress and wearout failure are characterized by a damage metric. In general, damage metrics for overstress failures are defined in terms of a pass or fail criteria. The damage metrics for wearout failures define how much useful life is consumed by an imposed environmental condition. Performance-tolerance failures may be associated with stress conditions that produce reversible changes in material properties. As a result, failures occurring at elevated stress levels may not occur during normal operation. Performance-tolerance failures are characterized by a performance metric. Failure occurs if the performance metric is outside of a specified tolerance range. Performance-tolerance failure mechanisms are usually associated with the failure of integrated circuits and not the CCAs (for example, excessive propagation delay in integrated circuits due at high temperature or excessive thermal transients due to inadequate diffusivity).

In the PoF approach, failure is related to the physical structure of the system under consideration and the environmental/operational loads. The failure modeling generally involves an interrelated two-step process. The first step is the evaluation of a stress metric. The stress metric, defined by the damage or performance metric, is generally a function of the stress condition, system geometry, and material properties (that is, the stress metric characterizes a physical situation). The second step is determining the damage metric, which is related to the material system or interface under consideration. The damage metric, using the stress metric as an input, determines the amount of useful life consumed by a physical situation. It is highly desirable to separate the stress metric and the damage metric, which allows easier solution to the problem. If the stress and damage metrics are closely coupled, their relationship can be confusing. In some cases, the stress metric is an artificial parameter that may be mistakenly treated as an actual stress value.

This section will focus on overstress and wearout failures that are more predominant in CCAs. These failure mechanisms can be categorized by the nature of the stresses that drive or trigger them. These categories include the following:

- *Mechanical.* Failures that result from elastic and plastic deformation, buckling, brittle and ductile fracture, fatigue-crack initiation and propagation, creep and creep rupture.

- *Thermal.* Failures that result from exceeding the critical temperatures of a component such as glass-transition temperature, melting point, or flash point.

- *Electrical.* Failures that include those due to electrostatic discharge, dielectric breakdown, junction breakdown, surface breakdown, surface and bulk trapping, hot electron injection.

- *Radiation.* Failures that may be caused by radioactive contaminants and secondary cosmic rays.

- *Chemical.* Failures that result from chemical environments which act as catalysts to corrosion, oxidation, or ionic surface dendritic growth.

This breakdown of failure mechanisms allows an analyst to consider stress metrics that must be evaluated to determine the potential for failure. To completely define the stress metric, the environmental/operational loads must be related to the geometry and materials under investigation. This can be achieved by categorizing failures in CCAs by common structures or building blocks. Common CCA structures include the following:

- Components (active and passive)
- Permanent interconnects
- Metallization
- Contacts

The failure information related to each structure is included in this section. Component failure is particular to the technology and physics of the specific devices and is beyond the scope of this chapter. This section will focus on failure mechanisms and models of permanent interconnects, PWB metallization, and contacts. Before addressing these specific failure mechanisms and model, the next section will describe generic damage models.

9.6.1   Damage models and metrics

Damage metrics approximate the probability of a failure occurring and are based on the physical conditions that produce failure. Damage is based on the assumption that failure results from irreversible degradation of the material which accumulates over time. The underlying assumption is that damage is additive, and this assumption is used to estimate the damage to permanent interconnects as a result of repeated stress reversals. The common damage metric for failures resulting from cyclic conditions is expressed as the ratio of applied (exposed) cycles to the estimated number of survivable cycles at the current exposure level. This is mathematically expressed as

$$DM = \frac{N_{applied}}{N_{life}} \qquad (9.54)$$

where $N_{applied}$ is the exposure time of the failure-inducing condition and N_{life} is the estimated time to failure for the failure-inducing condition.

If $N_{\text{applied}} = N_{\text{life}}$, the probability of failure is assumed to be one. Using this approach, failure can be evaluated for multiple environmental conditions. In this case, the damage metric is defined as

$$R = \sum_i \frac{N_{\text{applied}_i}}{N_{\text{life}_i}} \tag{9.55}$$

This ratio is referred to as Miner's ratio. Cycles to failure is commonly used for fatigue-related failures. For failure resulting from a sustained condition, such as creep rupture, the damage metric is generally defined as time to failure. The damage metric in this case is the application time over the estimated time to failure.

One of the inputs to the failure model is the stress metric. The term stress metric is introduced to distinguish it from the physical stress. In many cases, failure models blur the lines between actual physically measurable parameters and metrics that are used in modeling physical failures. The failure model is generally closely coupled with the method of characterizing the stress history.

9.6.2 Permanent interconnects

One of the greatest concerns related to CCA reliability is failure of package-to-board interconnects. For most modern packages, package-to-board interconnects provide a structural and an electrical interface between the component and the PWB. Solder is the most common method of interconnection. Failure mechanisms related to permanent interconnects include fatigue, creep rupture, mechanical overstress, intermetallic growth, and corrosion. Fatigue is generally considered to be the primary failure mechanism and can occur due to temperature cycling and/or mechanical vibration. While important, elevated temperature, chemical contamination, and moisture generally play a secondary role in precipitating interconnect failures. The fatigue failure of permanent interconnects will be addressed in the following sections.

9.6.2.1 Geometry and material characterization. Fundamental to permanent interconnect failure modeling is the understanding of the structures and the material used. In order to streamline production of the electronic system, electronic components are packaged in standard formats. Modern components have a multitude of package types or families. The standardization of package format is conducted by the Joint Electron Device Engineering Council (JEDEC) in conjunction with the Electronic Industry Alliance (EIA). JEDEC currently recognizes over 200 unique package styles, and new package formats are still being introduced. While there may be many similarities between package style, there are also important distinctions. The clearest

delineation of package styles is between insertion and surface-mount technologies. Insertion-mounted packages have leads that extend through the PWB. As the name implies, surface-mount packages are joined to a single surface and may be leadless or leaded. Most modern electronics use surface mountings due to their ease of assembly, higher input/output counts, and reduced land area. Insertion-mount technology, however, is still used extensively in consumer electronics and mixed technology boards are also being produced. Figures 9.14 and 9.15 show examples of typical surface-mount and insertion-mount interconnects, respectively.

Common insertion-mount packages include the dual in-line package (DIP), the pin grid array (PGA), can packages, and axial discretes (capacitors and resistors). Common surface-mount packages include leadless ceramic chip carriers (LCCC), plastic leaded chip carriers (PLCC), small outline j-leaded (SOJ) packages, small outline gull-wing packages (SOP), thin small outline packages (TSOP), leadless chip capacitors (LCC), and leadless chip resistors (LCR). New surface-mount packages include ball-grid arrays (BGA) and chip-scale packages (CSP). There are currently over 20 different CSP package formats. For surface-mount technology, components are mated to a PWB with solder in a reflow process. For insertion-mount technology, a wave-soldering process is typically employed. In both cases, a tin-lead alloy (solder) is generally used to create the bond.

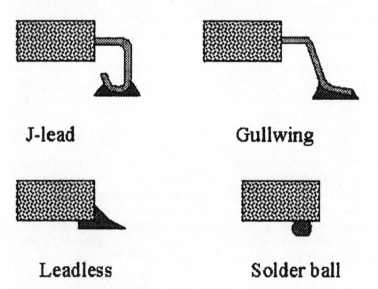

J-lead   **Gullwing**

Leadless   **Solder ball**

Figure 9.14 Illustration of typical surface-mount interconnects.

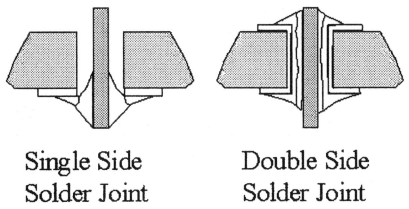

Single Side Solder Joint   Double Side Solder Joint

Figure 9.15   Illustration of fatigue-related failure sites for insertion-mount interconnects.

Eutectic or near eutectic tin-lead (60%Sn 40%Pb) is commonly used in CCAs. A variant of this alloy is a ternary alloy composed of 62%Sn36%Pb2%Ag, which is also used. Investigations into other lead-free systems have been conducted based on the environmental concerns related to lead. Initiatives are in progress to move toward lead-free electronics by the early part of the twenty-first century. These initiatives have spawned a search for replacements for tin-lead solder. Potential candidates include tin-silver, tin-indium-silver, tin-bismuth-silver alloys. At present, there is no clear leader.

Failure analysis of permanent interconnects requires a detailed understanding of the physical behavior of permanent interconnect materials. Under field loading conditions, it is important to understand the viscoplastic nature of solder. Several authors have addressed solder properties and behavior.[16–19] The melting point of eutectic or near eutectic solder is approximately 183°C. With temperature conditions in applications approaching 85°C or higher, the soldered interconnect is approximately one-half of its melting point. As a result, solder will creep at elevated temperatures and the stress in the interconnect will relax. This results in a hysteresis of the solder. Figure 9.16 shows the effect of the creep behavior in terms of an idealized stress-strain history. Since electronic products are subjected to temperature excursions, the creep behavior of solder is critical. *Creep* is defined as the time-dependent change in strain under a constant load and consists of an initial high strain rate followed by a steady-state strain rate. The Weertman steady-state creep law can be used to model the creep behavior of solder. For this model the creep strain is defined as

$$\gamma_{cr} = \gamma_0 \sigma^{1/n_c} \exp\left(\frac{-\Delta H}{T}\right) t \tag{9.56}$$

where $\gamma_0 = 6.82e_{-15}$ (lb/in²) $\frac{1}{n}$/s, $1/n_c = 6.28$, and $\Delta H = 8165.2$ for eutectic solder.[18]

In addition to the creep behavior, solder exhibits plastic flow under stress. This behavior can be modeled using the Ramsberg-Osgood relationship, which defines the strain as a function of elastic and plastic components as defined by

$$\varepsilon = \varepsilon_{el} + \varepsilon_{pl} = \frac{\sigma}{E} + \left(\frac{\sigma}{K}\right)^{1/n_p} \tag{9.57}$$

where σ = the equivalent stress
   E = the modulus of elasticity
   K = the Ramberg-Osgood constant
   n_p = the strain-hardening exponent

For eutectic solder, the yield stress is 4960 lb/in² and $E = 3.62e6$ lb/in², $K = 7025$ lb/in² and $np = 0.056$.[18]

The stress-strain history of a solder joint, depicted in Fig. 9.16, is directly related to the package, interconnect, and board. For leadless packages, the stress state may reach a plastic yield point during the course of a temperature excursion. If the system is maintained at the elevated temperature for any period of time, relaxation of the stress will occur within the soldered joint. On the downward side of the temperature cycle the reverse occurs. However, there may be little or no relaxation at low temperatures. The outer solid line in Fig. 9.16 depicts a hypothetical hysteresis loop for a leadless package. In the case of leaded packages, the stress may not reach the plastic yield state. In this case, relaxation occurs from somewhere below the yield state. As a result, the hysteresis loop is smaller as depicted by the dashed lines. Thus, leaded packages tend to have longer lives than leadless packages under similar temperature cycles.

τ **Hysteresis**

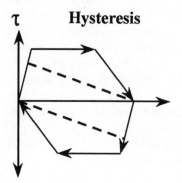

Figure 9.16   Idealized hysteresis loop for solder joints.

In order to forecast the overall life of a soldered interconnect, the stress-strain history and damage behavior of the interconnect material must be considered. As discussed previously, the PoF approach is a two-step process that involves evaluating a stress metric and then evaluating the damage metric. Researchers have proposed several stress metrics and damage laws for assessing the fatigue life of the permanent interconnects. These methods are based in stress, strain, and/or dissipative energy. The following paragraphs present several models and approaches for forecasting the fatigue life of soldered interconnects. These damage models include the stress approach (Basquin), the strain approach (Manson-Coffin), and the energy approach.

9.6.2.2 Stress approach. Fatigue life is often characterized in terms of the applied stress range. For many engineering materials, the fatigue life exhibits a linear relationship with the applied stress range when presented as a log-log plot. The relationship can be mathematically represented[20] as

$$N_1 \, \sigma_1^{\,b} = C \tag{9.58}$$

where C is the fatigue strength coefficient and b is the fatigue damage exponent. This relationship has a wide range of applicability. This model is most relevant for stresses within the elastic limits of the material. The stress metric for this damage model is the average cyclic stress range. Steinberg[21] demonstrated the use of this model to the fatigue life of permanent interconnects under both temperature cycling and vibration load conditions.

The problem with the Basquin damage model is that it does not capture the damage due to inelastic deformation. In general, solder undergoes inelastic deformations making the Basquin model questionable. The Basquin relationship can still be used, but it should be used with caution.

9.6.2.3 Strain approach. The strain-based damage models also make use of a power law form but use strain rather than stress as their input metric. The plastic strain life model[22,23] relates fatigue life to the plastic strain range. The damage model is given as

$$N_f = \frac{1}{2} \left(\frac{\Delta\gamma_p}{2\varepsilon_f} \right)^{1/c} \tag{9.59}$$

where c is the fatigue ductility exponent and ε_f is the fatigue ductility coefficient. For eutectic solder, ε_f is approximately equal to 0.625. The *fatigue ducility exponent* has been defined as a function of the mean cyclic temperature and the cycle frequency[16] and is expressed as

$$c = -0.442 - 6.0 \times 10^{-4}T_s + 1.72 \times 10^{-2}\ln(1 - f) \quad (9.60)$$

where T_s is the mean cyclic temperature and f is the cycle frequency. For this model, the cycle frequency is between 1 and 1000 cycles/day.

A more generalized strain range approach is to combine the elastic strain life (Basquin) and the plastic strain life (Coffin-Manson). This model relates the total strain range (elastic and plastic) to cycles to failure (damage metric) and is given as

$$\frac{\Delta\gamma}{2} = \frac{\Delta\gamma_e}{2} + \frac{\Delta\gamma_p}{2} = \frac{\sigma_f}{E}(2N_f)^b + \varepsilon_f(2N_f)^c \quad (9.61)$$

where $\Delta\gamma$ = the total strain range
$\Delta\gamma_e$ = the elastic strain range
$\Delta\gamma_p$ = the plastic strain range
E = the modulus of elasticity
σ_f = stress strength coefficient
b = the fatigue strength exponent
c = the fatigue ductility exponent
ε_f = the fatigue ductility coefficient

The advantage of the generalized Coffin-Manson equation is that it accounts for both the elastic and inelastic behavior of the material. A plot of the generalized Coffin-Manson relationship is presented in Fig. 9.17. As can be seen from Fig. 9.17, the elastic strain range dominates high cycle fatigue while the inelastic strain dominates low-cycle fatigue.

9.6.2.4 Crack propagation approach. The crack propagation approach is appealing since it represents the physics involved in the interconnect failure. Crack propagation assumes that there is a nucleation period followed by a crack growth phase. Cracks initially occur in the high stress concentration area of a solder joint. For gull-wing leads, the crack generally initiates at the heal where the lead bends to form the foot. Rather than shear loading, the normal (peeling) forces drive the crack formation and growth. Figure 9.18 shows a solder-joint fatigue crack for a gull-wing lead. The cracks can be seen as the dark lines through the solder. For BGA interconnects, cracks initiate and grow in the solder ball near the attached substrate. Cracks have been shown to form on the outer edge and move inward as can be seen in Fig. 9.19. For leadless interconnects, the crack initiates under the component at the pad-solder interface and grows outward into the solder fillet. Shear forces appear to drive leadless interconnect fatigue failures.

Classical fracture mechanics suggest that the crack initiation and growth is related to a stress intensity factor or the J-integral.

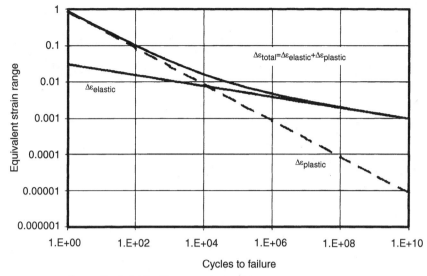

Figure 9.17 Generalized Coffin-Manson relationship.

Researchers examining soldered interconnects have also related crack initiation related to the strain range or viscoplastic strain energy. In experiments presented by Darveaux,[19] the cycles-to-crack initiation was modeled using the viscoplastic strain energy per cycle, ΔW. In this case, the crack initiation model is given as

$$N_o = C_i \Delta W^{b_i} \tag{9.62}$$

For 62%Sn36%Pb2%Ag, Darveaux[19] suggests the following material constants: $C_i = 7860$ and $b_i = -1.0$. In this same report, the crack growth was also related to the viscoplastic strain energy per cycle. In this case the crack growth model, defined as the area growth per cycle, was given as

$$\frac{dA}{dN} = C \, (\Delta W)^b \tag{9.63}$$

For 62%Sn32%Pb2%Ag, $C = 1.0e^9$ and $b = 1.19$. While the crack initiation and growth models provide a more realistic presentation of the failure of soldered interconnects, it presents a significant experimental challenge. The primary drawback stems from the fact that the acquisition of crack-related measurements is extremely difficult for specimens that match the geometries of actual solder joints. The measurement of bulk specimens may not provide relevant data.

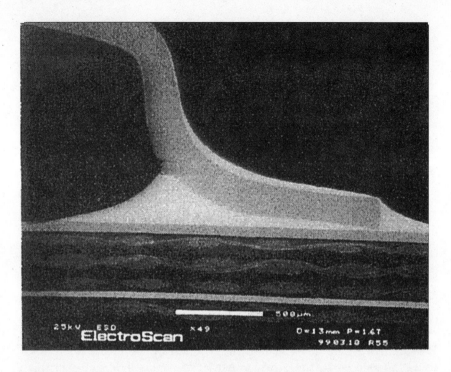

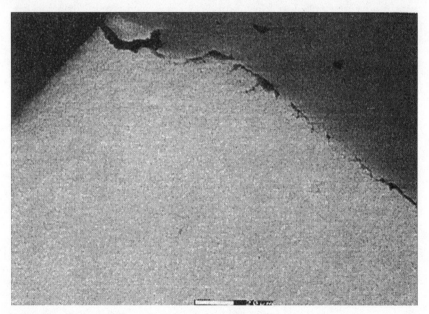

Figure 9.18 Fatigue failure of a gull-wing interconnect..

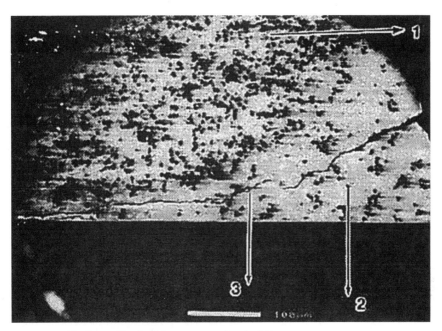

Figure 9.19 Broken solder ball interconnect due to fatigue.

9.6.2.5 Energy approach. The energy approach relates the failure of the interconnect to the energy dissipated during a stress cycle.[18] In this approach, the number of cycles to failure is proportional to the irreversible energy consumed at an interconnect joint. Like the strain range models, energy models assume the failure is a power law relationship with the hysteresis (cyclic) energy. The most basic model is given as

$$N_f = C\Delta W^m \tag{9.64}$$

where C and m are material properties and ΔW is the viscoplastic strain energy per stress cycle.

Analogous to the strain range partitioning approach, the need to separate the energies into rate-dependent curves results in the development of the energy partitioning approach.[18] In the energy partitioning approach, the energy dissipated during the stress cycle is partitioned into elastic potential energy, plastic work, and creep work. Each energy component is then related to the cycles to failure. This can be expressed as

$$\text{Energy} = U_e + W_p + W_{cr} = U_o N_{fe}^{b'} + W_{po} N_{fp}^{c'} + W_{co} N_{fc}^{d'} \tag{9.65}$$

where U_{eo} = the elastic coefficient
$\quad\quad b'$ = the elastic exponent
$\quad\ W_{po}$ = the plastic exponent
$\quad\quad c'$ = the plastic exponent
$\quad\ W_{co}$ = the creep coefficient
$\quad\quad d'$ = the creep exponent

A plot of the cyclic strain energy versus cycles to failure is presented in Fig. 9.20.

By examining the various fatigue life models, it is clear that the stress history of a solder interconnect is critical for forecasting the interconnect reliability. The stress develops in the solder interconnects as a result of temperature, vibration, and swelling due to moisture absorption. From the literature, it appears clear that temperature followed by vibration is the primary concern. Methods for approximating the stress-strain history, based on temperature and vibration induced loads, are presented in the following sections.

9.6.2.6 Temperature cycling. Temperature cycling of CCAs resulting from power cycling and exposure to uncontrolled environments directly affects

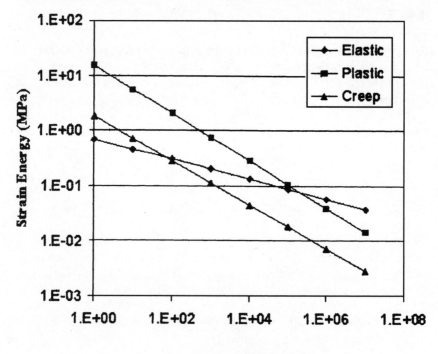

Figure 9.20 Strain energy versus fatigue life for 63%Sn37%Pb solder.

useful life. The concern associated with temperature cycling arises from the mismatch of thermal expansion rates between the various materials. The expansion rate of a material is characterized by a coefficient of thermal expansion (CTE), which may be measured experimentally using a thermomechanical analyzer (TMA). A list of CTE common packaging material and packages is provided in Table 9.10. For package-to-board interfaces, CTE mismatches may occur globally (package and board) and locally (lead and solder, solder and PWB pad). A depiction of problems related to temperature cycling and CTE mismatch is present Fig. 9.21. As a result of the CTE mismatches, the board, package, and interconnect undergo deformation. In general, the interconnect is the weak link and takes up most of the stress and consequently deforms the most. The stress-strain history in the solder joint is an important function of the package and joint geometry, as well as the applied temperature cycle. A discussion of different stress metrics used to characterize thermomechanical-induced fatigue is presented in the following paragraphs.

The stress range resulting from a temperature cycle is a simple metric that can be used to characterize the fatigue of a solder interconnect. The stress range may be obtained by FEA or by the application of basic mechanics of materials principles. This application of basic mechanics of materials principles allows an engineer to obtain a rudimentary understanding of the stress-strain condition within a soldered interconnect. An example of this approach will be discussed.

Global Mismatch   Local Mismatch

Figure 9.21 Thermomechanical load considerations.

TABLE 9.10   Typical Coefficient of Thermal Expansion (CTE) Values

Material	CTE (ppm/°C)
FR4 (in-plane)	15–20
FR4 (out-of-plane)	54
Plastic packages	20–25
Ceramic packages	5–9
PLCC	22
TSOP	6.7
Kovar	5.9
Solder	23
Polymide fiberglass	12–16

Figure 9.22 shows a surface-mount chip capacitor that is soldered to a PWB. By considering only in-plane forces and ignoring pad-to-solder expansion mismatch, the difference in expansion between the chip and the PWB can be expressed as

$$\alpha_c L_c \Delta T_c + \frac{P L_c}{E_c A_c} = \alpha_{pwb} L_c \Delta T_{pwb} + \frac{P_{pwb} L_{pwb}}{E_{pwb} A_{pwb}} + \frac{P_{solder} h_{solder}}{G_{solder} A_{solder}} \qquad (9.66)$$

where α_{pwb}, α_c = the CTE of the PWB and component
ΔT_{pwb}, ΔT_c = one-half of the temperature cycle range from the high to low temperature dwells for the PWB and component
E_{pwb}, E_c = the modulus of elasticity of the PWB and the component
A_{pwb}, A_c = the cross-sectional area of the PWB and the component
L_{pwb}, L_c = the half lengths of the PWB and the component
G_{solder} = the shear modulus of the solder
h_{solder} = load-bearing height of the solder joint
A_{solder} = the load-bearing area of the solder joint

From this expression, the average force in the interconnect due to a temperature excursion may be determined. A free-body analysis of the chip, the interconnect, and the PWB indicates that

$$P_c = P_{solder} = P_{pwb} = P \qquad (9.67)$$

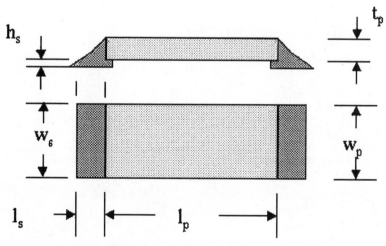

Figure 9.22 Leadless chip carrier diagram.

Based on this simplification, the force, P, acting on the interconnect can be expressed as

$$P = \frac{(\alpha_{pwb} L_c \Delta T_{pwb} - \alpha_{chip} L_{chip} \Delta T_{chip})}{\left\{ \left[\dfrac{L_{pwb}}{(E_{pwb} A_{pwb})} \right] + \left[\dfrac{h_{solder}}{(G_{solder} A_{solder})} \right] + \left[\dfrac{L_{chip}}{(E_{chip} A_{chip})} \right] \right\}} \tag{9.68}$$

From this result, we can see that the shear stress in the solder joint is

$$\tau_{solder} = \frac{P_{solder}}{A_{solder}} \tag{9.69}$$

A similar approach can be used to approximate the stress for other package styles. The estimated shear stress can be used to forecast the fatigue life by using a Basquin power law relationship for stress versus cycles to failure. As previously discussed, stress is not generally considered a good metric for evaluating the fatigue life of solder. For this approach, calibration factors are needed to obtain good agreement between field failure data and model outputs.

A simple model for estimating the fatigue life of interconnects, based on the cyclic strain range and the cyclic strain energy, was suggested by Engelmaier.[24] In this approach, the strain range is approximated by assuming that the expansion was completely taken up by the solder. In calculating this stress metric, transients and PWB warpage are ignored. The strain range may be approximated as

$$\Delta\gamma = \frac{L_D \Delta (\alpha \Delta T)}{h} \tag{9.70}$$

where L_D = the half the diagonal length of the package
$\Delta (\alpha \Delta T) = |\alpha_s (T_{s\_high} - T_{s\_low}) - \alpha_c (T_{c\_high} - T_{c\_low})|$
   α_s, α_c = the CTEs of the PWB and package
   T_{c_low} = package temperature at the low temperature dwell
   T_{c_high} = package temperature at the high temperature dwell
   T_{s_low} = PWB temperature at the low temperature dwell
   T_{s_high} = PWB temperature at low temperature dwell
   h = the height of the solder joint (or half the solder paste stencil thickness)

To handle leaded packages, strain energy is approximated by considering the stiffness of the lead. The energy is then converted back to a strain range by the introduction of an empirical constant. For a leaded package, the stress metric is defined as

$$\Delta\gamma = \frac{K [L_D \Delta (\alpha \Delta T)]^2}{(200 \ \text{lbf/in}^2) \, Ah} \tag{9.71}$$

where K is the diagonal flexural stiffness of the cornermost lead, A is the effective minimum load-bearing solder-joint area (two-thirds of the lead area projected to the pad), and the other parameters are the same as defined for the leadless case in Eq. (9.70).

The metrics presented in Eqs. (9.70) and (9.71) can be used to estimate the useful life of various packages. A high stress metric normally indicates a low expected life. From inspection, it can be observed that the stress metric is reduced by increasing the solder-joint height, minimizing the temperature expansion differential, or reducing the imposed temperature excursion. For the leaded case [Eq. (9.71)], the stress metric can also be reduced by increasing the load-bearing area of the solder joints, or decreasing the lead stiffness. For the stress metrics defined by Eqs. (9.70) and (9.71), the Manson-Coffin fatigue life relationship is used to forecast fatigue life [see Eq. (9.59)].

Strain energy is a useful stress metric for evaluating the fatigue life of soldered interconnects. While Eq. (9.71) provides an approximation of the strain energy, it does not truly capture the complexity of the stress-strain history in the solder joint. Two major problems are the idealized behavior of the solder joint and the assumed symmetry of the stress-strain history. To truly capture the complexity of the stress-strain history, the stress metric must evaluate the stress state in the soldered interconnect over time.

The time-based evaluation of the stress-strain history in a solder joint can be obtained by deriving the equilibrium equations for a component, interconnect, and board assembly, and then solving for the stress and strain instantaneously with respect to time. This approach was proposed and demonstrated for leadless chip resistors by Jih and Pao[25] and extended to leaded packages by Sundararajan et al.[26] In this approach, a two-dimensional free-body diagram of the package, interconnect, and board is developed. The stress in the assembly is assumed to arise from the in-plane CTE mismatch. A free-body diagram for a lead package is presented in Fig. 9.23. The equilibrium analysis yields

$$\gamma + \frac{\tau}{K} + \frac{\tau}{K_L} = \frac{(\alpha_c - \alpha_p) \, \Delta T \, L}{h_s} \tag{9.72}$$

where γ = the shear strain in the solder
τ = the shear stress in solder
α_p, α_c = the CTE of the PWB and package
h_s = the solder-joint height
L = half the length between the solder interconnect locations
ΔT = the temperature cycle range
K = a characteristic stiffness of the surface-mount assembly
K_L = the stiffness of the interconnect

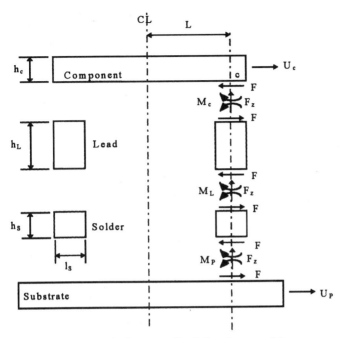

Figure 9.23 Free-body diagram of leaded package model.

The stiffness of the assembly is a function of the component and board stiffnesses. The stiffness of the interconnect can be determined by finite element analysis or by application of mechanics of materials principles. Differentiating Eq. (9.71), ignoring plastic yield and using a creep constitutive model for solder to account for creep, yields a first-order, time-dependent, nonlinear differential equation for shear stress in the critical solder joint:

$$\frac{d\tau}{dt} = \left[\frac{1}{K} + \frac{\tau}{K_L} + \frac{1}{\mu(T)}\right]^{-1} \left[\frac{(\alpha_c - \alpha_p)\,L}{h_s}\,\frac{dT}{dt}\right.$$

$$\left. + \frac{\tau}{\mu^2(T)}\frac{d\mu}{dT}\frac{dT}{dt} - \gamma_o \exp\left(\frac{-\Delta H}{T}\right)\tau\,|\tau|^n\right] \qquad (9.73)$$

The stress history in a solder joint can be approximated by solving Eq. (9.72). The strain history can be determined simultaneously by solving

$$\frac{d\gamma}{dt} = \frac{1}{\mu}\left(\frac{\tau_m - \tau_{m-1}}{\Delta t}\right) + \gamma_o \exp\left(\frac{-\Delta H}{T}\right)\tau^n \qquad (9.74)$$

This method was used to analyze a 9.0-mm^2 20-pin LCCC subjected to a −55 to 125°C temperature cycle. Deriving the assembly and lead constants, then solving Eqs. (9.73) and (9.74) using a differential equation solver yields the stress-strain history of the solder. This information is presented graphically in Fig. 9.24 in the form of a hysteresis loop. For comparison purposes, a finite element model of the same assembly was developed and analyzed. The resulting hysteresis plot is presented in Fig. 9.25. By examining these two figures, it can be seen that this approach provides a close approximation to finite element results. The two stress metrics derived from these results are potential elastic energy U_o and creep strain energy W_{cr}. These stress metrics are used in an energy partition relationship to forecast fatigue life. Strain range may also be determined from this analysis and used to estimate the value of the Manson-Coffin damage metric.

This section has provided methods for approximating stress-based, strain-based, and energy-based stress metrics. Each approach has its advantages and disadvantages. The major advantages include their representation of the actual physical situation and their speed of solution. A major limitation of these approaches is that they do not capture the more complex stress field that actually exists in the solder interconnect. For instance, none of these methods considers the mismatch of CTE between the solder and pad or solder and the lead. In addition, the stress metrics defined up to this point have assumed a uniform or linear distribution for stress and strain in the solder joint. As a result, calibration factors are often introduced to take into account uncertainties such as stress concentrations, plastic behavior, and manufacturing quality. It is relatively easy to criticize each of the previously defined stress modeling approaches because each has its own individual weakness. However, with proper calibration and rule-based guidance, these models may be successfully used for fatigue life forecasting and design trade-offs.

Hysteresis plot for LCCC

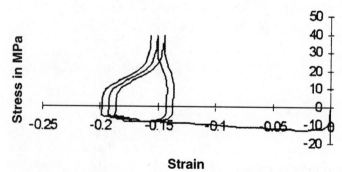

Figure 9.24   Stress-strain history using analytic model.

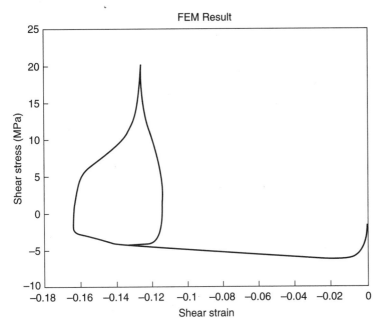

Figure 9.25 Finite element analysis stress-strain history.

As discussed previously, the driving mechanisms for solder-joint fatigue failure can be categorized into global and local thermal expansion mismatches. The global expansion mismatch results from the thermal expansion between the component and the substrate. The local thermal expansion mismatch results from the thermal expansion between the lead material and the solder, as well as the solder and the bond pad on the PWB. The relative contributions of the global and local mismatches to solder-joint failure vary significantly. While the global mismatch normally drives the failure, local mismatch for some lead materials (for example, kovar leads) can significantly influence the useful life. Material properties of the component, lead, solder, PWB, and lead compliance[27,28] must be considered in evaluating solder-joint fatigue. Closed-form stress metrics for fatigue analysis of surface-mount solder joints tend to oversimplify the complex stress states that occur in solder joints under cyclic thermal loading conditions.

To capture the complexity of the solder stress state, more sophisticated analytic methods are required. The most common approach is to use a nonlinear finite element analysis to examine the behavior of the solder interconnect under an imposed temperature cycle. Finite element analysis requires a skilled analyst and can be quite time consuming even on today's computer systems. The analysis is time consuming, especially if a design-of-experiments approach is required to determine the optimum design.

To address these problems, Ling and Dasgupta[29] proposed a simplified general-purpose, stress analysis model for both the global and local thermal expansion mismatch in surface-mount solder joints under cyclic thermal loading conditions. In this approach called the multiple domain Rayleigh ritz (MDRR) methodology, the solder joint is mapped in different domains based on the expected stress distribution. Colonies of nested subdomains are introduced where stress gradients are expected. The displacement fields are modeled using general polynomial-based functions. Polynomial displacement fields are superimposed within each nested subdomain in order to model large local displacement gradients. These enhanced displacement fields are carefully chosen in order to maintain continuity between neighboring domains. Continuity between neighboring materials at the lead/solder interface and PWB/solder interface are enforced as constraints using transformation techniques. Unlike conventional finite element modeling where nodal displacements are degrees of freedom, coefficients of the displacement polynomials are the degrees of freedom in MDRR.

An example of the domain mapping for a BGA interconnect is presented in Fig. 9.26. The reduction in complexity can be readily seen when compared to a finite element model, shown in Fig. 9.27. Since the enhanced displacement fields are only needed at selective regions where stress concentrations are expected, the degrees of freedom contained in a typical MDRR analysis are much less than a finite element model with similar accuracy. The benefits of the MDRR scheme include a simple, quick, and accurate solution of the thermal expansion mismatch problem in surface-mount solder joints, while eliminating the effort of time-consuming finite element model generation.

In the MDRR stress analysis, the elastic-plastic behavior of the solder is considered throughout loading and unloading cycles. The strain energy stored in the solder joint, lead, and PWB is calculated for each of the temperature increments by minimizing the potential energy. As a result, all the unknown degrees of freedom are solved and the incremental strain and stress are calculated. The strain is further categorized into its elastic, creep, and plastic components and is accumulated throughout the loading and unloading history, accordingly.

In comparing the MDRR approach with finite element modeling, it has both computational advantages and provides similar results. As an example of an MDRR analysis, consider a ceramic BGA with a CTE of 6.4 ppm/°C subjected to a 0 to 100°C temperature cycle. The stress profile of the solder ball at the maximum cyclic temperature calculated by MDRR is presented in Fig. 9.28 and by finite element analysis in Fig. 9.29. By comparing these figures, it can be seen there is close agreement between the two profiles.

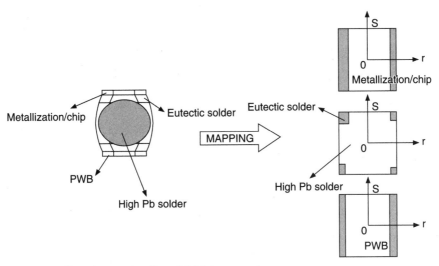

Figure 9.26 Domain mapping for a BGA interconnect.

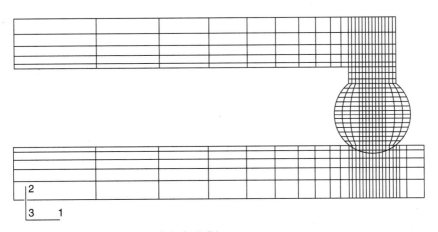

Figure 9.27 Finite element model of a BGA.

9.6.2.7 Vibration-induced stress cycling. As mentioned previously, fatigue failures may be induced by vibration and/or temperature cycling. Vibration loading is characterized by a high number of stress cycles with small amplitudes. Component interconnects are particularly prone to fatigue failure from vibration-induced loads. Failure of materials under vibration-induced loading is typically modeled by the Basquin power law.

For CCAs, test data show that most damage occurs at the fundamental resonant mode where board curvature and stresses are the highest. The stress metric in vibration is the curvature of the PWB,

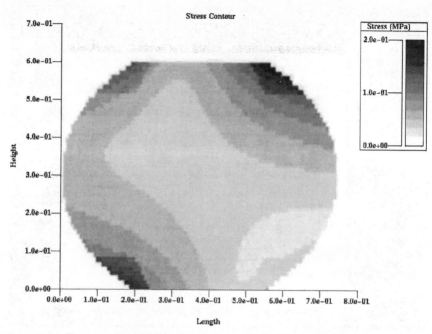

Figure 9.28   Stress profile in BGA solder ball as calculated by MDRR analysis.

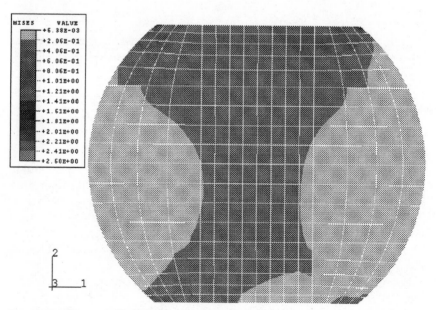

Figure 9.29   Stress profile in BGA solder ball as calculated by finite element analysis.

and under simple-simple loading the maximum curvature coincides with the maximum displacement. Based on these assumptions, Steinberg[11] developed a simple stress metric for vibration fatigue of permanent interconnects. In this approximation, the CCA is treated as a simple spring mass system. Based on this assumption, the maximum displacement (curvature) can be expressed as

$$Z = \frac{9.8G}{f_n^2} \qquad (9.75)$$

where

$$G = 3G_{\text{rms}} = \sqrt{\frac{\pi}{2} Pf_n Q} \qquad (9.76)$$

where P = the PSD at the natural frequency
f_n = the natural frequency
Q = the transmissibility

Calculations for the natural frequency and PSD are discussed in Sec. 9.5. The transmissibility is the amount of energy transferred through the supports to the printed wiring assembly. Researchers[11,15] indicate that the transmissibility (amplification) due to vibration is proportional to the square root of the natural frequency, $\sqrt{f_n}$ and is expressed as follows:

$$Q = A \sqrt{f_n} \qquad (9.77)$$

where A = 1/2 for f_n < 100 Hz
A = 1 for 200 < f_n < 300 Hz
A = 2 for f_n > 400 Hz

Amplification increases at lower temperatures, which is countered by the fact that natural frequency tends to drop as the temperatures increase. Low frequencies tend to result in higher curvatures and potentially higher stresses on the component interconnects.

The stress associated with vibration is produced by the maximum board curvature under component and inertia forces. Inertia forces become important for components with large masses, high centers of gravity, and/or weak board-to-package interconnects. For most components, inertia forces are relatively low and curvature-related stresses are dominant in producing failures. An empirical model for determining the maximum displacement at a component was proposed by Steinberg.[11] This model defines the critical displacement as

$$Z_c = \frac{0.00022B}{Chr\sqrt{L}} \qquad (9.78)$$

where B = the PWB length
C = the component factor
h = the board thickness

r = the position factor
L = is the component length

For random vibration, if the displacement is maintained below Z_c, the components are assumed to be able to survive approximately 20 million stress reversals. Based on this information, and using Eqs. (9.75) and (9.78), the fatigue life based on a vibration-induced load can be forecasted by using the Basquin fatigue relationship[11] as follows:

$$N_f = N_r \left(\frac{Z_c}{Z_o}\right)^b \tag{9.79}$$

For electronic components, it has been suggested that N_r = 20 million cycles and b is 6.4.[11] This method has been shown to yield good results. A more detailed approach would be to determine the stress based on FEA or the application of mechanics of material principles.

9.6.3 PWB metallization

Metallization on a CCA provides the signal, power, and ground interconnections for components. Reliability concerns for PWB metallization include thermal fatigue of plated-through holes, metal migration in the form of conductive filament formation, dendritic growth, and metal corrosion. These concerns are discussed in this section.

9.6.3.1 Plated-through holes. The complexity of modern electronics has led to the development of multilayer PWBs, where interconnections of signal and power between various layers is accomplished with plated-through holes (PTHs). A PTH is formed by drilling a hole through the laminated panel from which one or more PWBs will be extracted. The drilled hole is then plated using an electroless plating followed by an electrochemical plating. In general, it is desirable to have 1-mil-thick copper plating and a resistance of below 1 mΩ on the PTH.

The reliability of PTHs has been addressed by several researchers.[21,30–32] The failure mode of the PTH is an increased electrical resistance or opening due to cracking of copper plating. Cracks may be formed by fracture due to overstress or by fatigue when the CCA is subjected to temperature cycling. Overstress fractures are a particular concern during the exposure of the PWB to soldering temperatures above 200°C.[30,31] Temperature cycling during operation can also result in fatigue failure of the PTH.

Stress in the PTH is developed by thermomechanical deformations that occur during temperature excursions. The thermomechanical deformations result from a mismatch of the out-of-plane (z axis) CTE of the PWB and the PTH plating material. Expansion can produce barrel cracking of the PTH plating, separation of the PTH and surface metallization land interface, and separation of PTH and interior metallization interface. Figure 9.30 illustrates the effect of temperature cycling on a PTH.

Stress analysis of the PTH under temperature cycling must be conducted to evaluate the likelihood of failure. The stress metrics include the tensile stress in the barrel of the PTH, the stress at the PTH/land interface, and the strain range due to the temperature cycle. The stress and strain data can be determined by finite element analysis or by developing an analytic model.[21,30,33] The generalized Coffin-Manson equation is typically used to estimate the fatigue life of PTHs.

A simple approximation of the stress in the PTH can be obtained by applying basic principles of mechanics of materials. Assuming linear elastic material behavior, the force developed in a PTH subject to a temperature cycle can be defined as

$$P = \frac{\alpha_{pth}L_{pth}\Delta t_{pth} - \alpha_{pwb}L_{pwb}\Delta t_{pwb}}{\left\{ \dfrac{L_{pth}}{(A_{pth}E_{pth})} + \dfrac{L_{pwb}}{(A_{pwb}E_{pwb})} \right\}} \qquad (9.79)$$

where L_{pth}, L_{pwb} = the thickness of PTH and board
A_{pth}, A_{pwb} = the area of the board and plating
E_{pth}, E_{pwb} = the modulus of elasticity of the board and plating
α_{pth}, α_{pwb} = the CTEs of the PTH and board

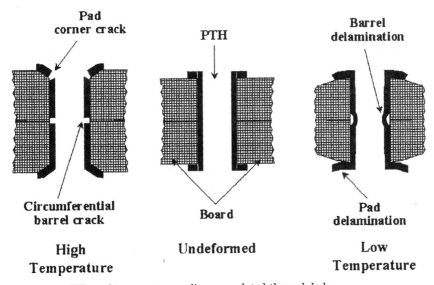

Figure 9.30 Effect of temperature cycling on a plated-through hole.

For board area, Steinberg[21] suggests using the length of the PTH versus the hole diameter. The tensile stress in the barrel is then calculated by

$$S_t = \frac{KP}{A_{\text{pth}}} \qquad (9.81)$$

where K is the stress concentration factor. The stress is then compared to the ultimate tensile strength or used in a Basquin fatigue relationship.

While the stress approximation provides some quantification of the physical situation, the strain range due to the temperature cycling is more appropriate. Models for approximating the temperature cycle strain range have been developed.[33,34] These models were based on approximating the strain range imposed by the thermomechanical expansion of the PTH and the multilayer PWB. Orien[34] chose to simplify this model by characterizing the multilayer PWB and PTH with effective CTE and modulus of elasticity. Bhandarkar et al.[33] chose to consider the individual plating materials used to create the PTH, as well as the individual layers used to construct the multiple layer PWB. From Bhandarkar et al., the total axial deformation of the PTH is defined as

$$\delta = \frac{\sum_i \alpha_{\text{pth}_i} \Delta T_{\text{pth}_i} E_{\text{pth}_i} A_{\text{pth}_i} - \left\{ \sum_j \alpha_{\text{pwb}_j} L_{\text{pwb}_j} \Delta T_{\text{pwb}_j} \middle/ \sum_j [L_{\text{pwb}_j} (A_{\text{pwb}_j} E_{\text{pwb}_j})] \right\}}{\left\{ 1 \middle/ \sum_j [L_{\text{pwb}_j} / (A_{\text{pwb}_j} E_{\text{pwb}_j})] + \sum_i A_{\text{pth}_i} E_{\text{pth}_i} \middle/ L_{\text{pth}} \right\}}$$

$$(9.82)$$

where α_{pth}, α_{pwb} = the CTE of the plating and PWB layers
  A_{pth}, A_{pwb} = the area of the plating and PWB layers
  E_{pth}, E_{pwb} = the modulus of elasticity of the plating and PWB layers
  L_{pth} = total length of the PTH
  L_{pwb} = the thickness of the layers
  i = the individual plating material
  j = the index for the individual layers

The total axial force in the plating is defined as

$$P = \frac{\delta - \sum_j \alpha_{\text{pwb}_j} L_{\text{pwb}_j} \Delta T_{\text{pwb}_j}}{\sum_j [L_{\text{pwb}_j} / (A_{\text{pwb}_j} E_{\text{pwb}_j})]} \qquad (9.83)$$

Equations (9.81) and (9.82) are used as stress and strain states in the PWB layers and the through-hole plating. The area used for the each layer is assumed to be equal. The value of A_{pwb} is defined as a function of material and geometric parameters listed previously with the addition of a calibration constant. The constant is evaluated by a closed-form equation. The function is introduced to account for pad radius, pad thickness, proximity of other PTHs, board thickness, and the radius of the PTH. The calibration coefficients were determined by a two-level, half factorial numerical design of experiments.

The damage metric for thermal fatigue of the PTHs is cycles to failure and the generalized Coffin-Manson equation was used to calculate the value. The axial strain in the plating was determined by dividing the total deformation by the length of the PTH. A numerical solution algorithm calculates the stress and strain states by incrementally changing the temperature to account for the inelastic plastic behavior of the plating and the property changes that result from exceeding the glass transition point of the board layer material. The results of the algorithm were validated by comparison with experimental data. The experimental data was taken from a comprehensive study of hundreds of PTHs with different configurations on polymide kevlar boards. These configurations were modeled using general-purpose FEA and the developed analytic model. The results of this analysis are presented in Fig. 9.31. As can be seen from the figure, the analytic model provides good correlation with experimental results.

9.6.3.2 Metallization corrosion. Corrosion is an electrochemical reaction that results in the deterioration of a material due to its reaction with the environment. Corrosion is best understood in terms of an electrochemical cell. An idealized electrochemical cell is presented in Fig. 9.32. An electrochemical circuit consists of two connected pieces of metal in a conducting solution or electrolyte. Features of an electrochemical circuit include:

- *Anode.* The metal is oxidized (ionizes) and corrodes.

- *Cathode.* Under corrosion, the metal ions react with other mobile ions to form gas, liquid, and/or solid by-products.

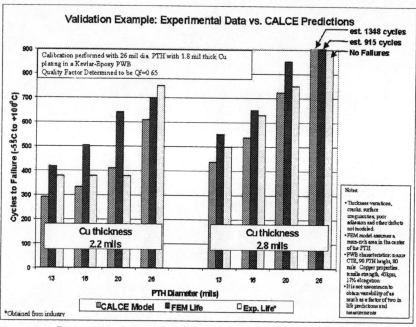

Figure 9.31 Comparison of experimental and numerical estimates of PTH fatigue life.

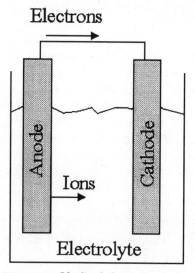

Figure 9.32 Idealized electrochemical cell.

- *Physical contact.* The anode and the cathode must be in electrical contact with each other.

- *Electrolyte.* The conductive medium that completes the circuit. This medium provides a means for metallic ions to travel from the anode and to accept electrons at the cathode. In many cases, water acts as the electrolyte.

.The metal acting as the anode undergoes oxidation and produces metal ions. For CCAs the most prevalent metal is copper and the reaction for copper is defined as

$$Cu \rightarrow Cu^{n+} + ne^-$$ (9.84)

Water located at the anode will also react, releasing oxygen and hydrogen ions. The reaction is defined as

$$H_2O \rightarrow \frac{1}{2} O_2 [\uparrow] + 2H^+ + 2e^-$$ (9.85)

At the cathode, water will react with mobile electrons as follows:

$$H_2O + e^- \rightarrow \frac{1}{2} H_2 [\uparrow] + 2OH^-$$ (9.86)

Metal ions will react with mobile electrons, and for copper, the reaction is as follows:

$$Cu^{n+} + ne^- \rightarrow Cu$$ (9.87)

In electroplating, an imposed voltage is used to drive the reaction, while in corrosion a natural potential is developed. This potential can result from the tendency of metals to give up electrons. Electromotive force (emf), a measure of this tendency, compares the electrode potential of the metal to that of a hydrogen electrode at ambient conditions. A list of electrode potentials for selected materials is provided in Table 9.11.

Corrosion may occur as a uniform attack or a galvanic attack. A uniform attack occurs when some regions of metal in an electrolyte are anodic to other regions. Under uniform attack, the anodic and cathodic regions shift over time. Unlike the uniform attack, in a galvanic attack the anodic and cathodic regions remain fixed. Galvanic reactions can occur when dissimilar metals are connected by an electrolyte. The tendency of a material to have an anodic reaction may be evaluated by examining its position in a galvanic series. A galvanic series lists materials based on their tendency to have an anodic reaction. Materials higher in the list react anodically to materials lower in the list. An abbreviated galvanic series for materials in seawater is presented in Table 9.12. As we can see from this table, solder is anodic to copper and will corrode.

TABLE 9.11   Electromotive Force (emf) (Adapted from Ref. 32)

	Metal	Electrode potential (V)
Anodic	$Ni \rightarrow Ni^{2+} + 2e^-$	-0.25
	$Sn \rightarrow Sn^{2+} + 2e^-$	-0.14
	$Pb \rightarrow Pb^2 + 2e^-$	-0.13
	$H_2 \rightarrow 2H^+ + 2e^-$	0.00
	$Cu \rightarrow Cu^2 + 2e^-$	$+0.34$
	$2H_2O \rightarrow O_2[\uparrow]+4H^+ + 2e^-$	$+1.23$
Cathodic	$Au \rightarrow Au^{3+} + 3e^-$	$+1.5$

TABLE 9.12   Galvanic Series in Sea Water[32]

Anodic	Magnesium
	Zinc
	Aluminum
	Cast iron
	50%Pb 50%Sn solder
	Lead
	Tin
	Cu-40% Zn brass
	Nickel-based Alloys (active)
	Copper
	Cu-30% Ni alloy
	Stainless steel
	Silver
	Gold
Cathodic	Platinum

In addition to dissimilar materials, galvanic reactions can occur between regions of high and low stress, called *stress corrosion.* Under stress corrosion, high stress areas tend to be anodic to lower stress regions. Corrosion of the high stress area weakens the material and cracks may form.

The lack of oxygen also increases the anodic reaction of metals. Because cracks and crevices have lower oxygen concentrations, these areas are particularly susceptible to corrosion. The corrosion of cracks by pitting can further accelerate fatigue failures. This type of galvanic attack is sometimes referred to as *concentration corrosion.*

Corrosion in CCAs can produce failures at multiple sites. Copper, which is used in most PWB designs, is extremely susceptible to corrosion from chlorides and sulfides. In addition, copper oxidizes relatively quickly. Surface corrosion can increase surface film resistance and produce by-products that accelerate wear. The effect of surface corrosion is discussed in more detail in Sec. 9.6.4. At present, the effect of corrosion on signal distortion is not well documented. Another corrosion failure risk in PWBs is the reduction of insulation resistance between adjacent conductors. Insulation resistance is lost by the formation of conductive bridges, and corrosion plays a significant role in this area. Failure under this condition may be attributed to dendrite growth or conductive filament formation.

9.6.3.3 Conductive filament formation and dendritic growth.

Conductive filament formation and dendritic growth are related to metallization corrosion. Each can result in the catastrophic failure of CCAs due to the sudden loss of insulation resistance between adjacent conductors. Both conductive filament and dendritic growth involve the migration of metal between adjacent conductors. This phenomenon has been observed to occur in CCAs.[35–37] Dendritic growth is a surface phenomenon where metal migrates between adjacent conductors. Figure 9.33 depicts separate paths for dendritic growth.

Conductive filament formation is more insidious because it occurs within the woven laminate. In conductive filament formation, a filament of metal grows along the fiber/resin interface in a PWB between adjacent conductors. This phenomenon has been readily observed during humidity/temperature testing at 85°C/85% relative humidity. Conductive filament paths are depicted in Figs. 9.34 through 9.36. In both conductive filament formation and dendritic growth, metal migration requires the presence of mobile metal ions, a migration path, and voltage potential. A sign of failure is loss of insulation resistance between adjacent conductors.

Interfacial failures resulting from dendritic growth are suspected to be caused by hydrolysis. In dendritic growth, a highly localized anodic reaction produces a plumelike structure that extends through the cover coat between the board surface and a region of metallization. The erupted area is the focus of the leakage current. A degradation of the cover coat follows, resulting in exposure of a cathodic conductor. Ultimately, failure results from a short circuit between the adjacent conductors. Surface failure paths may also result from blistering (delamination), which may also open a path for dendritic growth. Dendritic growth and blistering are thought to be results of faulty processing and improper handling.[35]

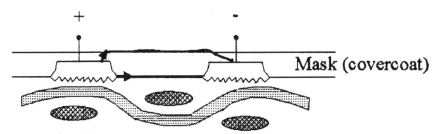

Figure 9.33 Dendritic grow between adjacent conductors.

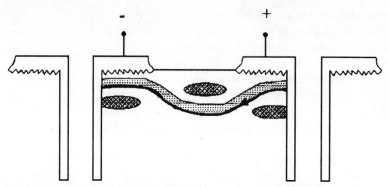

Figure 9.34   Conductive filament formation between adjacent PTHs.

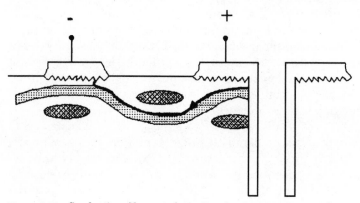

Figure 9.35   Conductive filament formation between a trace and an adjacent PTH.

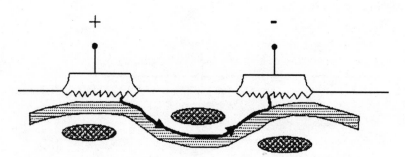

Figure 9.36   Conductive filament formation between adjacent traces.

Studies into conductive filament formation began in the mid-1970s.[35–40] Stress drivers included humidity, temperature, and voltage bias. In initial studies, the migration path was thought to occur due to the breakdown of the interface between the woven fiber and resin matrix due to thermomechanical loading. This hypothesis was supported by studies that showed a 35% reduction in time to failure observed for boards subjected to thermal shock (1 min at 260°C).[38] It was also found that some laminate materials were more resistant to this type of failure than others (for example, triazine showed a marked resistance to this type of failure[40]). Recent studies have indicated that better treatment of the glass fibers and high quality resin can be used to dramatically reduce the likelihood of conductive filament failures.

Another important finding is that the conductive filament formation failures do not require continuous bias.[40] Time to failure of unbiased conductor systems and continuously biased conductor systems did not show significant differences. This suggests that, given the opportunity, conductive filaments can form quickly.

The results of studies into conductive filament formation indicate that failure should be modeled as a two-step process. As a result, the time to failure may be defined as

$$t_f = t_1 + t_2 \tag{9.88}$$

where t_1 is the time to establish a migration path and t_2 is the time to form a conductive path between the adjacent conductors. Models for estimating the time to failure have been proposed by several researchers. From analyzing the collected data, Welsher et al.[40] suggested the following:

$$t_f = a\,(H)^b \exp\left(\frac{E_A}{RT}\right) + \frac{d\,l^2}{V} \tag{9.89}$$

where a, b, E_A, and d are material-dependent.

Tests conducted by Rudra and Jennings[41] suggested that the following empirical equation models the data:

$$t_f = \frac{af\,(1000\,kL)^n}{V^m\,(M - M_t)} \tag{9.90}$$

where L = the spacing between adjacent conductors
  V = the voltage bias between the conductors
  M = the moisture within the laminate
  M_t = the moisture threshold
  a = the function of the material, geometry, and laminated coating
  k = related to the conductor geometry
  n = the shape acceleration exponent
  m = the voltage acceleration exponent
  f = the multilayer connection factor

Failure (that is, growth) was found not to occur for moisture contents below a threshold level. The Rudra and Jennings study included the six-layer laminate test specimens constructed of FR-4, BT (bis-maleimide triazine), and CE (cyanate ester). Each specimen had two test areas that contained various metallization geometeries. In addition, test specimens were not coated, coated with a solder mask, or coated with a postcoat.

While fiber-resin delamination provides a migration path, recent studies[42] indicated that a more insidious path exists in the form of hollow fibers. The topic of hollow fibers will be discussed in Sec. 9.7, which addresses manufacturing defects. Dendritic growth and conductive filament formation can be controlled by proper design, material selection, and process control. This supply chain must be considered while addressing conductive filament formation concerns.

9.6.4 Contact and connector failures

Separable interconnects presents another reliability concern[43–45] for CCAs. In most cases, CCAs represent a replaceable item for electronic systems. Separable interconnects provide modularity in electronic systems and allow for field replacement and upgrades. Separable interconnects related to CCAs include sockets for packaged electronic devices, edge connectors for board to motherboard interconnects, and pin-in socket for wiring and peripheral interconnection. The advent of mobile electronics has also lead to the increased use of separable interconnects for peripheral devices.

The function of electrical connectors is to provide electrical interconnections with minimal electrical resistance. Contact resistance is defined by Holms law as

$$R_{\text{contact}} = R_{\text{cr}} + R_f \tag{9.91}$$

where R_{cr} is the constriction resistance and R_f is the film resistance. Constriction resistance, R_{cr}, is a function of the contact area and the resistivity of the mated contacts. The film resistance, R_f, is a strong function of the contact load. Oxide films can result in up to a 1000 times increase in the contact resistance.[44] For high-reliability applications, a gold surface finish is often used because it provides good electrical characteristics and resistance to corrosion.[43,44] The contact resistance for clean gold surfaces is typically below 1 mΩ.[43]

Connector reliability issues focus around the increase in contact resistance at the metal-to-metal interface. The contact resistance is a function of the materials, the geometries, and the contact area. Changes in the contact resistance can be attributed to a change in contact area and materials due to wear, stress relaxation, and corrosion.

9.6.4.1 Wear. Wear of connector contacts occurs due to the relative motion between two surfaces. This motion can be caused by mating cycles, vibration, or temperature cycling. The wear can occur by adhesion, abrasion, or brittle fracture. Adhesive wear occurs when there is transfer of metal. In adhesive wear, bonds form between the contacting members that are stronger than the cohesive strength of the metal which leads to transfer and wear as sliding continues. Abrasive wear results from the plowing of the surface by an opposing member that is substantially harder and is generally a result of contact misalignment. Finally, fracture wear occurs with brittle plating, especially when the substrate is easily deformed. The surface develops cracks during sliding which may result in loss of coating. Wear may be combatted by reducing the coefficient of friction between the contact surfaces. Lubricants, such as graphite, may be used to reduce friction and wear.

Wear reduces contact force, which increases contact resistance. Fretting wear can occur when contact surfaces under rapid motion wear through surface plating, which may cause intermittent failures.[21] Cracks form in the contact surfaces and oxides form as a result of local heating. The local oxide regions are nonconductive and result in intermittent opens. This problem is a particular concern for high vibration applications.

Wear may be model based on the applied normal force, relative motion, and material properties. Bayer et al.[46] developed an empirical model for estimating the depth of a scar caused by sliding. In this model the scar depth was given by

$$h = KP^m S^n \qquad (9.92)$$

where h is the scar depth; P is the contact force; S is the amount of sliding; and K, m, and n are related to the metallurgy. In his study, Bayer et al.[46] found that n was approximately 0.2 regardless of the metallurgy. The value of m was found to be inversely proportional to the metal thickness. The wear coefficient, K, was found to be proportional to the friction coefficient.

As mentioned previously, electrical contacts are often plated with gold because of its exceptional properties and resistance to corrosion. Due to the insertion/withdrawal cycles the connector undergoes during its life, the gold plating wears away and exposes the underplate or the base metal. When this occurs, the contact resistance increases due to film formation and corrosion on the nonnoble underplate or base metal. Failure is defined as the number of mating/unmating cycles required to penetrate and expose the nickel underlay. An example of wear on a gold-plated PWB contact is depicted in Fig. 9.37.

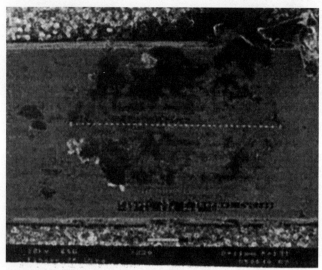

Figure 9.37 Wear due to temperature cycling.

9.6.4.2 Stress relaxation.

Stress relaxation of materials used to maintain the contact force in connectors is another area of concern. Contact resistance is a strong function of the normal force between the metal surfaces of a connector system. Based on the connector design, the contact force may be actively applied or be a part of the connector system. It is common for designs to use a cantilever spring to develop and maintain the normal force. Once mated, the spring system is under a continuous load, which provides the opportunity for materials to yield and thus reduces the contact force. The yielding of a material under a continuous load is termed *stress relaxation.* Copper alloys are used extensively in connectors because of their electrical conductivity, strength, and corrosion resistance.[47] The alloy system and heat treatment can produce variations in each alloy's ability to resist stress relaxation. A graph of several copper alloys is presented in Fig. 9.38.

In general, a high normal force is desired to provide a low and stable contact resistance. However, the normal force provided by the connector spring decreases with time and temperature due to stress and relaxation in the material of the spring. When stress relaxation occurs, the normal force load decreases, leading to a decrease in the area of the circular spot and an increase in the constriction resistance. In addition, the reduction in force may allow relative motion to occur between the contact surfaces. Thus, stress relaxation may permit fretting corrosion.

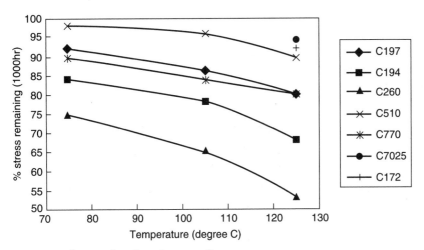

Figure 9.38 Stress relaxation of copper alloys.

9.6.4.3 Corrosion. Usually a copper alloy is used as the base metal for contact pins and springs. They are plated with precious metal to increase corrosion resistance. The gold contact plating typically will be in the range of 0.4 to 1.3 mm thick. However, despite having noble metal plating, corrosion is still a problem due to wear or the porosity in the plating. Porosity is a function of the plating process used, and for a given process the porosity is directly proportional to the thickness of the noble metal plating. When the plating is thicker, the porosity is lower. Porosity exposes the nonnoble underplate or base metal to the environment, which leads to its corrosion. These corrosion products creep out of the pores and spread over the noble metal plating. The corrosion process has been discussed previously in this section.

Corrosion affects contact resistance in several ways. One problem is that corrosion can attack at the periphery of any or all of the contact spots. This causes the constriction resistance to increase as the contact spot area and the spot distributions are reduced in size. Another related problem is that nonconductive corrosion products could form in the spaces between the contact spots. Corrosion products can produce intermittent failures when a contact spot comes in contact with a corrosion product, and then returns to its original position. Alternatively, the contact might ride up on the corrosion product and remain there. Finally, when the contact spot moves back, it might carry the corrosion product back with it. Increased contact resistance is a problem with corrosion products.

9.6.5 Summary of CCA Failure Mechanism and Models

In this section, common CCA failure mechanisms and models have been discussed. To completely address reliability, the impact of manufacturing and assembly must be understood. Manufacturing and assembly may introduce defects which could reduce the life of the CCA. Even a perfectly designed CCA can be unreliable if manufacturing defects cause early life failure. The next section discusses manufacturing and assembly defects.

9.7 Reliability Issues Related to Manufacturing and Assembly Defects in CCAs

The manufacture and assembly process for PWBs and CCAs can represent several reliability concerns. The concern related to manufacturing and assembly is the introduction of defects. The implication of these irregularities to long-term service reliability is crucial. Most defects fall into three major categories: intrinsic material defects (including improper material properties and cracks), improper geometry (including incorrect size, incorrect thickness), and improper location (including misregistration errors).[48] The presence of defects can increase the susceptibility of a CCA to potential failure mechanisms. Defect magnitudes can be addressed in terms of their effects on time to failure or on performance parameters. While the basic geometry, material, and operation parameters determine a nominal design, the worst-case defect magnitudes determine the defect-related reliability risk. While some defects exist in all CCAs within the limits of the statistical process control, the allowable defect magnitudes can be derived based on the life-cycle stress profile and mission life requirements.

Efficient application of screens can detect defects introduced by the manufacturing and assembly processes. Although it should be noted, some defects may not be screened. Defect types and locations are needed to design effective product screens. CALCE Electronics Products and System Center has a detailed on-line resource[48] for defects related to the assembly of PWBs. Cost-effective screens are determined by assessing the effects of the worst-case defect magnitudes during accelerated tests. The effectiveness of a screen is directly related to the defect for which the screen is designed. Broad or improperly applied screens may not catch intended defects and may reduce the useful life of the CCA. Some screens induce damage to the CCA, including coupon testing and stress screening. Coupon testing allows for destructive testing of samples subjected to the same fabrication process. Stress screens typically subject the CCA to a defined amount of temperature cycling and vibration. Although frequently used, stress screens are costly and have questionable effectiveness. Visual, x-ray, and infrared inspection offer a nondamage-inducing evaluation process.

To understand defects, it is important to understand the fabrication process of a PWB. A PWB is generally constructed by laminating dielectric cores together. The dielectric cores typically consist of an encased woven glass fabric which is then encased in an epoxy resin and sandwiched between two copper foils. To save materials and time, the dielectric cores come in standard rectangular sizes (typically 18 by 24 in) and are sometimes referred to as "panels." As a result, multiple PWBs are typically laid out as a panel. A circuit is formed by removing unwanted copper on the panels by subjecting the treated panels to an etching path. For PWBs with multiple layers, panels are stacked together with a resin prepreg inserted between the individual panels with a copper foil placed above and below the prepreg at the top and bottom of the stack. This stack is sometimes referred to as a "blank." Multiple blanks separated by aluminum are put together to form a book, and the layers are laminated in an evacuated elevated temperature press. Once laminated, the blank is drilled, processed, and etched. The individual PWBs are then extracted from the processed blank in a depanelizing process. After further post-processing, the PWBs are ready for the assembly process.

Under good process control, PWBs and assemblies rarely exhibit production-related failures.[49] When failures occur, they can generally be traced to defects or deficiencies in materials or the manufacturing process. Throughout this section, defects related to PWB fabrication and the production of PWB assemblies will be discussed as they relate to different steps in the manufacturing and assembly process. The defects related to CCAs can be categorized by physical structures that occur in circuit card assemblies. These categories include PWB laminates, plated-through holes (PTH), and soldered interconnects. Defects related to each of these structures are discussed in the following sections.

9.7.1 PWB laminates

Reliability concerns related to laminate typically are related to loss of insulation resistance, reduction of mechanical strength, and dimensional stability. Defects occurring in PWB laminates include

- Delamination
- Hollow fibers
- Laminate voids
- Plating slivers

These defects are discussed in this section.

In terms of PWBs, *delamination* is the separation of dissimilar materials that are bonded together to form the board. Delamination can occur between the resin and fiber bundles or the resin and conductive metallization. Delamination can often be observed as a change in color in the delaminated area. Figure 9.39 shows a cross section of the delamination of the resin from the fiber under magnification. The delamination can be seen as shadow images around fibers at the bottom of the fiber bundle. Delamination of the fiber bundle and resin may occur due to the starvation of epoxy in the glass cloth layers, contaminants present during the coating process, and/or incomplete resin curing. The CTE of epoxy resin is approximately 12 times as large as the CTE of the glass fibers. As a result, epoxy resin/glass fiber separation may arise from mechanical stresses due to the mismatch in the properties of epoxy resin and glass fibers. Temperature extremes encountered under reflow soldering present a particular concern. While generally lower, temperature cycling during usage may also be a concern. In addition, the difference in moisture absorption of glass fibers and epoxy resin can cause interfacial stresses. Metallization delamination can result from the board fabrication process due to excessive pressure and temperature or insufficient bonding of the laminate resin. An example of delamination between plated metal and epoxy fiber core is depicted in Fig. 9.40.

Measling and crazing are two forms of delamination where the glass fibers separate from the resin at the weave intersection. Both measling and crazing appear as white spots or "crosses" below the surface of the base material and may appear locally or over a large area. Measling differs from crazing in that it is thermally induced, rather than mechanically induced (crazing). Other factors that influence measling and crazing include moisture in the laminate and improper curing of the laminate.

Depending on the application, there are different board-acceptance criteria for measling and crazing. Ordinarily, measling and crazing are considered to be merely cosmetic defects with no documented board failures due to either measling or crazing. Crazing is generally considered unacceptable, because the almost continuous epoxy separation decreases dielectric strength. Some studies consider measling unacceptable in high humidity usage, because of the possibility for growth of metallic dendrites through the insulator, seriously reducing the dielectric strength. Gross delamination is cause for rejection, because if the resultant cavities fill with moisture or electrolyte, reduced insulation resistance and/or mass-soldering blowholes can occur. Delamination also provides an opportunity for conductive filament formation (CFF).

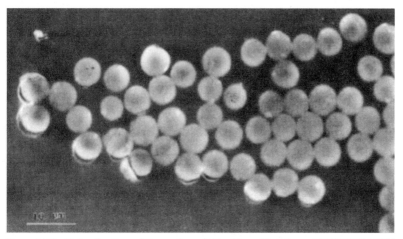

Figure 9.39 Delamination of the resin and the fiber. (Courtesy of CALCE EPSC.)

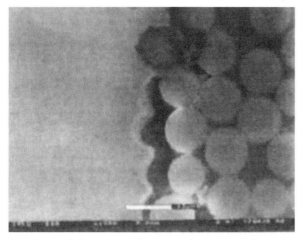

Figure 9.40 Delamination between the resin and copper. (Courtesy of CALCE EPSC.)

As discussed in Sec. 9.6.3, hollow fibers may also provide an opportunity for conductive filament formation to occur. Hollow fibers are assumed to arise from the fiber fabrication process. This fabrication process consists of four fundamental steps: conversion of raw materials into molten glass, fiber drawing, fabric weaving, and resin coating. In the drawing process, molten glass is fed through a metal plate with tiny holes (bushing) and cooled to form thin glass fibers. Hollow fibers are suspected to form during the glass formation process when gas bubbles become trapped in the molten glass as it is drawn through the bushing. Gas bubbles may arise from trapped air or impurities in the raw materials used to form the glass. If the trapped gas bubbles do not cause the fiber to break, the entrapped bubbles form capillaries in the fiber.

Researchers at CALCE have described a process for identifying hollow fibers by examining the exposed bare glass bundle matrix of a laminate.[50,51] In this process, a sample is excised from the laminate and the surrounding resin is burnt off the sample. The edges of all sides of the test specimens are sealed to prevent wicking (capillary action of a fluid into a hollow fiber), and placed in refracting oil overnight. The oil is selected to have a refractive index that matches the glass fiber. Light, which is directed onto the sample, travels freely until it hits a hollow fiber (air), and the change in the refractive index at the fiber/air boundary will be partially reflected. The use of a microscope with a camera attachment to identify the hollow fibers is recommended, despite the fact that hollow fibers are visible to the naked eye.

The presence of hollow fibers is a particular concern in the formation of fine-line printed circuit boards, laminated multichip modules, and plastic ball-grid arrays used in high reliability applications. In these applications, even a short hollow fiber can connect two conductive elements. Since the presence of hollow fibers is likely to increase the chance of conductive filament formation failure, the use of materials with any hollow fibers is not recommended. Further reading on hollow fibers can be found in the references.[52-56]

Laminate voids are defects that occur in multilayer PWB lamination and appear as voids or separations within the laminate material. Generally resulting from thermal stress heat contraction of the epoxy, voids may be adjacent to inner-layer copper foil or a PTH barrel. Void formation may arise from improper curing or flow of the resin, air entrapment, insufficient prepreg, and resin outgasing. Voids may be detected through visual inspection when the defect is gross or very close to the surface of the board. A vertical cross section of the PTH is frequently required to reveal these defects.

Plating slivers are etching defects occurring when unsupported overhang on the surface of the PWB fractures and results in small metallic particles or slivers. An excessive amount of either etch undercutting or edge overhang increases the probability of slivering during subsequent board assembly procedures and during use. Most PWB specifications require that all tin/lead–plated boards, based on epoxy glass, be reflowed to eliminate the tin/lead overhang along the conductor edges or pad rims. In general, there should be no evidence of overhang after etching.

The probability of slivering can be determined by transverse sectioning through representative circuit traces. These traces are examined using an optical microscope to determine the amount of etch undercutting or conductor overhang present. It is also possible to determine the extent of conductor overhang or etch undercutting by comparing the conductor width and the master line width on the 1/1 master pattern at the same location on the printed wiring diagram. Examples of techniques used to determine the amount of undercutting or overhang on PWB circuit traces can be found in the *Electronic Materials Handbook*.[57] Plating slivers are considered to be major defects because they can bridge the dielectric between traces and result in short circuits or dielectric breakdown.

9.7.2   PTH defects

Plated-through holes (PTHs) provide electrical interconnectivity between selected layers of CCAs consisting of more than a single layer. PTHs are normally created by a drilling process. During the drilling process, the blanks are pinned in stacks on tooling plates and drilled on multispindled drill machines. The computerized drill program determines hole locations, and automatically changes drill sizes when a drill size completes its path. To ensure complete drilling and proper alignment with the inner layer pads, the holes may be examined through x-ray and visual inspection. After these inspections, any burrs or resin smear covering inner layer connects are removed. This process slightly roughens the hole and allows for subsequent plating. The hole is plated by copper deposition under an electroless process. The thickness of the electroless layer is typically 80 to 100 millionths of an inch thick. After a number of processing steps, the barrel of the hole is electroplated with copper to a thickness of 1 mil. This is generally followed by plating of either tin or tin/lead. Potential defects in this process resulting from the drilling and desmearing process[58] include

- Etchback problems
- Misregistration
- Nail heading
- Plating folds and nodules
- Plating cracks at the PTHs
- Plating voids and thickness variation
- Resin smear
- Wicking

These defects are discussed in this section.

A desmearing process is used to clean the hole and remove resin that has covered interior metallization resulting from the drilling operation. The desmearing process can result in etchback or the removal of the laminate resin and the woven glass fabric that is used to reinforce the resin from the unplated hole. The desmearing procedure can cause the edges of the internal copper foil to project into the hole. Normally the plating of the hole makes contact with three surfaces of the internal foil: the top of the foil, the vertical edge, and the bottom surface. The proper etchback can increase mechanical strength and ensure good interconnection between the inner layer foil and the copper wall plating.

Excessive etchback of base material can result in unacceptable, nonuniform plating in the hole which does not meet the specified requirements. Inadequate etchback can result from improper curing of the laminate, hardening of the resin smear, insufficient agitation of the desmearing bath, improper bath temperature, and imbalance of chemicals in the desmearing bath. Reliability concerns with etchback include cracking of metallization and exposure of metal in the PTH.

Misregistration is the maximum amount of variation between the centerlines of all terminal pads within one PTH and occurs during board fabrication. It is directly attributed to problems involved with the production of the artwork and their materials, the setup procedures during lamination, or with the dimensional instability of the laminate used. Improper or careless registration of the artwork, improper assembly of blanks, loss of dimensional stability, and worn or deformed holes in the artwork may result in misregistration. Misregistration may be detected by x-ray inspection of the PWB. The amount of misregistration is determined by calculating the difference between centerlines of terminal pads that are shifted to extreme positions. The consequences of misregistration can be very serious, because violation of the minimum clearance could induce a possible short.

Folds and nodules in the plating of a through-hole present a quality control concern. These structures can be caused by rough or improper drilling, loose fibers, and/or insufficient cleaning. Incomplete curing of the resin may also cause folds and nodes. Nodules alone may not be cause for rejection, but they can cause other problems. For example, when a pin gauge is inserted into the holes to verify hole sizes, the nodules can be broken loose, creating voids. Additionally, insertion-mount interconnects may not allow the detection of voids caused by the dislodging of nodules. Later, outgasing of the laminate may occur during soldering. Figure 9.41 shows an example of plating folds where the plating, which is shown as a shade in this image, is very uneven and has numerous bulges.

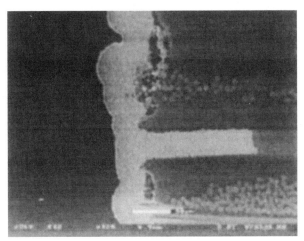

Figure 9.41   Example of plating folds. (Courtesy of CALCE EPSC.)

Nail heading is another indication of poor drilling. Nail heading results in flaring of inner layer metallization and is attributed to inadequate cutting of the metal by the drill bit. Dull or worn drill bits and improper drill speeds may cause this defect. The mechanical deformation of the metallization causes hardening and increases the risk of cracking. Figure 9.42 shows a microsection of a PTH in which the internal pads develop a cross section corresponding to a nail with a conical head. In this image, cracks in the copper are visible. For multilayer boards, nail heading can increase the internal copper to one or one-half times the thickness at the hole edge. Nail heading is usually not serious in itself, but in extreme cases, internal pads on multilayer boards may be ripped out by the drill.

In addition to nail heading, resin smearing is often found on the internal PTH structure. Resin smear is formed when the heat generated by drilling of a PTH causes the resin of the board to flow and then harden on top of the exposed copper layers. Resin smear may be caused by improper drill speeds and/or incompletely cured laminates. Smearing presents a reliability concern related to plating. Resin smearing and copper oxide formation may cause interconnection separation and/or marginal interconnections on the inner layer copper foil that contacts the plated copper. Partial and marginal interconnects may fail or be weakened further by the soldering process. Resin smearing can be revealed during microsectioning a suspect PTH. The vertical and horizontal microsectioning of test coupons may be conducted to assess the potential for inner layer disconnections due to resin smearing. If any irregularities are detected between the land and the copper

Figure 9.42 Example of nail heading. (Courtesy of CALCE EPSC.)

plating, it is likely that a problem exists. However, if there is no visible spacing between the plating and the land, the interconnection is acceptable. The impacts of resin smear are very serious on multilayer boards, because it prevents a reliable plating connection between the inner layers and the hole wall. If there is a separation at the interconnection, the PWB will not function electrically, or it may exhibit an intermittent electrical "open." Any interconnection separation that is detected could result in a failure. In Figure 9.43, the vertical cross section shows a resin smear separating the layer foil from the plating.

Cracking plated copper at the knee of the PTH or within the PTH barrel can compromise the reliability of a CCA. Since the CTE of the copper plating and the resin system in the PWBs can differ by a factor of 13, stress is exerted on the plated copper in the plated-through holes in the z axis. Models for approximating this stress were presented in Sec. 9.6. Stress cracks can result from inadequate ductility and poor bonding between inner layer metallization and the electroplated copper. Cracks, which support the development of open circuits during the wave soldering operation, can readily be seen in the microsection after the board has been subjected to thermal stress testing. Figure 9.44 shows examples of cracks in the PTH plating.

Voids and variation in plating thickness are an additional concern with PTHs. Voids and plating thickness variation can result from lack

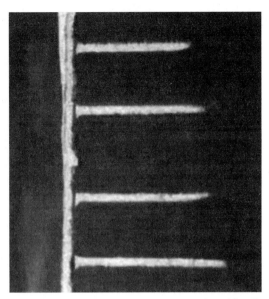

Figure 9.43 Example of resin smear and nail heading. (Courtesy of CALCE EPSC.)

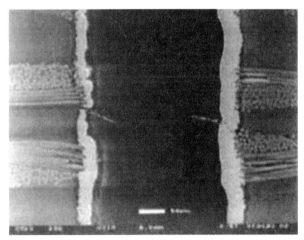

Figure 9.44 Cross section of a via with plating cracks. (Courtesy of CALCE EPSC.)

of proper agitation in the electroless deposition line, insufficient current or time in the plating bath, and/or the presence of contaminants or air bubbles in the hole during the plating process. While the presence of small voids does not necessarily present a reliability problem, voids can represent stress concentration areas. In addition, an isolated area with a thickness of 0.8 mil on a typical minimum thickness of 1 mil of electrolytic copper may be considered a reliability risk. The presence of excessive plating voids may cause open circuits and roughness in thin plating, resulting in outgasing of the laminate during the soldering operation. The impacts of insufficient copper plating thickness can be serious if a specific amount of metal is required to carry the functioning current. Furthermore, the difference in z expansion between the epoxy and the copper may cause the thin plating to crack. Figure 9.45 shows severe variations in copper plating thickness, resulting in open circuits.

The extension of copper from a PTH along the fibers generated by manufacturing defects is referred to as *wicking*. Wicking can occur along delaminated regions of a fiber bundle and may be caused by etchback or improper control of chemical cleaning. Wicking can be serious if it extends sufficiently to deter the dielectric strength or cause internal resistance breakdown between PTHs. Further, wicking provides a convenient starting point for conductive filament formation testing criteria on final fabricated board and inner layer testing. Figure 9.46 shows an example of wicking of copper into the fiber bundle.

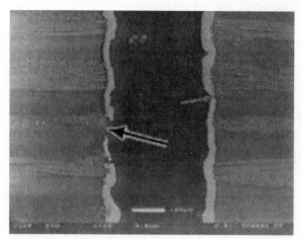

Figure 9.45 Example of plating thickness variation. (Courtesy of CALCE EPSC.)

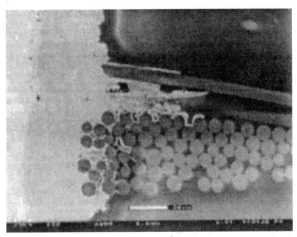

Figure 9.46 Example of copper wicking into the fiber bundle. (Courtesy of CALCE EPSC.)

9.7.3 Solder joint defects

As discussed in Sec. 9.6, cyclic forces caused by CTE mismatches between the package and the board material and vibration are known to produce wearout failures in soldered interconnects. Defects reducing the joint strength, therefore, are of particular concern and are often related to the "solderability" of the bonding surfaces (for example, PWB pad, component lead, or component pad). Solderability (that is, the ability of leads to form a strong solder joint) is dependent on the wetting angle of the lead/solder joint pair, the lead geometry (for example, shape coplanarity), the solder temperature, the solder process, and the metallurgy of the bonding surfaces. A more detailed discussion of the soldering process may be found in Ref. 59. Oxides and corrosion on bonding surfaces, intermetallic consumption of bonding surface coatings, and impurities and particulate contamination on the bonding surfaces can reduce solder strength. Corrosion creates soldering problems since solder only forms a strong bond with pure metal or alloys—not a corrosive residue or oxide layer. Adequate cleaning of the leads is required if corrosion products are present on the lead. Improper treatment can result in poor solder-joint quality. Defects related to forming solder joints include [49,57,60,61]:

- Cold solder joints
- Dewetting
- Excessive intermetallics

- Nonwetting
- Solder balling
- Solder bridging
- Solder voids
- Tombstoning

Defects related to soldering are discussed in the remainder of this section.

Cold solder joints are poorly formed joints that often possess a grayish or porous appearance. Cold joints are caused by an improper reflow process resulting from insufficient heat, inadequate cleaning, or excessive impurities in the solder. Cold joints are mechanically weak and produce early failure due to overstress and fatigue.

Dewetting is a lack of adhesion between solder and the board or package (lead) metallization. Dewetting is caused by a pulling back of the solder on the surface into irregular mounts, high reflow temperatures, and prolonged dwell during the reflow process.[57] These mounds can vary from barely noticeable to having metal exposed between them. Dewetting involves a gas evolution during exposure of the parts being soldered to the molten solder. The gas results from thermally degraded organic material or the release of water from inorganic material. Water vapor at soldering temperature is highly oxidizing and may cause oxidation of the molten solder film surface or subsurface interface. Gas evolution may also be seen where exposed intermetallics exist.

Intermetallics on the lead, which may form between the lead base and the finish, can degrade solderability by consuming the solderable element of the lead coating. Hot-dipping of leads and other high temperature failure accelerators will speed up this phenomenon. An especially susceptible lead base–solder combination is a non-nickel undercoated copper alloy base with a tin or tin-lead solder. Copper-tin intermetallics are not solderable. In addition, the amount of solderable tin is reduced because of the chemical reaction, thereby reducing the solder plating thickness. Nickel undercoatings reduce this intermetallic formation. In addition to the soldering process, intermetallics can occur as a result of prolonged exposure to high temperatures. Intermetallics can reduce the strength of the interconnect and may produce voids in the solder. The gross effect of intermetallic growth is depicted in Fig. 9.47. In this case, intermetallics resulted in voiding due to the application of a high tin solder.

Misaligned solder joints are inaccurate positioning of the package interconnects (lead/or bond pads) over the board bond pads where the solder joint forms. Misaligned solder joints can occur due to

Figure 9.47 Solder separation due to intermetallic formation. (Courtesy of CALCE EPSC.)

insufficient tack in the solder paste, poor lead formation, and improper package placement. Solder paste tack is affected by temperature and humidity. Reliability issues with respect to misalignment are related to weak joints which can produce early failures due to fatigue and overstress.

Nonwetting is the case where there is no adhesion between all or part of an interconnect surface and can result from the presence of a physical barrier between the base metal and the solder or a temperature that is too low to allow the solder to wet metallurgically.[57] There is no intermetallic bond between the solder and the base metal in nonwetting. In addition, a surface deposit caused by epoxy or paint on the lead may prevent melting or alloying. An oxidized coating of sufficient thickness may also prevent the melting or alloying. The effective solder area is reduced and also introduces film defects that can act as strong stress risers, which could turn into fracture-initiation sites. The heel area of the joint is most susceptible to cracking because of nonwetting.

Solder balling is the formation of extraneous small solder balls due to lack of cohesion of the solder during the joint formation process. The presence of oxides on the soldering surfaces and/or inadequate or excessive drying of solder paste can cause solder balling. Solder balling can produce short circuits due to undesired bridging.

Solder bridging is the formation of unintended interconnections between adjacent pads and/or leads. Excessive solder paste and inadequate solder masking can cause bridging. Bridging produces short-circuit failures by creating unintended conductive paths between adjacent conductors.

Solder voids are the holes and recesses that occur in solder joints. Voids can be caused during the solder-joint formation by inadequate reflow time, gas evolution from process contaminants, improper solder paste compositions, and interconnect metallization. Voids can also occur because of the growth of intermetallics due to material aging. Failure concerns related to voids are reduced strength, increased stress concentration, and reduced fatigue life.

Starved solder joints are joints that have a lower-than-normal volume of solder between intended interconnection terminals. Starved solder joints can be caused by component lift, insufficient coplanarity of component, insufficient paste, and improper preheating. Gross failure can be observed as open joints. Reliability issues are related to reduced mechanical strength and reduced fatigue life. Figure 9.48 depicts a starved joint resulting from the component lift.

Tombstoning is the flipping of a package onto one side during the reflow process. Tombstoning can occur as a result of uneven past deposition and improper preheating of the assembly. Tombstoning may result in a gross disconnect for electronic terminations on a package. An illustration of tombstoning is presented in Fig. 9.49.

Icicling is another defect related to soldering that may degrade the reliability of a PWA. *Icicles* are excessive solder in vias or through-holes that forms a point when it cools. Dross and contamination are the two main causes of icicles. In wave soldering, contaminants can cause defects in the solder. The contaminants are the result of the dissolution of plating materials, such as zinc and cadmium, from the surrounding hardware. *Dross* is the formation of oxides in the solder. Dross increases the surface tension in the solder. Icicles can cause solder bridging between pads and lead to board shorts.

9.7.4 Summary

In this section, defects related to the manufacture and assembly process for PWB assemblies have been presented. Defects may result in low yield or cause potential reliability concerns in field product. Although defects are generally low, designers and analysts should have a general understanding of the type and locations of defects. This understanding is particularly important when developing a screening process for PWB assemblies. The selection of qualified manufacturers

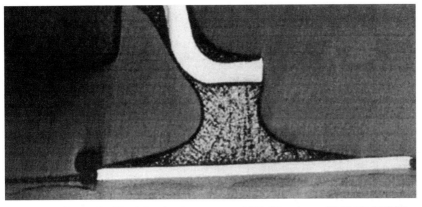

Figure 9.48 Starved solder joint due to component lift. (Courtesy of CALCE EPSC.)

Figure 9.49 Illustration of tombstoning.

and review of standard production procedures can aid engineers in reducing and eliminating defects.

9.8   Example 1: Military-Designed Circuit Board

To illustrate how to perform a PoF analysis on CCAs, three case studies are presented. This section will be devoted to the analysis activities involved in performing a PoF analysis for a military-designed CCA. Section 9.9 will discuss the analysis of a Vera Module European (VME) processor CCA, and Sec. 9.10 will present CCAs used in a military environment. Although the environments described in these three sections are harsh, the same approach is valid for more benign environments.

In this study, the CalcePWA software[62] was used to determine possible failure mechanisms due to operating and environmental conditions on three CCAs in a control module of a military radio.[63] Subsequent physical testing was performed to confirm the failure assessment made by CalcePWA software. A photo of the military radio can be seen in Fig. 9.50.

9.8.1 Description of the CCAs and environment

The control module consists of three CCAs, an aluminum frame, and two aluminum backplanes. Each circuit board has six layers and contains commercial and military components. Components are both ceramic and plastic and use surface-mount and through-hole attachment technology. The three CCAs are encased in an aluminum frame with CCA no. 1 and CCA no. 3 having a bonded aluminum backplane. A diagram of the three CCAs and their housing is depicted in Fig. 9.51.

The three CCAs contained 50 microcircuits, 7 connectors, 22 inductors, 44 semiconductors, 241 capacitors, 222 resistors, and 4 miscellaneous parts. Despite differences in layout, a number of the same parts were used on the CCAs. The bare boards were constructed of BT laminates and contained six signal/power/ground planes. The total thickness of boards CCA no. 1 and CCA no. 3 were approximately 37 mils and the thickness of board CCA no. 2 was approximately 58 mils. The boards were 5 in square with cutouts on the corners. The board layers were modeled by alternating copper and BT layers with the thickness of the copper approximately 1 mil and the BT varied between 5 and 7 mils.

The control module CCAs were modeled and analyzed in the CalcePWA software. To facilitate import of the data, the electronic design file generated by Zuken-Recal design system was imported through the CalcePWA import facility. Further data entry and manipulation were required after the initial import process, since the electronic design system did not carry all of the information necessary to perform the PoF analysis. For example, detailed lead geometry, lead

Figure 9.50 Military radio.

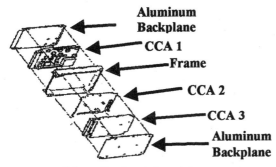

Figure 9.51   Schematic of control module.

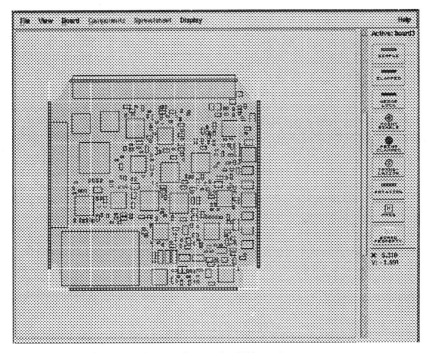

Figure 9.52   Vibration boundary conditions for CCA no. 3.

materials, package materials, detailed package dimensions, laminate materials, and laminate thicknesses were not specified in the electric design file. Figure 9.52 shows the layout for CCA no. 3 in the control module as modeled in the software.

While the model was being developed, a test plan was constructed to allow a one-to-one comparison of experimental and simulation results. Based on the radio's anticipated military environment, temperature and vibration stress conditions were selected to induce failures. The level and duration of the vibration test were based on the product qualification test procedure and included random vibration at $0.04g^2/\text{Hz}$, between 20 to 2000 Hz, for 4 h. The temperature test consisted of cycling between -50 and $95°C$ with 15-min dwells at the temperature extremes. A complete temperature cycle lasted approximately 2 h.

9.8.2 Vibration analysis

In order to estimate the local stresses within each CCA, a vibration analysis of each CCA was performed. Inputs to the software included a detailed set of boundary conditions, the applied power spectral density (PSD) curve from the test specifications, and a damping factor of 0.05 (normal default value). From examining the assembly, it was observed that CCA no. 1 and CCA no. 3 were supported completely along their edges. In addition, the support appears to be fairly rigid. As a result, the first vibration model defined the outer edges as clamped supports. A more conservative approximation was later used and the supports were changed to simple supports. CCA no. 2 was supported by several screws, and simple point supports were used to conservatively approximate their support. CCAs no. 1 and no. 3, including the aluminum backplane, each weighed approximately ⅛ lb and had a rigidity of approximately 185 lb-in. CCA no. 2 weighed less than ⅒ lb and had a rigidity of approximately 77 lb-in. Figure 9.52 depicts the screen-captured image of the boundary conditions for CCA no. 3.

The analysis results indicated that the first mode frequency of each board was above 500 Hz. Figure 9.53 depicts the first mode shape as well as the first three calculated natural frequencies for CCA no. 3. Based on the input PSD, a simulated one-sigma board deflection is depicted in Figure 9.54 for CCA no. 3. For this analysis, the weight and stiffening effects of the components were ignored, which provided a conservative result since the components provided more stiffness than weight in this particular situation. The relatively high first mode frequency results suggest that vibration-induced fatigue did not pose a significant reliability risk.

9.8.3 Thermal analysis

As discussed previously, components and board temperatures are important inputs to the failure assessment process. For this study, a thermal analysis was performed on each CCA to identify component temperatures and the thermal distribution throughout each board due to component power dissipation and the effectiveness of the cooling paths. Inputs to the thermal analysis module included component,

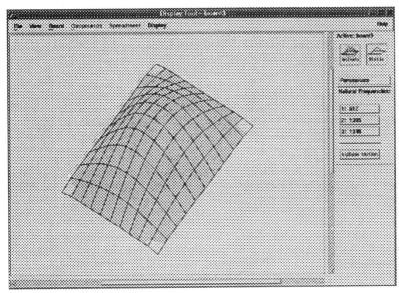

Figure 9.53   First fundamental mode shape for CCA no. 3.

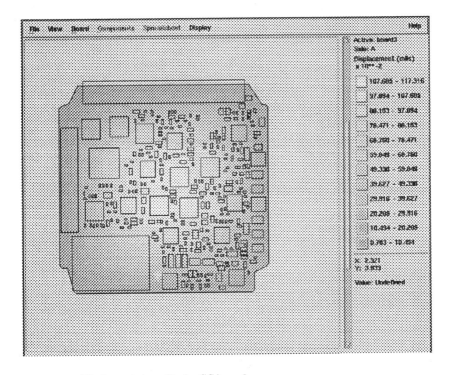

Figure 9.54   Displacement profile for CCA no. 3.

PWB, and cooling information. Package dimensions, power dissipation, junction-to-case thermal resistance, and material properties were taken from the manufacturer's datasheets.

In developing the thermal analysis models, the metal housing which supported each board along its edges was assumed to be the primary heat sink. Several analyses were performed for each board to determine the temperature at different operating conditions. Figure 9.55 depicts the applied thermal boundary conditions for one of the thermal analysis runs performed on CCA no. 3. Similar boundary conditions were used for the other CCAs under investigation.

Simulation results provided temperature distributions within the board layers, as well as junction, case, and board temperatures for each component. Figures 9.56 and 9.57 depict the temperature profile of the first material layer and the component temperature for CCA no. 3, respectively. Due to the relatively small heat dissipation of the components on each card, the board and component temperatures were only slightly higher than the applied boundary condition temperature. Discussions with the radio module vendor confirmed that the calculated CCA temperatures closely matched the applied boundary conditions. This observation simplified the failure mechanism analysis.

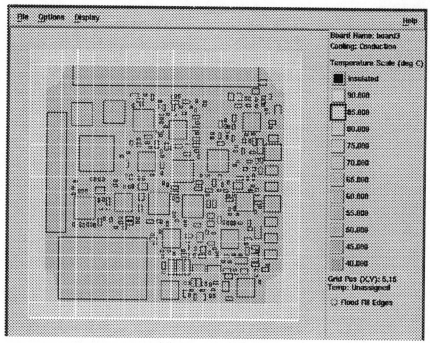

Figure 9.55 Thermal boundary conditions for CCA no. 3.

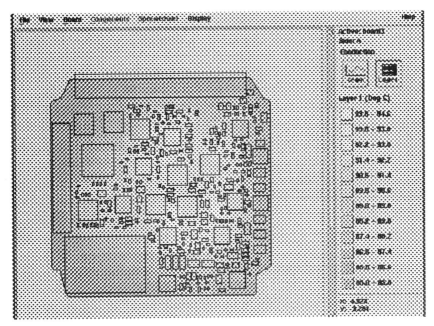

Figure 9.56   Temperature profile for CCA no. 3.

9.8.4   Failure mechanism analysis

The failure analysis of each CCA was conducted to identify potential failure sites. The following failure mechanisms and sites were examined: conductive filament formation between metal conducts, PTH failure due to thermal-induced fatigue, and component interconnect fatigue due to temperature cycling and vibration.

PTH barrel fatigue was analyzed to predict the fatigue life of PTH plating when subjected to repeated thermal cycles. To achieve a conservative solution, only the worst-case PTH was modeled. The worst-case PTH was 12 mils in diameter, 1 mil plating of copper plating, and 500 mils from the closest other PTH. Figure 9.58 shows the model of the PTH. The failure analysis was based on the planned physical tests evaluated for a ΔT of 145°C and predicted life greater than 20 years. The long life predicted by the model indicated that failure due to barrel fatigue did not pose a significant problem.

As discussed in Sec. 9.6, component interconnect failure due to fatigue is one of the primary wearout failures observed in CCAs. Inputs for the interconnect failure models include the component position, component size, interconnection format, interconnection material, and component material. Component information was extracted from the CCA design models. Board curvature at the component,

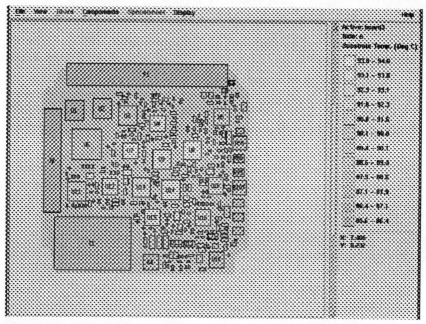

Figure 9.57   Component temperature.

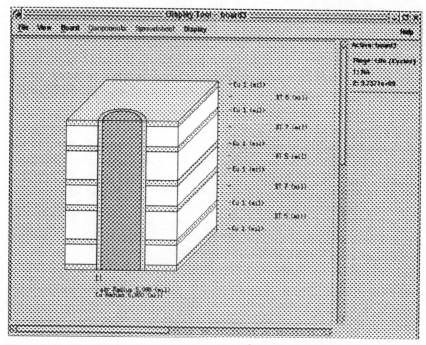

Figure 9.58   Plated-throughhole model in CalcePWA.

component case temperature, and board temperature from the thermal and vibration analysis results were used as inputs to this failure assessment. Other information, such as temperature cycle dwell times and frequency of vibration exposure, was directly entered to perform the analysis.

The random vibrational fatigue failure model was applied against each of the three CCAs with an excitation level $0.04g^2$/Hz. Figure 9.59 presents a graphical display of the vibration-induced fatigue model results for CCA no. 3. From Fig. 9.59, it can be seen that some of the microcircuits located near the high curvature region of the board have lower times to failure. However, the results of the vibrational fatigue model indicated that no significant problems existed for a vibration loading of $0.04g^2$/Hz. Similar observations were made when examining the results for CCA no. 1 and CCA no. 2. Again, no significant vibration problems were indicated.

The thermal-induced fatigue model was applied for each of the three CCAs with an imposed ΔT of 145°C. The results exposed a potentially serious failure problem for certain parts on CCA no. 2 and no. 3. Figure

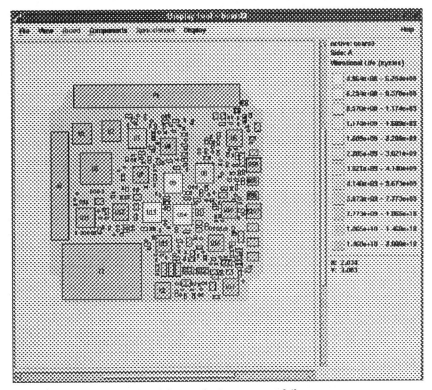

Figure 9.59   Vibration-induced fatigue for interconnect failures.

9.60 provides a graphical depiction of thermal fatigue analysis results for CCA no. 3. From Fig. 9.60, it can be seen that eight 20-pin leadless chip carriers (LCCs) have estimated life predictions below 900 temperature cycles. On CCA no. 2, the failure analysis indicated that one LCC would have a life below 1000 temperature cycles. Based on the projected field-service environment, these results indicated that failures will occur in less than 7 years. Other packages have greater lifetimes and should not influence the CCA's life.

9.8.5 Experimental test results

From the failure analysis results, solder fatigue due to temperature cycling was the dominant failure mechanism. Therefore, the final test program incorporated a 2-h thermal cycle between −50 and 95°C with 15- to 20-min dwells at the extreme temperatures. Electrical functionality was verified at 1-week intervals using existing production equipment. Modules that passed the functional test were returned to the temperature testing chambers, while failed modules were subjected to root-cause failure analysis. Repairable modules were fixed and returned to the temperature chamber.

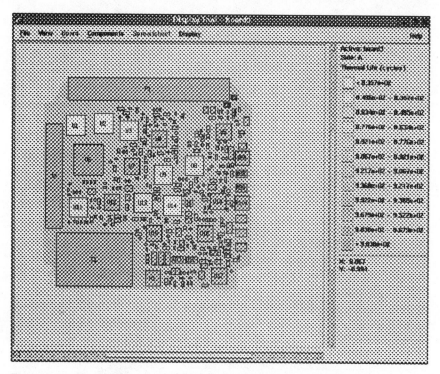

Figure 9.60   Thermal fatigue failure analysis of solder joints for CCA no. 3.

Although the analysis showed that vibration-induced fatigue was not a major failure mechanism, the modules were subjected to 10 h of random vibration at $6.10G_{rms}$ to verify the simulation result. The modules were not powered during the vibration test. The vibration test was broken into four segments and the modules were functionally tested at the end of each segment.

Ten modules containing the three CCAs were used in the physical tests. As expected, no failures occurred during the vibration tests. For the thermal tests, the 10 modules averaged slightly over 2000 h. Observed failures included solder-joint failure of ceramic leadless chip carriers (LCC), failure of electrical coils due to conformal coating, shorting of a voltage regulator due to manufacturing defects, lead failure of connectors due to thermal cycling, device failure of a hybrid, and solder-joint failure of a J-leaded device. Of the precipitated failures, the LCC solder-joint fatigue was clearly the dominant failure. Eight of the 10 modules experienced at least one LCC solder-joint failure, while the two modules that did not exhibit failure had less than 900 cycles. Physical examination of these two modules revealed that the solder joints were beginning to crack. The mean cycles to failure for the LCCs were 967 compared to the predicted value of 900 h.[64] The other failure mechanisms were isolated incidents associated with manufacturing problems. The test results closely matched the failure analysis results.

9.8.6   Cost saving from this PoF analysis

As a result of this study, a design change was implemented in this military radio which eliminated the LCC from the control module. By eliminating these premature failures from the military radio's design, a potential of 90,000 CCA repairs were avoided during the expected 20-year lifetime of the 5000 radios to be fielded. The repair, transportation, inventory, and logistics costs associated with these 90,000 CCA repairs were estimated to be $27 million.[65] By performing this failure assessment on the control model and eliminating the dominant failure mechanism, $27 million in costs to the U.S. military was avoided. In addition to the significant cost savings, the availability of these radios was greatly increased. Although this PoF assessment saved substantial repair costs, it could have saved redesign and rework costs if performed early in the design process.

9.9   Example 2: VME Processor Circuit Card Assembly

This section presents a case study of a PoF reliability assessment of a commercial VME processor CCA.[66–67] The purpose of the assessment was to determine if a commercial VME processor CCA could withstand a harsh military environment. Vibrational, thermal, and failure analyses were performed on the CCA using the CalcePWA software.[62]

9.9.1 Description of CCA and environment

The CCA assessed was a VME single-board computer based on a 33-MHz MC68040 microprocessor with 32 Mbytes of dynamic RAM on a mezzanine board. Dimensions of the main board are 9.2 × 6.3 in (IEEE VME standard), and the dimensions of the mezzanine board are 5.4 × 5.6 in. Figure 9.61 shows the commercial main board. The boards were constructed of FR-4 and modeled as alternating layers of copper and epoxy fiberglass. The VME CCA is the main processor card in an Army radar processing system, composed of radar display and

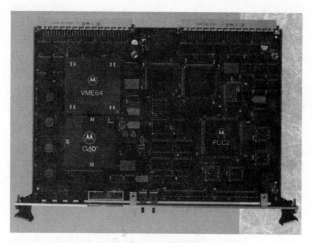

Figure 9.61   Commercial VME CCA.

Figure 9.62   Military vehicle and shelter.

processing equipment housed in a shelter mounted on a high mobility multipurpose wheeled vehicle (HMMWV). The HMMWV and shelter are shown in Fig. 9.62.

The thermal management for the CCA consists of forced air blown over the card from an environmental control unit. The maximum PSD for the transverse (out-of-plane) axis in the processor card specification is $0.02g^2$/Hz for operation and $0.1g^2$/Hz for qualification tests. Test measurements on the HMMWV and shelter yielded similar values for the maximum PSD.

9.9.2 Thermal analysis

Inputs to the thermal analysis included component, board, and cooling data. For the commercial CCA package dimensions, power dissipation, junction-to-case thermal resistance, and material properties were taken from the manufacturer's datasheets. For this analysis, only active devices were considered; thermal contributions from passive devices were considered negligible. The dissipated power of the integrated circuits was taken from the manufacturer's datasheets. The PWB dimensions are in accordance with the IEEE VME standard. Contributions from the PWB traces, due to their insignificant thermal conductivity, were not considered. Components were placed on the PWB in accordance with the manufacturer's specifications. Cooling was modeled as forced convection with an airflow rate of 0.44 lb/min at a 25°C ambient air temperature.

Case temperatures calculated from the thermal analysis ranged from 36 to 97°C. Further results showed that heat dissipated into the PWB from the components caused a temperature profile ranging from 28 to 83°C based on forced-convection cooling in the military application. This large thermal gradient was due to one or two components that were very hot compared to the majority of components. Most components had temperatures close to the lower values. Figure 9.63 shows the thermal profile of the PWB. Component data are highlighted in the window. Some component temperatures were found to be greater than their rated value of 70°C, which could cause either the microcircuit to not operate properly or a reliability failure of the microcircuit. Further analysis showed that increasing the airflow across the components reduced the temperatures below their rated value. The case temperature for each microelectronic package and the board temperature at all locations on the board were inputs to the failure analysis.

9.9.3 Vibrational analysis

Table 9.13 shows the board properties input into the vibrational module. The worst case approximation of one layer of epoxy fiberglass was used.

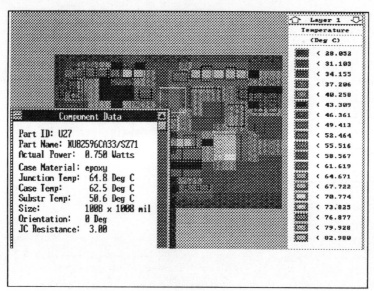

Figure 9.63   Thermal profile of the board.

The CCA is supported vertically by metal slots as it is inserted into the box with two tab connectors along the backplane. Additional support is provided by two lock-down clamps once the card is slid into place, along with two metal supports for attachment of the mezzanine card on the clamped edge. The mezzanine board is supported on the four corners by screws clamped down to the VME CCA, which in turn provides additional support. These supports are the boundary conditions for the vibrational finite element analysis and are shown in Fig. 9.64. The results of the vibrational analysis revealed a first natural frequency for the commercial processor board of 109 and 224 Hz for the mezzanine board. These results were used to perform a solder-joint fatigue analysis.

9.9.4   Failure mechanism analysis

The failure analysis of the CCA was performed on potential failure sites. For this study, the following failure mechanisms and sites were examined: PTH failure due to thermal-induced fatigue and component-interconnect fatigue due to temperature cycling and vibration.

To achieve a conservative solution, only the worst-case PTH was modeled, as shown in Fig. 9.65, with a worst-case ΔT of 100°C. The PTHs on this board were known to have a quality problem of plating folds and nodules, which were caused by rough drilling or rough copper plating. The PTH had a 14-mil drill size and was plated with 1 mil of copper. Figure 9.66 shows a cross section of the PTH and the rough drilling can be seen. This knowledge of poor quality drilling was factored into the analysis. The barrel fatigue life prediction was 15 years in the usage

TABLE 9.13   PWB Properties for Example 2

	Computer board	Mezzanine board
Layers	1	1
Length (in)	9.187	5.404
Width (in)	6.299	5.600
Thickness (in)	0.063	0.063
Density (lb/in^3)	0.078	0.057
Rigidity (lb-in)	60.712	60.712

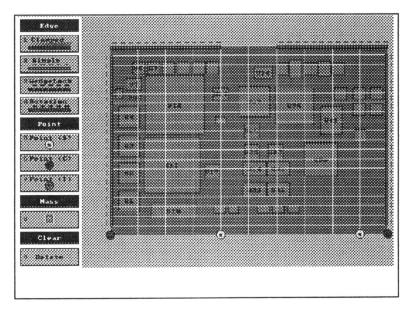

Figure 9.64   Commercial main board vibration setup.

environment. This lifetime was less than the desired life of 20 years, but could be increased by improving the poor quality of the PTHs.

Inputs for the solder-joint fatigue analysis include the diagonal length, second-level interconnect, and case material. From the thermal analysis, the case and substrate (PWB under the component) temperatures for each component were input to the solder-joint fatigue analysis. Other inputs to the solder-joint fatigue analysis from the commercial CCA are shown in Table 9.14.

Results of the solder-joint fatigue analysis include on/off cycles-to-failure, thermal fatigue life (h), vibration fatigue life (h), and combined fatigue life (h). Tables 9.15 and 9.16 provide the five lowest microelectronic-package solder-joint fatigue lifetimes for the main and mezzanine cards, respectively. Other packages have greater lifetimes and should

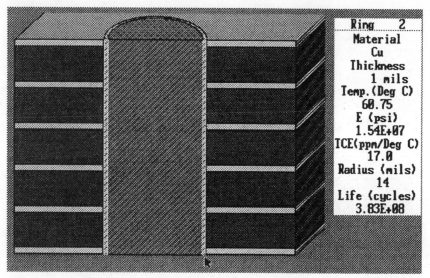

Figure 9.65 Plated-through hole model in CalcePWA.

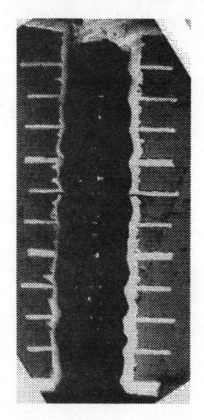

Figure 9.66 Cross section of the PTH with poor quality plating.

TABLE 9.14   Inputs to the Solder-Joint Fatigue Analysis (Example 2)

Input	Value	Source
Power-off temperature (°C)	25.00	System environment
Vibration (h/d)	1.000	System environment
On-off (cycle/day)	2.000	System environment
Natural frequency	109.0, 224.0	Vibrational analysis
Dwell time (min)	360.0	System environment
Fatigue ductility	0.325	Material constant
Effective CTE x(ppm/°C)	17.600	Board material prop.
Effective CTE, y(ppm/°C)	17.600	Board material prop.
Maximum PSD (g^2/Hz)	0.02	System environment

TABLE 9.15   Solder-Joint Fatigue Life for the Main Card (Example 2)

Part no. ID	Thermal life (h)	Vibrational life (h)	Combined life (h)
U34	$2.0E+10$	$1.4E+05$	$9.9E+04$
U27	$2.6E+07$	$1.6E+05$	$1.1E+05$
U41	$2.0E+10$	$1.7E+05$	$1.2E+05$
U24	$3.4E+09$	$2.1E+05$	$1.5E+05$
U32	$2.5E+09$	$3.0E+05$	$2.1E+05$

TABLE 9.16   Solder-Joint Fatigue Life for the Mezzanine Card (Example 2)

Part no. ID	Thermal life (h)	Vibrational life (h)	Combined life (h)
U64	$1.8E+07$	$6.0E+05$	$4.0E+05$
U52	$1.8E+07$	$6.0E+05$	$4.0E+05$
U30	$1.8E+07$	$6.0E+05$	$4.0E+05$
U17	$1.8E+07$	$6.0E+05$	$4.0E+05$
U48	$1.8E+07$	$6.0E+05$	$4.0E+05$

not influence the CCA's life. The solder-joint life of each component is shown graphically in Fig. 9.67.

9.9.5   Analysis conclusions

Results from the vibration analysis included the board fundamental frequencies and displacements. The first natural frequency was a relatively low 109 Hz and was cause for concern for vibration-related failure mechanism. Junction, case, and substrate temperatures were calculated based on component power dissipation, component materials, and circuit card cooling methods. Some component temperatures were over their rated values, which could result in operational problems or reliability failures. The analysis also uncovered a potential

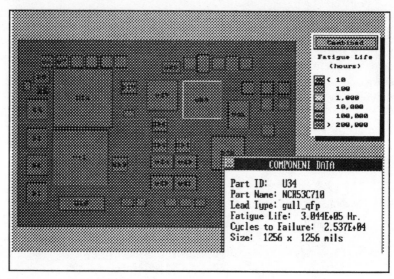

Figure 9.67   Solder-joint fatigue life estimates.

problem with fatigue in the PTHs, which was caused by poor manufacturing quality.

The solder-joint fatigue from thermal stress was minimal because of the small CTE mismatch between the components and the board, which caused the thermal life of the solder joints to be greater than 10^6 h. The life of the solder joint was greatly influenced by the vibrational stresses. The solder-joint life of the components on the mezzanine board tended to be greater than on the main card, due to the high first natural frequency of the mezzanine card (224 Hz) as opposed to the main card (109 Hz). The solder-joint life of the lowest component on the mezzanine board was 400,000 h, or approximately 45 years, while the solder joint life of the lowest components of the main board was approximately 100,000 h, or 11 years. While this estimate of life was less than the 20-years desired life, a decision was made to purchase the commercial CCA because the next generation processor will probably replace this card before failures occur. The commercial CCA cost approximately $7000 while a ruggedized version cost $19,000. For each processor CCA used, a $12,000 cost savings would be realized. The total cost savings based on using the commercial CCA instead of the ruggedized CCA was on the order of $1 million for only the first 10 systems.

9.10   Example 3: Circuit Card Assembly in a Military Environment

This section presents a case study of a PoF reliability assessment of electronic CCAs used in a harsh vibration environment. The electronics in this system form a highly automated system that can handle target location data, automatically prepare the messages required, and integrate the radio suite communicating this data. Figure 9.68 shows the tracked vehicle. Several CCAs in a tracked-vehicle system were assessed to determine possible failure mechanisms due to operating and environmental stresses and to evaluate their impact on useful life.[68–69] PoF reliability analyses were performed for six circuit CCAs, including a video processor, power filter, processor, interface, serial I/O, and backplane CCAs. The CCAs were evaluated using the CalcePWA software.[62]

9.10.1 Vibration analysis

Inputs to the vibration analysis module included the CCA supporting boundary conditions (shown in Fig. 9.69), CCA weight, the worst-case power spectral density (PSD) curve from the subsystem specifications (shown in Fig. 9.70), and a typical CCA damping factor of 0.01. Boundary conditions for each CCA were selected by using engineering drawings to determine clamp and support locations. Because they invoke a worst-case vibrational response condition, simple point and edge supports were used to model the fastening effects on the boards. For each board, several vibrational analysis models were developed based on simple or clamped support structures and the extent of components on the board being modeled as stiffeners. Components, deemed of significant size to add a stiffening effect to the board, were

Figure 9.68   Tracked vehicle.

modeled as stiffeners on the board by calculating an effective elastic modulus for each of their respective finite elements, which took into account the combined stiffness of the component and board.

Vibration analyses are not only dependent on structural board boundary conditions, but are also heavily dependent on the weight of the assembly and the type of input vibration. The weight of the assembly was provided by the CCA producer, and the vibration was considered random due to the nature of the displacement caused by tracked vehicles on paved and off-road surfaces. The worst-case random vibration PSD curve was taken from the electronics specifications. A graphical representation of the modal analysis results of the processor CCA can be seen in Fig. 9.71, and the tabular results of the vibration analysis of all the CCAs can be seen in Table 9.17.

The contractor also performed environmental stress screening (ESS) surveys on the CCAs, which measured the CCA's first natural frequencies. These measured frequencies were compared to the frequencies calculated in this analysis. For all of the CCAs, except for the interface CCA, the simulated natural frequency was lower than the

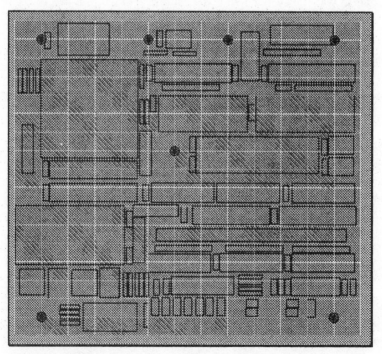

Figure 9.69   Processor CCA boundary conditions with simple point supports and stiffened elements.

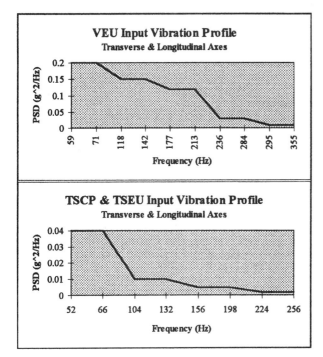

Figure 9.70   Input PSD curves.

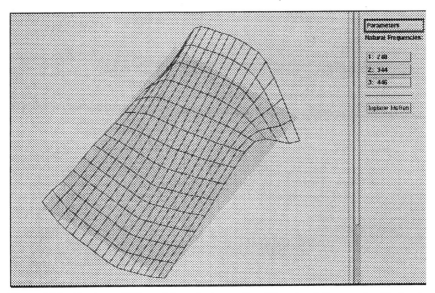

Figure 9.71   Processor CCA modal analysis results with natural frequencies (first natural frequency mode shape shown).

TABLE 9.17   Vibration Analysis Results (Example 3)

CCA	First natural frequency CalcePWA results (Hz)	Second natural frequency CalceWA results (Hz)	Third natural frequency CalcePWA results (Hz)	First natural frequency ESS results (Hz)
Video processor	372	404	419	510
Power filter	226	352	469	230
Processor	248	344	446	265
Interface	303	666	729	265
Serial I/O	261	540	801	N/A
Backplane	417	604	1134	410

ESS measured frequency. Since a lower natural frequency indicates a greater amount of damage, the ESS measured value was used in the fatigue analysis for the interface CCA. This allows for a conservative solder-joint fatigue life assessment for the Interface CCA.

9.10.2   Thermal analysis

A thermal analysis was conducted on the CCAs to identify component and board temperatures due to the power dissipation of components. Component power values for this analysis were provided by the manufacturer and thermal boundary conditions (that is, conduction paths) were obtained from engineering drawings. Package dimensions, material properties, and junction-case resistance values were taken from the component manufacturer's datasheets. For this study, a worst-case convection path gap of 500 mils was assumed for each board. Simulations were conducted at three different ambient temperatures, of which the minimum and maximum were identified in the corresponding design specifications. The minimum and normal ambient temperature used for the thermal analysis for all of the CCAs were -32 and 25°C, respectively. The maximum ambient was 52°C for all CCAs except the video processor and power filter CCA, which had a maximum ambient of 71°C. From these initial and boundary conditions, junction, case, and board temperatures were simulated for each component and board location. Results identified the "hot" spots on each board and graphically displayed the thermal distributions throughout the board layers (shown in Fig. 9.72). The case temperature for each microelectronic package and the temperature at all locations on the board were inputs to the solder-joint and plated-throughhole fatigue analysis tools. See Table 9.18 for the thermal analysis results.

9.10.3   Failure analysis

For solder-joint fatigue analyses, the operational PSD was taken to be $0.06g^2/Hz$. Also, it is important to note that only vibration cyclic stresses

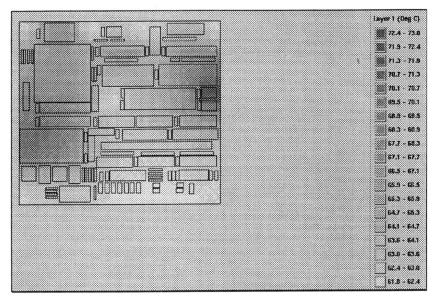

Figure 9.72   Processor CCA board level temperature distribution (°C).

TABLE 9.18   Thermal Analysis Results (Example 3)

CCA	Ambient temperature (°C)	Maximum board temperature (°C)	Maximum case temperature (°C)	Maximum junction temperature (°C)
Video	71	86	143	170
processor	25	43	97	125
	−32	−14	41	67
Power	71	81	82	87
filter	25	38	39	42
	−32	−21	−20	−15
Processor	52	73	117	123
	25	46	90	95
	−32	−10	33	38
Interface	52	64	70	75
	25	38	44	49
	−32	−19	−13	−9

were considered in most of these CCA solder-joint fatigue analyses. This stems from the use of through-hole components, which are generally not affected by thermal cycling stresses. Other inputs to the solder-joint fatigue analysis, as represented by the Processor CCA input parameters, are shown in Table 9.19.

TABLE 9.19   Inputs to Solder-Joint Fatigue Analysis for the Processor CCA (Example 3)

Input	Value	Source
Power-off temperature (°C)	25.00	System environment
Vibration (h/d)	1.000	System environment
On-off (cycle/day)	1.000	System environment
Natural frequency (Hz)	248.0	Vibrational analysis
Dwell time (min)	60.0	System environment
Fatigue ductility	0.325	Material constant
Effective CTE, x(ppm/°C)	17.43	Board material prop.
Effective CTE, y(ppm/°C)	17.43	Board material prop.
Maximum PSD (g^2/Hz)	0.06	System environment

For these solder-joint fatigue analyses, the first five components to fail and their cycles to failure were recorded on each assembly. Then for each assembly, assuming that the system will experience a random vibration input of 200 h/years, the fatigue life (years) was calculated for the five subject components. These components consisted of dual in-line packages, resistors, capacitors, converters, and filters. Connector fatigue results were not considered, because they have mechanical connections to the board in addition to the solder-joint connections, which would prevent significant solder-joint fatigue damage.

The results of the failure assessment indicate that, under the anticipated conditions, the CCA should last well beyond the 25-year required life. This long life estimate was due to the use of through-hole components, which tend to be more resistant to vibrational fatigue than their surface-mounted counterparts. An example of these results for the processor CCA can be found in Table 9.20.

Using the thermal analysis results, the PTH fatigue life was simulated under repeated thermal cycles. To achieve a conservative solution for this study, the worst-case PTH was modeled. This model consisted of a 26-mil drill diameter hole, a PTH spacing of 500 mils (0.5 in), and a 1-mil (0.001-in) copper plating thickness. Using these worst-case PTH conditions, and assuming a maximum change in temperature of 100°C and one on/off cycle/day, worst-case PTH fatigue lives were determined in terms of temperature cycles for each CCA board configuration. The results of this analysis indicated no PTH reliability problems.

9.10.4   Analysis conclusions

From the vibrational analysis it was seen that the board displacements are relatively small, due to natural frequencies that were all greater than 200 Hz. The use of through-hole components reduced the likelihood of fatigue-related problems. Although the thermal analysis showed some high board and component case temperatures, especially

TABLE 9.20   Processor CCA Solder-Joint Fatigue Results (Example 3)

Part no. ID	Life (cycles)	Life (years)
C14	$6.9E+8$	>30
U22	$4.3E+09$	>30
U10	$1.1E+10$	>30
C18	$1.3E+09$	>30
C27	$1.4E+09$	>30

for the video processor CCA, the PTH analysis did not show any failures associated with the high temperature cycle gradients. Both the solder-joint fatigue and the PTH reliability analyses, for each subject CCA, showed fatigue lives of greater than 30 years. Therefore, in this application and under these environmental stresses, one would not expect to observe failures in the solder joints or PTHs during the useful life of the components. The high temperatures in the component case and junction temperatures, 140 and 160°C, respectively, on the video processor CCA, were found to be the results of conservative power estimates by the contractor. This conclusion was confirmed by the contractor's thermal ESS survey results for the video processor CCA.

This analysis was performed after the CCA designs were completed, and the results initiated some minor adjustments to the designs of some of the assemblies. Once again, early application of PoF methods could easily have identified potential problems prior to design completion.

9.11   Summary

The application of the PoF approach leads to reliability modeling and assessment that uses knowledge of failure mechanisms and results in robust design and manufacturing practices. In this chapter, the concept of PoF and its role in reliability assessment of CCA have been presented. Concerns raised by common environmental conditions have been reviewed, and methods for evaluating stress with a CCA have been introduced. After the discussion on environments and stress analysis, failure mechanisms and models related to CCAs were presented. In addition to evaluating design reliability, manufacturing and assembly defects associated with CCAs were described. PoF has been applied to actual CCA design and has resulted in increased reliability, decreased design and testing costs, and reduced operating and logistics costs. Three case studies that use the PoF approach have been reviewed. This PoF approach goes beyond merely characterizing the thermal performance and vibration behavior of the system. A detailed understanding of how stresses produce failures, and an identification of where failures are likely to occur, allow designers to strengthen their design and managers to evaluate the potential risks.

Early detection and elimination of potential failures result in greater customer satisfaction, improved reliability and warranties, and increased competitiveness. In the PoF approach, models of root-cause failure processes or mechanisms are employed to qualify and quantify the likelihood of failure. Input to these models includes parameters that have been found to influence structural and operational integrity of a system. These parameters are associated with physical geometry, material properties, as well as the environment and operational characteristics that lead to electrical, chemical, thermal, and mechanical stresses and lead ultimately to failure. Research into physical failure mechanisms is subjected to scholarly peer review and published in the open literature. The failure models are validated through experimentation and replication by multiple researchers.

Acknowledgments

The authors would like to thank the personnel at U.S. Army Materials Systems Analysis Activity (AMSAA), Aberdeen, Maryland, and the CALCE Electronic Products and Systems Center at the University of Maryland for their help in creating this chapter. The authors would especially like to thank Dr. Donald Barker, Pradeep Sharma, Keith Rogers, Lynn Anderson, Randy Wheeler, Barry Hum, Michael Deckert, and Dr. Michael Cushing for their valuable inputs. Finally, the authors would like to thank Dr. Charles Harper for his patience.

References

1. B. W. Boehm, "A Spiral Model of Software Development and Enhancement," *IEEE Computer,* vol 21, no. 5, 1998, pp. 61–72, 1998.
2. P. McCluskey, M. Osterman, and M. Pecht, "Software for Reliability Assessment of Electronic Components, Circuit Cards and Equipment," *Proceedings of the Second International Symposium on Electronic Packaging Technology,* Shanghai, China, December 9–12, 1996., pp. 472–483.
3. W. Weibull, "Statistical Design of Fatigue Experiments," *ASME Journal of Applied Mechanics,* vol. 19, no. 1, March 1952, pp. 109–113.
4. N. Mann, R. Schafer, and N. Sigpurwalla, *Methods for Statistical Analysis of Reliability and Life Data,* John Wiley, New York, 1974.
5. M. Modarres, *What Every Engineer Should Know about Reliability and Risk Analysis,* Marcel Dekker, New York, 1993.
6. W. Nelson, *Accelerated Testing, Statistical Models, Test Plans, and Data Analysis,* John Wiley, New York, 1990.
7. P. Lall, M. Pecht, and E. Hakim, *Influence of Temperature on Microelectronics and System Reliability: A Physics of Failure Approach,* CRC Press, New York, 1995.
8. F. P. McCluskey, R. Grzybowski, and T. Podlesak, *High Temperature Electronics,* CRC Press, New York, 1996.
9. M. Pecht, *Handbook of Electronic Package Design,* Marcel Dekker, Inc., New York, NY, 1991.
10. "Test Method Standard for Environmental Engineering Considerations and Laboratory Tests," U.S. Department of Defense Military Standard 810E, 1989.

11. D. S. Steinberg, *Vibration Analysis for Electronic Equipment,* John Wiley, New York, 1991.
12. M. Pecht, L. Nyguyen, and E. Hakim, *Plastic Encapsulated Microelectronics,* John Wiley, New York, 1994.
13. P. R. Engel, T. Corbett, and W. Baerg, "A New Failure Mechanism of Bond Pad Corrosion in Plastic Encapsulated ICs under Temperature, Humidity, and Bias Stress," *Proceedings of the 33rd Electronic Components Conference,* 1986, pp. 127–131.
14. *CalcePWA Explanation and Validations, A Guide for CALCE Software Users,* CALCE EPRC, University of Maryland, College Park, Md., 1996.
15. J. Sloan, *Design and Packaging of Electronic Equipment,* Van Nostrand Reinhold, New York, 1985.
16. W. Engelmaier, "Fatigue Life of Leadless Chip Carriers Solder Joints during Power Cycling," *IEEE Transactions CHMT,* vol. CHMT-6, September 1983, pp. 232–237.
17. H. Solomon. "The Solder Joint Fatigue Life Acceleration Factor," *Transactions,* ASME, JEP, vol. 113, June 1991, pp. 186–190.
18. A. Dasgupta et al., "Solder Creep-Fatigue Analysis by an Energy-Partitioning Approach," *Transactions,* ASME JEP, vol. 114, June 1992, pp. 152–160.
19. R. Darveaux and K. Banerji, "Constitutive Relations for Tin-Based Solder Joints," *IEEE Transactions on CHMT,* vol. 15, no. 6, December 1992, pp. 1013–1024.
20. O. H. Basquin, "The Exponential Law of Endurance Tests," *ASTM Proceedings,* vol. 10, 1910, pp. 625–630.
21. D. Steinberg, *Cooling Techniques for Electronic Equipment,* 2d ed., John Wiley, New York, 1991.
22. L. F. Coffin, Jr., "A Study of the Effects of Cyclic Thermal Stresses on a Ductile Metal," *Transactions,* ASME, vol. 76, 1954, pp. 931–950.
23. S. S. Manson, "Fatigue: A Complex Subject-Some Simple Approximations," *Experimental Mechanics,* no. 5, 1965, pp. 193–226.
24. W. Engelmaier, "Generic Reliability Figures Of Merit Design Tools For Surface Mount Solder Attachments," *IEEE Transactions CHMT,* vol. 16, no. 1, February 1993, pp. 103–112.
25. E. Jih and Y. Pao, "Evaluation of Design Parameters for Leadless Chip Resistor Solder Joints," *Transactions,* ASME, JEP, vol. 117, June 1995, pp. 94–99.
26. R. Sundarajan et al., "Semi-Analytic Model for Surface Mount Solder Joints," ASME IMECE, 97-WA/EEP-12, Dallas, Tex., November 1997.
27. J-P Clech et al., "A Comprehensive Surface Mount Reliability Model Covering Several Generations of Packaging and Assembly Technologies," *Transactions IEEE CHMT,* vol. 16, no. 8, December 1993, pp. 949–960.
28. A. Dasgupta, S. Ling, and S. Verma, "A Generalized Stress Analysis Model for Fatigue Prediction of Surface Mount Solder Joints," *Advances in Electronic Packaging,* ASME, EEP-vol. 4-2, 1993, pp. 979–986.
29. S. Ling and A. Dasgupta, "A Nonlinear Multi-Domain Stress Analysis Method for Surface-Mount Solder Joints," *Journal of Electronic Packaging,* vol. 118, no. 2, June 1996, pp. 72–79.
30. M. Oien, "Methods for Evaluating Plated-Through-Holes Reliability," *Proceedings of 14th Annual IEEE Reliability Physics Symposium,* 1976, pp.129–131.
31. D. Barker et al., "Transient Thermal Stress Analysis of Plated Through Holes Subjected to Wave Soldering," *Transactions,* ASME, *Journal of Electronic Packaging,* vol. 113, no. 2, June 1991, pp. 149–155.
32. D. Askeland, *The Science and Engineering of Materials,* PWS Engineering, Boston, 1984.
33. S. Bhandarkar et al., "Influence of Design Variables on Thermomechanical Stress Distributions in Plated Through Hole Structures," *Transactions,* ASME, *Journal of Electronic Packaging,* vol. 114, no. 1, March 1992, pp. 8–13.
34. M. Oien, "A Simple Model for Thermo-Mechanical Deformation of Plated-Through-Holes in Multilayer Printed Wiring Boards," *Proceedings of 14th Annual IEEE Reliability Physics Symposium,* 1976, pp.121–128.
35. R. Delaney and J. Lahti, "Accelerated Life Testing of Flexible Printed Circuits, Part II: Failure Modes in Flexible Printed Circuits Coated with UV-Cured Resins,"

Proceedings of the 1976 International Reliability Physics Symposium, 1976, pp. 114–117.

36. P. Boddy et al., "Accelerated Life Testing of Flexible Printed Circuits, Part I: Test Program and Typical Results," *Proceedings of the 1976 International Reliability Physics Symposium,* 1976, pp. 108–114.

37. Bi-Chu Wu, M. Pecht, and D. Jennings, "Conductive Filament Formation in Printed Wiring Boards," *1992 IEEE/CHMT International Electronics Manufacturing Technology Symposium,* 1992, pp. 74–79.

38. J. Lahti, R. Delaney, and J. Hines, "The Characteristic Wearout Process in Epoxy-Glass Printed Circuits for High Density Electronic Packaging," *Proceedings of the 17th Annual Reliability Physics Symposium,* 1979, pp. 39–43.

39. D. Lando, J. Mitchell, and T. Welsher, "Conductive Anodic Filaments in Reinforced Polymeric Dielectrics," *Proceedings of the 17th Annual Reliability Physics Symposium,* 1979, pp. 51–63.

40. T. Welsher, J. P. Mitchell, and D. J. Lando, "CAF in Composite Printed Circuit Substrates: Characterization, Modeling, and a Resistant Material," *18th Annual Proceedings on Reliability Physics,* 1980, pp. 235–237.

41. B. Rudra and D. Jennings, "Tutorial Failure-Mechanism Models for Conductive-Filament Formation," *IEEE Transactions on Reliability,* vol. 43, no. 3, September 1994, pp. 354–360.

42. A. Shukla et al., "Hollow Fibers in Woven Laminates," *Printed Circuit Fabrication,* vol. 20, no. 1, January 1997, pp. 30–32.

43. R. Martens, M. Osterman, and D. Haislet, "Design Assessment of a Pressure Contact Connector System," *Circuit World,* vol. 23, no. 3, April 1997, pp. 5–9.

44. W. Reyes et al., "Factors Influencing Thin Gold Performance for Separable Connectors," *Transactions on Components, Hybrids, and Manufacturing Technology,* vol. CHMT 4, no. 4, December 1981, pp. 499–508.

45. R. Bayer and Gregory, "An Engineering Approach to Vibration Induced Wear Concerns of Electronic Contact Systems," ASME, EEP-vol. 14-1, *Advances in Electronic Packaging,* 1993, pp. 525–536.

46. R. Bayer, E. Hsue, and J. Turner, "A Motion-Induced Sub-Surface Deformation Wear Mechanism," *Wear,* no. 154, 1992, 193–204.

47. R. Mroczkowski, *Electronic Connector Handbook,* McGraw-Hill, New York, 1998.

48. CALCE Web Site, *http://www.calce.umd.edu,* reviewed October 1998.

49. R. Tummla and E. Rymaszeewski, *Microelectronics Packaging Handbook,* Van Nostrand Reinhold, New York, 1989.

50. A. Shukla et al., "Hollow Fibers in PCB, MCM-L and PBGA Laminates May Induce Reliability Degradation," *Circuit World,* vol. 23, no. 2, 1997, pp. 5–6.

51. A. Shukla et al., Hollow Fibers in Woven Laminates," *Printed Circuit Fabrication,* vol. 20, no. 1, January 1997, pp. 30–32.

52. Z. Tian and S. R. Swanson, "The Fracture Behavior of Carbon/Epoxy Laminates Containing Internal Cut Fibers," *Journal of Composite Materials,* vol. 25, no. 7, November 1991, pp. 1427–1444.

53. J. Z. Wang and D. F. Socie, "Failure Strength and Damage Mechanisms of E-Glass/Epoxy Laminates under In-Plane Biaxial Compressive Deformation," *Journal of Composite Materials,* vol. 27, no. 1, 1993, pp. 40–480.

54. A. Shukla, M. Pecht, J. Jordon, K. Rogers, and D. Jennings, *Circuit World,* vol. 23, no. 2, 1997, pp. 5–6.

55. "IPC-CC-110 Guidelines for Selecting Core Constructions for Multilayer Printed Wiring Board Applications," IPC-CC-110, Institute for Interconnecting and Packaging Electronic Circuits, Lincolnwood, IL, January 1994.

56. N. Patel, V. Rohatgi, and L. James Lee, "Micro Scale Flow Behavior and Void Formation Mechanism During Impregnation Through a Unidirectional Stitched Fiberglass Mat," *Polymer Engineering and Science,* vol. 35, no. 10, May 1995, pp. 837–851.

57. *Electronic Materials Handbook,* vol. 1, *Packaging.* ASM International. ASM International, Materials Park, Ohio, 1989.

58. M. Oien, "Methods for Evaluating Plated-Through-Holes Reliability," *Proceedings of 14th Annual IEEE Reliability Physics Symposium,* 1976, pp. 129–131.

59. M. Pecht, *Soldering Processes and Equipment,* John Wiley, New York, 1993.
60. Kester Solder Technical Notes, *http://www.metcal.com/kester/,* reviewed February 1999.
61. D. L. Millard, "Solder Joint Inspection," *Electronic Materials Handbook Packaging,* vol. 1: *Packaging,* ASM International, Materials Park, OH, pp. 735–739.
62. *CalcePWA Release 1.3 User's Manual,* CALCE EPRC, University of Maryland, College Park, Md., 1996.
63. M. Osterman and T. Stadterman, "Failure-Assessment Software for Circuit-Card Assemblies," *Proceedings of the Annual Reliability and Maintainability Symposium,* January 1999, pp. 269–276.
64. "Reliability Assessment Process Improvement Demonstration (RAPID)," Contract No: F33615-96-D-5302, Delivery Order 041, Subtask: 3.3, Prepared for ESC/DIT, GRCI Inc., Fairborn, OH, 1998.
65. E. Grove et al., Draft of "Army Materiel Command/Training and Doctrine Command Guide for Reliability Quantitative Requirements," U.S. Army Materiel Systems Analysis Activity, Aberdeen Proving Ground, Aberdeen, Md., 1998, pp. 38–44.
66. T. Stadterman et al., "Design-Reliability Assessment of Commercial Circuit Cards Based on Failure-Mechanism Modeling," *Proceedings for the Second International Society of Science and Applied Technology International Conference on Reliability and Quality in Design,* Orlando Fla., March 1995, 8–10.
67. T. Stadterman et al., "Use of Physics of Failure for Reliability Assessment in Support of Commercial Circuit Card Use in Military Systems," *Proceedings for the Institute of Environmental Sciences Symposium,* April 30–May 5 1995, pp. 57–62.
68. R. Wheeler and T. Stadterman, "Physics of Reliability Analysis of Circuit Card Assemblies in the Bradley Fire Support Team (BFIST) Vehicle," *Proceedings of ARDEC Predictive Technology Symposium,* November 1997.
69. M. Osterman, T. Stadterman, and R. Wheeler, "CAD/E Requirements and Usage for Reliability Assessment of Electronic Products," *Proceedings of the American Society of Mechanical Engineers INTERpack'97 Conference,* June 1997.

Status of IPC Standardization Programs*

A.1 ANSI Approved

IPC-2221	Generic Standard on Printed Board Design—Chair, Lionel Fullwood, Wong's Kong King International
IPC-2222	Sectional Design Standard on Organic Printed Boards—Chair, Lionel Fullwood, Wong's Kong King International
IPC-4130	Specification and Characterization Methods for Nonwoven "E" Glass Materials—Chair, Dennis Lockyer, Crane and Co.

A.2 Published/Pending ANSI Approval

J-STD-002A‡	Solderability Tests for Component Leads, Terminals, and Wires—Chair, Dave Hillman, Rockwell
J-STD-012§	Implementation of Flip Chip and Chip Scale Technology—Chair, Ray Prasad, Ray Prasad Consultancy Group
J-STD-013†	Implementation of Ball Grid Array and Other High Density Technology—Chair, Ray Prasad, Ray Prasad Consultancy Group
IPC-EG-140† Amendment 2	Specification for Finished Fabric Woven from "E" Glass for Printed Boards—Chair, Joel Murray, Clark Schwebel

*With permission, *IPC Review,* Mike Buetow, ed., Institute of Printed Circuits, Northbrook, Ill., April 1999.

IPC-CF-152B‡	Composite Metallic Material Specification for Printed Wiring Boards—Chair, Rolland Savage, Gould
IPC-FC-234‡	Pressure Sensitive Adhesives Assembly Guidelines for Single-Sided and Double-Sided Flexible Printed Circuits—Chair, Terry Shepler, Sheldahl
IPC-D-279‡	Design Guidelines for Reliable Surface Mount Technology Printed Board Assemblies—Chairs, Nick Virmani, NASA, and Ed Aoki, Hewlett-Packard
IPC-D-356A‡	Bare Board Electrical Test Information in Digital Form—Chair, Duane Delfosse, ECT Circuitest
IPC-QL-653A‡	Qualification of Facilities that Inspect/Test Printed Board Components and Materials—Chair, Bob Neves, Microtek Laboratories
IPC-CC-830A†	Qualification and Performance of Electrical Insulating Compound for Printed Board Assemblies—Chair, John Waryold, Humiseal
IPC-2223§	Sectional Design Standard for Flexible Printed Boards and Assemblies—Chair, Russ Griffith, Tyco
IPC-2224‡	Sectional Standard for Design of PWBs for PC Cards—Chair, Leonard Roach, Lucent Technologies
IPC-2225‡	Sectional Design Standard for Multichip Modules (MCM-L) and MCM-L Assemblies—Chair, Kaz Hirasaka, Eastern Company.
IPC-2511‡	Generic Requirements for Implementation of Product Manufacturing Description Data and Transfer Methodology—Chair, Harry Parkinson, Compaq Computer
IPC-3406‡	Guidelines for Electrically Conductive Surface Mount Adhesives—Chair, Richard Thompson, Loctite
IPC-3408‡	General Requirements for Anisotropically Conductive Adhesive Films—Chair, Richard Thompson, Loctite
IPC-4101§	Specification for Base Materials for Rigid and Multilayer Boards—Chair, Erik Bergum, Polyclad
IPC-6013§	Specification for Printed Wiring, Flexible and Rigid-Flex—Chair, Roy Keen, Rockwell, and Larry Dexter, Advanced Circuit Technology

IPC-6015†   Performance Specification for Organic Multichip
                     Module   Structures   (MCM-L)—Chair,   Kaz
                     Hirasaka, Eastern Company
IPC-6018‡   Microwave End Product Board Inspection and
                     Test—Chair, John Kelly, Motorola GTSG

A.3 Published

J-STD-0351†   Acoustic Microscopy for Non-Hermetic Encapsula-
                     ted Electronic Components
IPC-QE-605A   Printed Board Quality Evaluation Handbook—
                     Chair, Ron Thompson, NSWC-Crane
IPC/JPCA-6202   Performance Guide Manual for Single- and
                     Double-Sided Flexible Wiring Boards

A.4 Pending Reaffirmation

IPC-SM-8172‡   General Requirements for Dielectric Surface
                     Mounting Adhesives

A.5 Pending DOD Adoption

IPC-6011†   Generic Performance Specification for Printed
                     Boards—Chair, Phil Hinton, Hinton "PWB"
                     Engineering
IPC-6012   Qualification and Performance Specification for
                     Rigid Printed Boards—Chair, Phil Hinton,
                     Hinton "PWB" Engineering

A.6 Second Interim Final

IPC-SM-782A   Surface Mount Design and Land Pattern
                     Standard—Chair, Vern Solberg, Tessera
IPC-2512†   Sectional Requirements for Implementation of
                     Administrative Methods for Manufacturing Data
                     Description—Chair, Harry Parkinson, Compaq
                     Computer
IPC-2513‡   Sectional Requirements for Implementation of
                     Drawing Methods for Manufacturing Data
                     Description—Chair, Harry Parkinson, Compaq
                     Computer
IPC-2514‡   Sectional Requirements for Implementation of
                     Printed Board Manufacturing Data Description—
                     Chair, Dana Korf, Hadco

IPC-2515‡	Sectional Requirements for Implementation of Bare Board Product Electrical Testing Data Description—Chair, Bob Neal, Hewlett-Packard
IPC-2516‡	Sectional Requirements for Implementation of Assembled Board Product Manufacturing Data Description—Chair, John Minchella, Celestica
IPC-2517‡	Sectional Requirements for Implementation of Assembly In-Circuit Testing Data Description—Chair, Bob Neal, Hewlett-Packard
IPC-2518‡	Sectional Requirements for Implementation of Bill of Material Product Data Description—Chair, Harry Parkinson, Compaq Computer
IPC-2521‡	Customer PCB Data Quality Rating—Chair, Dana Korf, Hadco
IPC-6012A†	Performance Specification for Rigid Printed Boards—Chair, Lisa Greenleaf, Teradyne
IPC-9502‡	PWB Assembly Soldering Process Guideline for Non-IC Electronic Components—Chair, Dave Nicol, Lucent Technologies

A.7   Interim Final

J-STD-003A	Solderability Tests for Printed Boards—Chair, Dave Hillman, Rockwell
J-STD-020A†	Moisture/Reflow Sensitivity Classification of Plastic Surface Mount Devices—Chair, Steve Martell, Sonoscan
J-STD-026†	Semiconductor Design Standard for Flip Chip Applications
J-STD-028‡	Performance Standard for Flip Chip/Chip Scale Bumps
J-STD-033†	Packaging and Handling of Moisture Sensitive Non-Hermetic Solid State Surface Mount Devices—Chair, Steve Martell, Sonoscan
IPC-CH-65A†	Guidelines for Cleaning of Printed Boards and Assemblies—Chair, Frank Cala, Church & Dwight
IPC-MF-150G†	Metal Foil for Printed Wiring Applications—Chair, Dave McGowan, DRM Consulting
IPC-FC-241D†	Flexible Metal-Clad Dielectrics for Use in Fabrication of Flexible Printed Wiring—Chair, Clark Webster, Precision Diversified Industries Inc.
IPC-A-600F†	Acceptability of Printed Boards—Chair, Floyd Gentry, Sandia National Laboratories

IPC-2141†	Controlled Impedance Circuit Boards and High Speed Logic Design—Chair, Andy Burkhardt, Polar Instruments
IPC-2221 Amendment 1	Generic Standard on Printed Board Design—Chair, Lionel Fullwood, Wong's Kong King International
IPC-4104†	Qualification and Conformance of Materials for High Density Interconnection Structures and Microvias—Chair, Ceferino Gonzalez, DuPont
IPC-4121†	Guidelines for Selecting Core Constructions for Multilayer Printed Wiring Board Applications—Chair, Doug Sober, isolaUSA
IPC-4411‡	Specification and Characterization for Nonwoven Para-aramid Reinforcement—Chair, David Powell, DuPont
IPC-6016†	Qualification and Performance Specification for High Density Interconnect Structures—Chair, Rolf Funer, Funer Associates
IPC-9191†	General Requirements for Statistical Process Control Implementation—Chair, Nick Watts, Intel
IPC-9503‡	Moisture Sensitivity Classification for Non-IC Components—Chair, David Nicol, Lucent Technologies

A.8   Official Representative Proposal

J-STD-001C†	Acceptability of Soldered Electrical and Electronic Assemblies—Chair, Jeff Koon, Raytheon Systems, and Teresa Rowe, AAI Corp.
J-STD-027†	Mechanical Outline Standard for Flip Chip or Chip Scale Configurations
J-STD-0291†	Test Methods for Flip Chip of Chip Scale Products
J-STD-030†	Qualification and Performance of Flip Chip Underfill Materials
J-STD-032†	Performance Standard for Ball Grid Array Bumps and Columns
IPC-SC-60A†	Post Solder Solvent Cleaning Handbook—Chair, Bill Kenyon, GCPC
IPC-FC-231D†	Flexible Base Dielectrics for Use in Flexible Printed Wiring—Clark Webster, Precision Diversified Industries Inc.

IPC-FC-232D†	Adhesive Coated Dielectric Films for Use as Cover Sheets for Flexible Printed Wiring and Flexible Bonding Films—Chair, Clark Webster, Precision Diversified Industries Inc.
IPC-A-610C†	Acceptability of Electronic Assemblies—Chair, Tino Gonzalez, ACME Consulting
IPC-7071†	General Requirements for Component Mounting—Chair, Peggi Blakley, NSWC Crane
IPC-7072†	Sectional Requirements for Through-Hole Component Mounting
IPC-7073†	Sectional Requirements for Standard Surface Mount Technology Component Mounting
IPC-7075†	Sectional Requirements for High Pin Count Area Array Component Mounting
IPC-7078†	Sectional Requirements for Flip Chip Component Mounting (Direct Chip Attach)
IPC-7525	Guidelines for Stencil Design—Chair, Bill Coleman, PhotoStencil
IPC-7911†	Barcode Label Performance Specification—Chair, Vicki Heideman, Brady USA
IPC-7912	Calculation of Defects per Million Opportunities (DPMO) in Electronic Assembly Acceptance Inspection—Chair, Fritz Byle, Rockwell Automation
IPC-9193†	SPC Implementation Requirements Applied to Printed Board and Interconnection Structure Manufacture—Chair, Nick Watts, Intel
IPC-9199†	SPC Quality Rating—Chair, Nick Watts, Intel
IPC-9501A	PWB Assembly Process Simulation for Evaluation of Electronic Components (Preconditioning IC Components)—Chair, Dave Nicol, Lucent Technologies

A.9   Working Draft

J-STD-002B	Solderability Tests for Component Leads, Terminals, and Wires—Chair, Dave Hillman, Rockwell
J-STD-004A	Requirements for Soldering Fluxes—Chair, Laura Turbini, Georgia Tech
IPC-HDBK-005	Soldering Pastes Handbook—Chair, Andy Mackie, Praxair
J-STD-005A	Requirements for Soldering Pastes—Chair, Andy Mackie, Praxair

J-STD-006A	Requirements for Electronic Grade Solder Alloys and Fluxed and Non-Fluxed Solid Solders for Electronic Soldering Applications—Chair, Dave Dodgen, Multicore Solders
J-STD-031†	Mechanical Outline Standard for Ball Grid Arrays and Other High Density Technology
IPC-T-50G†	Terms and Definitions—Chair, Pete Gilmore, Morton Electronic Materials
IPC-D-316A†	High Frequency Design Guide—Chair, Suzanne Seymour, Taconic Plastics
IPC-D-317B	Design Guidelines for Electronic Packaging Utilizing High Speed Techniques—Chair, Tom Bresnan, Multek
IPC-SS-605A	Printed Board Quality Evaluation Slide Set—Chair, Ron Thompson, NSWC Crane
IPC-2226†	Design Standard for High Density Array or Peripheral Leaded Component Mounting Structures—Chair, Foster Gray, PC Interconnects
IPC-2227	Design Standard for Advanced IC Interconnection Mounting Structures
IPC-2315†	Design Guide for HDI Interconnect and Microvias—Chair, Happy Holden, TechLead
IPC-2401	Performance Specification for Production Phototools—Chair, Chaye Bookey, Praegitzer Industries
IPC-2431	Performance Specification for Phototool Materials—Chair, Chaye Bookey, Praegitzer Industries
IPC-2615‡	Printed Board Dimensions and Tolerances—Chair, John Sabo, Rockwell Automation
WHMA/IPC-3710	Acceptability of Electronic Wire Harnesses and Cables—Chair, Ralph Hersey, Hersey & Associates
IPC-4101A†	Specification for Base Materials for Rigid and Multilayer Boards—Chair, Erik Bergum, Polyclad
IPC-4103†	Specification for Plastic Substrates, Clad or Unclad, for High Speed/High Frequency Interconnection—Chair, Mike Norris, Rogers
IPC-5001	Fabrication Process Control Handbook—Chairs, Dan Nelson, Coates ASI, and Ceferino Gonzalez, DuPont
IPC-7074	Sectional Requirements for Fine Pitch and High Pin Count Component Mounting

IPC-7076	Sectional Requirements for Chip Scale and Chip Size Component Mounting
IPC-7077	Sectional Requirements for Wire Bonding Bare Chip Component Mounting
IPC-7095	Design and Assembly Process Implementation for BGAs—Chair, Ray Prasad, Prasad Consulting
IPC-9261	Calculation of Defects per Million Opportunities (DPMO) in Electronic Assembly Operations—Chair, Fritz Byle, Rockwell Automation
IPC-9701	Qualification and Performance Test Methods for Surface Mount Solder Attachments—Chair, Vern Solberg, Tessera

A.10   Test Method Proposals

2.3.25C	Detection and Measurement of Ionizable Surface Contaminants
2.3.25.1	Ionic Cleanliness Testing of Bare PWBs
2.3.37B	Volatile Content of Adhesive Coated Dielectric Films
2.3.42	Identification of Solder Mask Products Using Fourier Transform
2.4.22C	Bow and Twist
2.5.5.7A	Characteristic Impedance and Time Delay of Lines on Printed Boards
2.6.14C	Resistance to Electromigration, Polymer Solder Masks
2.6.14.1	Electromigration Resistance Test 2.6.24 Junction Stability Under Environmental Conditions
2.6.26	DC Current Induced Thermal Cycling Test

---

†ANSI PIN submitted.

‡ANSI BSR-8 submitted.

§Final ANSI approval pending.

Index

ABOUT THE EDITOR

Charles A. Harper is President of Technology Seminars, Inc., of Lutherville, Maryland, and is widely recognized as one of the leaders in the electronic packaging industry. He has authored over a dozen well-known books in the field, and is among the founders and past presidents of the International Microelectronics and Packaging Society. He is series editor of the well-known McGraw-Hill Electronic Packaging and Interconnection Series. Mr. Harper is a graduate of The Johns Hopkins University School of Engineering, where he has also served as adjunct professor.